Biological Mechanisms in the Treatment of Neuropsychiatric Diseases

Biological Mechanisms in the Treatment of Neuropsychiatric Diseases

Guest Editors

Randall D Marshall
Frank S. Menniti

MDPI

Basel • Beijing • Wuhan • Barcelona • Belgrade • Novi Sad • Cluj • Manchester

Guest Editors
Randall D Marshall
Sophrosyne Pharmaceuticals
Chicago, IL
USA

Frank S. Menniti
BioPharmaWorks, LLC
Groton, CT
USA

Editorial Office
MDPI AG
Grosspeteranlage 5
4052 Basel, Switzerland

This is a reprint of the Special Issue, published open access by the journal *Cells* (ISSN 2073-4409), freely accessible at: https://www.mdpi.com/journal/cells/special_issues/927IJRP8A3.

For citation purposes, cite each article independently as indicated on the article page online and as indicated below:

Lastname, A.A.; Lastname, B.B. Article Title. *Journal Name* **Year**, *Volume Number*, Page Range.

ISBN 978-3-7258-8173-4 (Hbk)
ISBN 978-3-7258-8174-1 (PDF)
https://doi.org/10.3390/books978-3-7258-8174-1

Contents

About the Editors

Randall D Marshall

Randall D Marshall, Chief Medical Officer at EuMentis Therapeutics and Interim Chief Medical Officer at Sophrosyne Therapeutics, earned his MD at Johns Hopkins University and completed his psychiatry residence at Columbia University, where he attained Associate Professor status and was PI, Co-PI, or Advisor to ~30 research grants funded through NIH, nonprofits and industry. After the 9/11 attack on New York City, he founded and led a multi-year program providing evidence-based mental health diagnosis and treatment training to thousands of professionals in the Greater New York area over several years through a consortium of medical institutions. Dr. Marshall has published more than 140 peer-reviewed publications, book chapters and books across a wide range of subjects, including trauma-related disorders, psychiatric consequences related to the 9/11 attack in New York City, translational studies for novel therapies, treatment of multiple anxiety disorders, and the development of novel outcome measures in rare diseases. After leaving academia to work in biotechnology, Dr. Marshall has contributed to or led pre-clinical, translational and clinical programs across multiple CNS disorders in both neurology and psychiatry, in hepatology and pediatric oncology, and in rare disease, developing treatments for depression, treatment-refractory depression (multiple programs), neuropathic pain, insomnia, ADHD, alcohol use disorder, pantothenate kinase-associated neurodegeneration, cerebrotendinous xanthomatosis, Parkinson's disease, intestinal failure associated liver disease, and lymphatic malformations. He also co-founded Protara Therapeutics, now a public company listed on NASDAQ. At present, he works primarily as a Chief Medical Officer or medical consultant at start-up or early-stage biotechnology companies developing novel treatments for a range of neuropsychiatric conditions.

Frank S. Menniti

Frank S. Menniti, is a neuropharmacologist with BioPharmaWorks, LLC and a Ryan Research Professor of Neuroscience and Adjunct Professor, Department of Biomedical and Pharmaceutical Sciences, in the College of Pharmacy, University of Rhode Island. He was a Founder and formerly Chief Science Officer of MindImmune Therapeutics, Inc., a drug discovery company focused on developing immunological modulators to treat diseases of the mind and brain. Prior to MindImmune, Dr. Menniti was Founder and Chief Scientific Officer of Mnemosyne Pharmaceuticals, Inc., developing subunit-selective NMDA receptor modulators. Mnemosyne developed GluN2B NAMs for depression, with Novartis fully acquiring this asset as well as a second Mnemosyne asset, an NMDA receptor PAM for schizophrenia, through the purchase of Cadent Therapeutics. Prior to founding Mnemosyne, Dr. Menniti was a Research Fellow in the CNS Discovery group at Pfizer, Inc., in Groton, CT, and a principal researcher in developing the scientific rationale for use of NMDA receptor GluN2B antagonists for stroke, neuropathic pain, Parkinson's disease, and depression and in the clinical development of the prototype GluN2B antagonist CP-101,606. He was also involved in developing the therapeutic utilities of phosphodiesterase inhibitors for neuropsychiatric disease. This includes the identification of PDE10A inhibitors for schizophrenia and Huntington's disease, with the first PDE10A inhibitor to reach Phase II development, the first PDE9A inhibitor into Phase II testing for the treatment of Alzheimer's disease and the first PDE5A inhibitor into Phase II testing to improve functional recovery after stroke. After receiving a Ph.D. in Pharmacology from the University of North Carolina, Chapel Hill, in 1987, Dr. Menniti served as a Staff Fellow in the laboratory of Dr. James W. Putney at the National Institute of Environmental Health Sciences, participating in research elucidating the fundamentals of intracellular calcium signaling.

Preface

Diseases affecting the nervous system are the leading cause of overall disease burden in the world, affecting 3.4 billion people. Although research in basic neuroscience has thrived for several decades, interest in applied neurosciences—particularly developing new therapies—has generally lagged behind in recent years; however, we have seen a resurgence of commitment to therapeutics development. The urgent need for progress in the understanding of molecular, circuit-related, and genetic causal factors for serious neurological and psychiatric conditions is once more attracting the considerable governmental and private investments needed to progress an empirically founded idea into a successful therapeutic.

This Special Issue of *Cells* focuses on cellular and mechanism studies relevant to therapeutics development, including pre-clinical studies of novel treatments, translational studies crossing from preclinical to clinical data, and literature syntheses related to pathological processes and potential novel therapeutic strategies. Advances in the understanding of the molecular mechanisms of neuropsychiatric diseases have ushered in a renaissance of interest in both rare and common conditions and the development of increasingly relevant translational models to inform rational therapeutic development. Research into the molecular mechanisms of effective treatments increases our understanding of disease processes and can lead to even more effective therapies.

With this selection of ongoing work in the field, we hope to encourage an optimistic mindset toward the possibility that the following decade will see the creation of many new and better therapies to improve the lives of millions of people with these conditions.

Randall D Marshall and Frank S. Menniti
Guest Editors

cells

MDPI

Article

Investigation of the Roles of the Adenosine A(2A) and Metabotropic Glutamate Receptor Type 5 (mGlu5) Receptors in Prepulse Inhibition and CREB Signaling in a Heritable Rodent Model of Psychosis

Anthony M. Cuozzo, Loren D. Peeters, Cristal D. Ahmed, Liza J. Wills, Justin T. Gass and Russell W. Brown *

Department of Biomedical Sciences, James H. Quillen College of Medicine, East Tennessee State University, Johnson City, TN 37614, USA; cuozzoa@etsu.edu (A.M.C.); johnsonc15@etsu.edu (C.D.A.)
* Correspondence: brownl@etsu.edu

Abstract: The metabotropic glutamate receptor type 5 (mGlu5) and adenosine A(2A) receptor form a mutually inhibitory heteromer with the dopamine D2 receptor, where the activation of either mGlu5 or A(2A) leads to reduced D2 signaling. This study investigated whether a mGlu5-positive allosteric modulator (PAM) or an A(2A) agonist treatment could mitigate sensorimotor gating deficits and alter cyclic AMP response element-binding protein (CREB) levels in a rodent neonatal quinpirole (NQ) model of psychosis. F0 Sprague–Dawley rats were treated with neonatal saline or quinpirole (1 mg/kg) from postnatal day 1 to 21 and bred to produce an F1 generation. F1 offspring underwent prepulse inhibition (PPI) testing from postnatal day 44 to 48 to assess sensorimotor gating. The rats were treated with mGlu5 PAM 3-Cyano-N-(1,3-diphenyl-1H-pyrazol-5-yl) benzamide (CDPPB) or A(2A) agonist CGS21680. Rats with at least one NQ-treated parent showed PPI deficits, which were alleviated by both CDPPB and CGS21680. Sex differences were noted across groups, with CGS21680 showing greater efficacy than CDPPB. Additionally, CREB levels were elevated in the nucleus accumbens (NAc), and both CDPPB and CGS21680 reduced CREB expression to control levels. These findings suggest that targeting the adenosinergic and glutamatergic systems alleviates sensorimotor gating deficits and abnormal CREB signaling, both of which are associated with psychosis.

Keywords: mGlu5; A(2A); dopamine; CREB; prepulse inhibition; psychosis

1. Introduction

Adenosine is a nucleoside and neuromodulator that plays a critical role in proper brain function, contributing to sleep regulation, neuroprotection, and mediation of the effects of drug use [1]. Adenosine regulates the release of neurotransmitters at the synapse level via A(1) and A(2A) receptors. A(1) receptors are largely enriched in glutamatergic terminals and exert an inhibitory effect [2], whereas A(2A) receptors are facilitatory and responsible for the integration of dopaminergic and glutamatergic neurotransmission [3]. Thus, the A(2A) receptor presents itself as a possible therapeutic target for the treatment of psychiatric disorders. Previous work in our laboratory has shown that administration of the A(2A) agonist CGS21680 alleviates sensorimotor gating deficits in rats given neonatal quinpirole (NQ) [4], which results in increased dopamine D2 receptor sensitivity throughout the animal's lifetime. Aberrant sensorimotor gating is a biomarker of schizophrenia (SZ) that was established via prepulse inhibition (PPI) testing in a clinical population [5].

The most widely accepted and implemented theory of SZ pathology is based on DA hyperactivity within the mesolimbic pathway of the brain [6]. This idea is supported by the prescription of D_2 antagonists as antipsychotics, as well as DA-enhancing drugs inducing delusions and hallucinations that are associated with SZ [7]. However, it is likely that other neurotransmitters and neuromodulators contribute to the pathophysiology of SZ. Adenosine is a modulator of glutamate and DA signaling, and diminished function of the adenosine system can produce a hyperdopaminergic state [8]. A(2A) deficiency in the striatum has been previously linked to an enhanced response to amphetamines, and A(2A) deficiency in the hippocampus produces N-methyl-D-aspartate receptor (NMDAR) hypoactivity that is associated with learning deficits [9].

Glutamate is the most abundant excitatory transmitter in the brain and is heavily involved in mechanisms of neuroplasticity and synaptic reorganization [10]. Metabotropic glutamate receptor type 5 (mGlu5) is a modulatory receptor that plays a role in cognition, reward processing, and synaptic plasticity [11]. The mGlu5 receptor modulates NMDA receptor-mediated signaling and has been implicated in the pathology of SZ [12]. Previous work in our laboratory has demonstrated that the administration of mGlu5-positive allosteric modulator (PAM) CDPPB attenuates sensorimotor gating deficits in NQ-treated rats and restores striatal D_2 signaling to control levels [13]. In summary, the mGlu5 receptor has been identified as a significant therapeutic target for symptom treatment in psychiatric disorders.

The glutamatergic hypothesis of SZ posits that NMDAR hypofunction in the prefrontal cortex and hippocampus contributes to observed symptomology [14]. The mGlu5 receptor, in particular, has been implicated via genome-wide association studies (GWASs), revealing reduced availability and function of the receptor [15]. Several models of psychosis are induced via NMDAR hypofunction through the administration of ketamine, MK-801, or phencyclidine (PCP). MK-801 has been previously shown to impair cognition and spatial memory in rodents [16]. These findings overwhelmingly suggest that the glutamatergic system is disrupted in SZ; thus, it is a necessary therapeutic target for treating the full range of symptoms and insults in the disorder.

The current standard for SZ treatment is founded in D_2 antagonism, but these compounds produce significant side effects that include motor impairment, metabolic dysregulation, and tardive dyskinesia [17]. Attenuating hypersensitive D2 signaling while reducing the occurrence of adverse effects is imperative to improving treatment outcomes in this population. The A(2A)-D_2-mGlu5 heteromer is abundant in the striatum, and synergistic actions between the A(2A) and mGlu5 subunits have previously been suggested [18]. Stimulation of either of these subunits results in an overall reduction in DA signaling, as the heteromer is mutually inhibitory [19]. In the present study, we analyzed both CGS21680, an A(2A) agonist, and CDPPB, a PAM, towards mGlu5. We hypothesize that these compounds work to restore hypofunction within the adenosinergic and glutamatergic systems while also producing a net reduction in D_2 signaling.

In order to model the enhanced hyperactivity of the D_2 system in SZ, our laboratory has developed a heritable rodent model of psychosis that is induced by neonatal treatment with quinpirole, a D_2-like receptor agonist [20]. Recently, our laboratory established the heritability of this NQ model by breeding pairs of NQ- or NS-treated rats. The resulting offspring (F1 generation) is untreated and, remarkably, demonstrates enhanced dopamine D2 signaling and presents with the same phenotypical sensorimotor gating deficit as F0 founders [21]. We have previously shown several consistencies between the NQ model and clinical data in SZ [22,23]. Interestingly, we have also demonstrated an enhanced drug reward sensitivity and behavioral response to nicotine [22], which is based on overwhelming epidemiological data demonstrating the vulnerability of this population towards cigarette smoking [24]. Additionally, F1 rats that are the offspring of two NQ-treated founders demonstrate a 26-fold enrichment of the nicotine addiction Kyoto Encyclopedia of Genes and Genomes (KEGG) pathway in the infralimbic cortex (IfL) compared to controls [25].

PPI of the acoustic startle response is an operational measure of sensorimotor gating that reflects the ability of an individual to filter out unnecessary environmental stimuli [26]. Systemic administration of D2-like agonist quinpirole has been previously shown to induce disruptions of PPI [27]. Deficits in PPI in psychosis have been observed across numerous studies, and the administration of second-generation antipsychotics has been shown to normalize these deficits [28]. PPI is highly translatable between animals and humans and possesses significant cross-species measures that can enhance understanding of the pathophysiology of SZ and related psychoses [29]. Thus, we evaluated the efficacy of CGS 21680 and CDPPB to reduce sensorimotor gating deficits in a heritable model of psychosis.

The transcription factor CREB is heavily involved in gene expression that affects downstream neural circuit function and synaptic plasticity [30]. Within psychosis, deficits in CREB have been previously observed in the dorsolateral prefrontal cortex and cingulate gyrus [31]. Changes in CREB within the NAc have been linked to alterations in gating between emotional stimuli and behavioral responses [32]. In previous work in the F1 generation model, we reported an increase in the cortisol synthesis pathway via RNAseq analysis of the dorsal striatum, which has an abundance of D2-like receptors [21]. This finding is consistent with an enhanced stress response, which plays an important role in the manifestation of psychotic symptoms [33]. The association of CREB with emotional reactivity within the context of PPI can greatly affect the ability to filter out irrelevant sensory stimuli [34]. This enhanced emotional reactivity-induced deficit is hypothesized to be involved in the PPI deficits observed within the NQ model. Given this, in the present study, we analyzed protein levels of CREB via ELISA in the NAc to establish changes in CREB within our model and assess the effects of CDPPB and CGS21680 on this emotional reactivity marker.

2. Materials and Methods

2.1. Experiment 1: Subjects

For this experiment, a total of 185 Sprague–Dawley animals were used as subjects, with 100 subjects that were males and a total of 85 that were females. These animals were a product of a total of 22 breeder pairs in which male/female rats were neonatally treated with saline (NS) or quinpirole HCl (NQ; 1 mg/kg). All breeders and their offspring were housed in the AALAC-accredited animal colony at the East Tennessee State University Quillen College of Medicine. A total of 1–2 males/females were used per litter to control for within-litter variance All animals were fed regular chow and tap water ad libitum and were kept on a 12/12 light/dark cycle. All animals were tested during the light portion of their daily cycle. All procedures reviewed and approved by the University Committee on Animal Care at East Tennessee State University, which conforms to the NIH Guide on the Care and Use of Animals. Across all behavioral analyses, females and males are presented in separate figures. This was done, in part, because there were significant sex differences in many analyses but also based on the number of groups analyzed.

2.2. Experiment 1: Drugs

The drugs used in experiment 1 were Quinpirole HCl (Product# Q102) at a dose of 1 mg/kg and adenosine A(2A) agonist CGS 21680 (Product #199137) at doses of 0.03 mg/kg and 0.09 mg/kg. Both drugs were ordered from Sigma-Aldrich, Inc. (St. Louis, MO, USA). Quinpirole was mixed in 0.9% saline. The dose of quinpirole was chosen based on past work, which has shown a dose of 1 mg/kg i.p. administered from postnatal day (P)

1 to 21 results in an increase of dopamine D2 receptor sensitivity throughout the animal's lifetime [35]. CGS 21680 (Product # C141) was mixed in a 1:1 mixture of 2% dimethyl sulfur oxide (DMSO) and 0.9% NaCl. The doses of CGS 21680 were based on past work that has shown these doses to be effective in alleviating PPI deficits in F0-generation NQ-treated rats [4]. In experiment 1, both drugs were given through an intraperitoneal (i.p.) route of administration.

2.3. Experiments 1 and 2: Research Design

F0-generation breeders, which were the offspring of males and females ordered from Inotiv, Inc. (Indianapolis, IN, USA), produced the F1-generation offspring, which were used as subjects. The day of birth for the F0-generation rats was counted as postnatal day (P)0. These animals were neonatally i.p. administered saline or quinpirole (1 mg/kg) once daily from P1 to 21, and neonatal treatments followed our past work [21]. These animals were raised to P60 with occasional handling by experimenters. Breeding commenced on approximately P60, with one male and one female per cage. The F1-generation offspring of F0-generation breeders were used as subjects. F1-generation animals were untreated. There were four crosses used: two breeders that were both neonatally treated with saline (MSxFS), two breeders that were both neonatally treated with quinpirole (MQxFQ), a male breeder that was neonatally treated with quinpirole bred with a female neonatally treated with saline (MQxFS), and a male breeder that was neonatally treated saline bred with a female that was neonatally treated with quinpirole (MSxFQ).

Only F1-generation animals that were the offspring of at least one NQ-treated founder were given CGS 21680 or CDPPB. The rationale for this is that the focus of this study is to analyze whether these drugs are effective in alleviating PPI deficits in animals expressing increased DAD2 receptor sensitivity. The focus was not on whether these drugs had any effect on sensorimotor gating in controls. Furthermore, adding this factor would result in a four-factor study, which could result in a statistically significant four-way interaction, which is an uninterpretable statistical effect.

Finally, we did not systematically analyze the estrous cycle stage in females because the number of animals would have to be dramatically increased 3- to 4-fold to analyze the impact of the stage of the estrous cycle on sensorimotor gating. Adding this factor would increase the number of factors and complicate the research design.

2.4. Experiments 1 and 2: Sensorimotor Gating Apparatus

The Startle Monitor II apparatus and software (Kinder Scientific, Poway, CA, USA) were used for all sensorimotor gating testing across experiments 1 and 2. Within the apparatus, subjects were placed into an enclosed dome (height = 8 cm) that was attached to a stainless-steel platform (11 cm wide × 15 cm long). This platform was mounted in a sound-attenuating chamber (28 cm tall × 30 cm wide × 36 cm depth). Subject responses were recorded in Newtons (N) within a 250 millisecond (ms) timeframe following presentation of a stimulus. The apparatus was calibrated during testing based on the weight of the animal in the enclosure. Each subject was tested within the same apparatus across all five days of behavioral analysis.

2.5. Experiments 1 and 2: Prepulse Inhibition (PPI) Methods

PPI behavioral methodology is based on previous work completed by our laboratory [36]. Rats began testing on P44 and completed testing on P48 for a total of 5 days. In experiment 1, subjects were i.p. administered saline and 0.03 mg/kg or 0.09 mg/kg CGS21680 10 min prior to placement in the PPI apparatus. In experiment 2, subjects were subcutaneously (s.c.) administered saline and 10 mg/kg or 30 mg/kg CDPPB 20 min prior to testing. In each daily test, the first 5 min consisted of a habituation period in which white noise (70 dB) was the only stimulus. All animals were then administered 60 total trials consisting of 20 pulse, 30 prepulse, and 10 no-stimulus trials. The pulse trial consisted of a 120 dB startle pulse administered alone to produce a startle response. The prepulse trial was an auditory stimulus 3, 6, or 12 dB above the 70 dB white noise (73, 67, and 82 dB, respectively), followed by the 120 dB startle pulse 250 ms later. The stimulus trial consisted of presentation of stimulus. The startle response of the subject was measured in Newtons (N) via software provided by Kinder Scientific and averaged over the five test days for all trials.

2.6. Experiments 1 and 2: Cyclic AMP Response Element Binding Protein (CREB) ELISA

All animals were given one more injection of all drug treatments to be consistent with past work [4]. Approximately 24 h following this final injection, on P50, rats were live-decapitated, and brain tissue was harvested, flash-frozen in ice-cold isopentane, and immediately stored at −80 °C. The NAc was dissected using tissue punches that were placed into pre-weighed 1.5 mL Eppendorf tubes and stored for future analysis. Tissue was homogenized using a Fisher Scientific (Atlanta, GA, USA) Dismembrator 5000 at an amplitude of 50% for 3 s per tissue punch. Prior to homogenization, RIPA buffer with protease and phosphatase inhibitors (all from Sigma-Aldrich: phenylmethylsulfonyl fluoride and product numbers P5726, P0044, and P8340) were added to each sample tube. Following homogenization, the samples were centrifuged twice at 4 °C. CREB was analyzed via an ELISA kit from MyBioSource.com (San Diego, CA, USA; Catalog No.: MBS762390), and the protocol was closely followed. In brief, the provided standard was diluted in a serial dilution from 10 ng/mL to 0.156 ng/mL and run in duplicate. Then, 100 µL per well of homogenized sample was added and run in duplicate. The plate was incubated for 90 min at 37 °C, then washed twice. Subsequently, 100 µL of the biotin-labeled antibody working solution was added to each well (100 µL) and incubated for 60 min at 37 °C. The plate was washed three times, and 100 µL of the HRP–streptavidin conjugate working solution was added to each well and incubated for 60 min at 37 °C. The plate was washed five times, after which 90 µL of TMB substrate solution was added and incubated for 20 min. After this incubation, 50 µL of stop solution was added, and the plate was immediately read at 450 nm on a BioTek elx800 Plate Reader (BioTek, Winooski, VT, USA). The resulting data were calculated according to the given standard curve and converted to pg of CREB protein/mg of tissue.

2.7. Experiment 2: Subjects

For this experiment, a total of 186 Sprague–Dawley animals were used as subjects, with 91 subjects that were males a total of 95 that were females. These animals were a product of a total of 22 breeder pairs in which male/female rats were neonatally treated with saline (NS) or quinpirole HCl (NQ; 1 mg/kg). Notably, animals in the MSxFS group are the same subjects that were used in experiment 1. Identically to experiment 1, all breeders and their offspring were housed in the AALAC-accredited animal colony at East Tennessee State University Quillen College of Medicine with a 12/12 light/dark cycle. A total of 1–2 males/females were used per litter to control for within-litter variance, had ad libitum access to food and water, and were tested during the light portion of their daily cycle.

2.8. Experiment 2: Drugs

In experiment 2, quinpirole HCl was used again, in addition to CDPPB. CDPPB was ordered from EAG Laboratories (San Diego, CA, USA). CDPPB was dissolved in 10% Tween-80 from Sigma-Aldrich, Inc., and mixed with 0.9% NaCl at doses of 10 mg/kg and 30 mg/kg. Quinpirole was given via i.p. injection in experiment 2, with CDPPB administered s.c. The doses of CDPPB were chosen based on past works that have shown that these doses are behaviorally effective in rats [37], as well as in F0-generation rats given NQ [13].

2.9. Experiments 1 and 2: Statistical Analysis

For statistical analysis of PPI, a three-way ANOVA was used as the primary statistic, with sex, group, and decibel of prepulse used as the repeated measures. Group was a combination of the founder pair (MQxFQ, MQxFS, MSxFQ, or MSxFS) and the drug that was given (saline, CGS 21680, or CDPPB). The rationale for analysis in this manner was two-fold. First, CGS 21680 or CDPPB was not administered to the MSxFS control group, simply because the study was not focused on analyzing the effects of either of these two drugs alone on sensorimotor gating in controls. Rather, the study was focused on utilizing these drugs to alleviate sensorimotor gating deficits in F1-generation offspring of NQ- or NS-treated founders. Secondly, including drug treatment as a fourth factor in the ANOVA would significantly complicate the statistical analysis and interpretation. For analysis of the startle response, only a two-way ANOVA was used; because there was not a repeated measure, we averaged the startle over the days of testing. For all post hoc analyses, the Bonferroni test was used to analyze any group differences ($p = 0.05$). Across both experiments, there was an N of 7–12 males/females per group that were behaviorally tested. For CREB analyses, there was an N of 6–7 males/females per group.

3. Results

3.1. CGS21680–MQxFQ PPI and Acoustic Startle

Prepulse inhibition (%) is presented as a function of the decibel of prepulse and group in Figure 1A (females) and Figure 1C (males). The number of animals per group for females was MSxFS-S = 9, MQxFQ-S = 10, and 6 each for MQxFQ-0.03 and MQxFQ-0.09. For males, the number of animals was MSxFS-S = 10, MQxFQ-S = 11, MQxFQ-0.03 = 6, and MQxFQ-0.09 = 8. A three-way ANOVA revealed significant main effects of sex ($F(1,58) = 5.5, p < 0.022$; group $F(3,58) = 46.22, p < 0.001$) and group × dB of prepulse interaction ($F(6,116) = 2.45, p < 0.029$). Females and males were analyzed separately based on the main effect of sex. Overall, females demonstrated significantly improved performance as compared to males. In the separate analysis of females, there was a significant main effect of group ($F(3,27) = 23.41, p < 0.001$) and a significant two-way group × decibel of prepulse interaction ($F(6,54) = 3.34, p < 0.007$). In the separate analysis of males, a two-way ANOVA revealed a significant main effect of group ($F(3,31) = 24.42, p < 0.001$). For both females and males, at all dB levels of prepulse, MSxFS, MQxFQ-0.03, and MQxFQ-0.09 groups all performed equally and were all significantly improved over the MQxFQ group given vehicle. In other words, both doses of the A(2A) agonist improved performance in MQxFQ rats. Acoustic startle is presented in Figure 1B (females) and Figure 1D (males). No effects on acoustic startle were revealed.

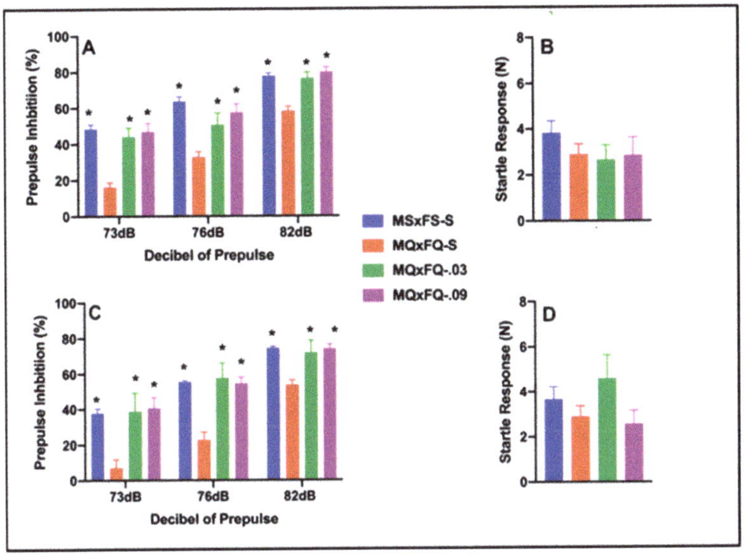

Figure 1. Prepulse inhibition (%) is presented as a function of MQxFQ groups in (**A**) (females) and (**C**) (males). In females and males, the MSxFS, MQxFQ-0.03, and MQxFQ-0.09 groups were equivalent and all significantly greater in PPI than the MQxFQ-S group (indicated by *, $p < 0.05$). Acoustic startle is presented as a function of MQxFQ groups in (**B**) (females) and (**D**) (males). There were no significant differences between groups.

3.2. CGS21680–MQxFS PPI and Acoustic Startle

Prepulse inhibition (%) is presented as a function of decibel of prepulse and group in Figure 2A (females) and Figure 2C (males). The number of animals per group for females was MSxFS-S = 9, MQxFS-S = 10, MQxFS-0.03 = 8, and MQxFS-0.09 = 7. For males, the number of animals was MSxFS-S = 10, MQxFS = S = 12, and 6 in MQxFS-0.03 and MQxFS-0.09. A three-way ANOVA revealed a significant main effect of drug ($F(3,60) = 17.9$, $p < 0.001$), a significant two-way interaction of drug ×dB of prepulse ($F(6,120) = 2.64$, $p < 0.019$), and a significant three-way interaction of sex × drug × dB of prepulse ($F(6,120) = 2.50$, $p < 0.026$). We then analyzed the two sexes separately. In females, a two-way ANOVA revealed a significant main effect of group ($F(3,30) = 4.95$, $p < 0.007$) and a significant two-way interaction of group × decibel of prepulse ($F(6,60) = 3.25$, $p < 0.008$). In females, at 73 and 76 dB, no difference between MQxFS-0.03 and vehicle were observed, but MQxFS.09 and MSxFS-S were greater than MQxFS-S. No group differences were observed in females at 82 dB. In males, a two-way ANOVA revealed a significant main effect of group ($F(3,30) = 16.82$, $p < 0.001$). In males, at all three dB levels, the MSxFS, MQxFS-0.03, and MQxFS-0.09 groups all performed equally and were all significantly improved over the MQxFS group given vehicle. In MQxFS, the 0.03 mg/kg dose of CGS 21680 was not effective in females at the lower decibel levels, and deficits in female MQxFS did not demonstrate as severe of PPI deficits as MQxFQ.

Acoustic startle is presented in Figure 2B (females) and Figure 2D (males). A two-way ANOVA revealed a significant main effect of group ($F(3,60) = 4.39$, $p < 0.008$). The MSxFS group was significantly higher than the MQxFS-0.03 group. Thus, the lower dose of CGS 21680 significantly decreased acoustic startle in the MQxFS group.

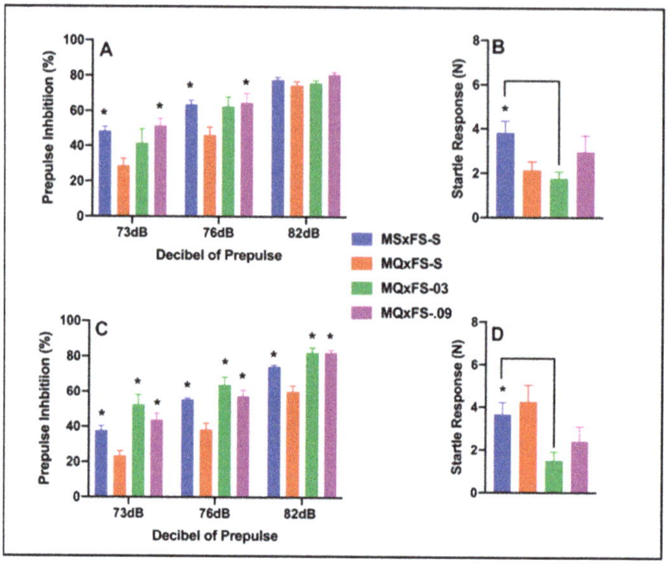

Figure 2. Prepulse inhibition (%) is presented as a function of MQxFS groups in (**A**) (females) and (**C**) (males). In females, at 73 and 76 dB, the MSxFS-S and MQxFS-0.09 groups were equivalent and both significantly greater in PPI than the MQxFS-S group (indicated by *, $p < 0.05$). There were no effects at 82 dB. In males, the MSxFS, MQxFS-0.03, and MQxFS-0.09 groups were equivalent and all significantly greater in PPI than the MQxFS-S group across all decibels of prepulse (indicated by *, $p < 0.05$). Acoustic startle is presented as a function of MQxFQ groups in (**B**) (females) and (**D**) (males). The MSxFS-S group was significantly greater than the MQxFS-0.03 group in both females and males.

3.3. CGS21680–MSxFQ PPI and Acoustic Startle

Prepulse Inhibition (%) is presented as a function of decibel of prepulse and group in Figure 3A (females) and Figure 3C (males). The number of animals per group for females was nine in groups MSxFS-S and MSxFQ-S and six in groups MSxFQ-0.03 and MSxFQ-0.09. For males, the number of animals was MSxFS-S = 10, MSxFQ = S-S = 9, and 6 each in MSxFQ-0.03 and MSxFQ-0.09. A three-way repeated-measures ANOVA revealed a significant main effect of group ($F(3,51) = 12.78$, $p < 0.001$) and significant two-way interactions of sex × decibel of prepulse ($F(2,102) = 3.21$, $p < 0.045$) and group × decibel of prepulse ($F(6,102) = 4.30$, $p < 0.002$). Based on the significant two-way interaction of sex × decibel of prepulse, females and males were analyzed separately. In females, a two-way repeated-measures ANOVA revealed a significant main effect of group ($F(3,24) = 3.83$, $p < 0.023$). In females, at 73 dB and 76 dB, the MSxFS was greater than MSxFQ-S; however, there were no differences between groups given either 0.03 or 0.09 mg/kg CGS 21680 and MSxFS-S or MSxFQ-S at this level of prepulse. At 82 dB, there was no significant group difference. In males, a two-way repeated-measures ANOVA revealed a significant main effect of group ($F(3,27) = 14.90$, $p < 0.001$) and a significant interaction of group × decibel of prepulse ($F(6,54) = 5.08$, $p < 0.001$), and MSxFQ-S was significantly below all other male groups at all three decibels of prepulse. Therefore, CGS 21680 was effective in alleviating PPI deficits in MSxFQ males at all decibels of prepulse.

Acoustic startle is presented in Figure 3B (females) and Figure 3D (males). There were no significant effects on startle.

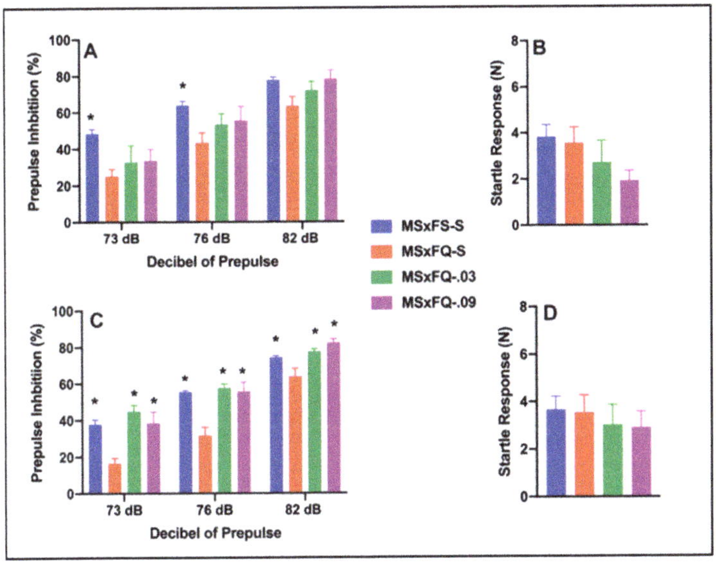

Figure 3. Prepulse inhibition (%) is presented as a function of MSxFQ groups in (**A**) (females) and (**C**) (males). In females, at 73 dB and 76 dB of prepulse, the MSxFS was greater than MSxFQ-S (indicated by *, $p < 0.05$); however, there were no differences between groups given either 0.03 or 0.09 mg/kg CGS 21680 and MSxFS or MSxFQ-S at this level of prepulse. At 82 dB, there were no significant group differences. In males, the MSxFS, MSxFQ-0.03, and MSxFQ-0.09 groups were equivalent and all significantly greater in PPI than the MSxFQ-S group across all decibels of prepulse (indicated by *, $p < 0.05$). Acoustic startle is presented as a function of MQxFQ groups in (**B**) (females) and (**D**) (males). There were no significant differences between groups.

3.4. CDPPB–MQxFQ PPI and Acoustic Startle

Prepulse inhibition (%) is presented as a function of decibel of prepulse and group in Figure 4A (females) and Figure 4C (males). The number of animals per group for females was MSxFS-S = 9, MQxFQ-S = 10, and 6 in each of MQxFQ-10 and MQxFQ-30. For males, the number of animals was MSxFS-S = 10, MQxFQ-S = 11, and 6 each in MQxFQ-10 and MQxFQ-30. A three-way repeated-measures ANOVA revealed significant main effects of sex ($F(1,56) = 4.7$, $p < 0.034$) and group ($F(3,56) = 25.74$, $p < 0.001$), as well as a significant two-way interaction of group × dB of prepulse $F((6,112) = 2.26$, $p < 0.042$). Based on the significant main effects of sex, females and males were analyzed separately. In females, a two-way repeated-measures ANOVA revealed a significant main effect of group ($F(3,28) = 9.61$, $p < 0.001$) and a significant two-way interaction of group × decibel of prepulse ($F(6,56) = 2.61$, $p < 0.027$). At 73 and 76 dB, MSxFS was equivalent to MQxFQ-10 and MQxFQ-30 and demonstrated improved PPI compared to MQxFQ-S. However, the MQxFQ-10 group was improved compared to MQxFQ-S, but MQxFQ-S was equivalent to MQxFQ-30. At 82 dB, groups MSxFS, MQxFQ-10, and MQxFQ-30 were all significantly greater than MQxFQ-S. It appears the lower dose of 10 mg/kg CDPPB was more effective in females, but the 30 mg/kg dose did not improve PPI until the decibel of prepulse reached 82 dB. In males, a two-way repeated-measures ANOVA revealed a significant main effect of group ($F(3,28) = 20.71$, $p < 0.001$). At all three levels of prepulse, MSxFS-S and MQxFQ-30 were significantly improved compared to MQxFQ-S, but MQxFQ-S did not significantly differ from MQxFQ-10. Therefore, the 30 mg/kg CDPPB but not the 10 mg/kg dose was effective in alleviating PPI deficits in males.

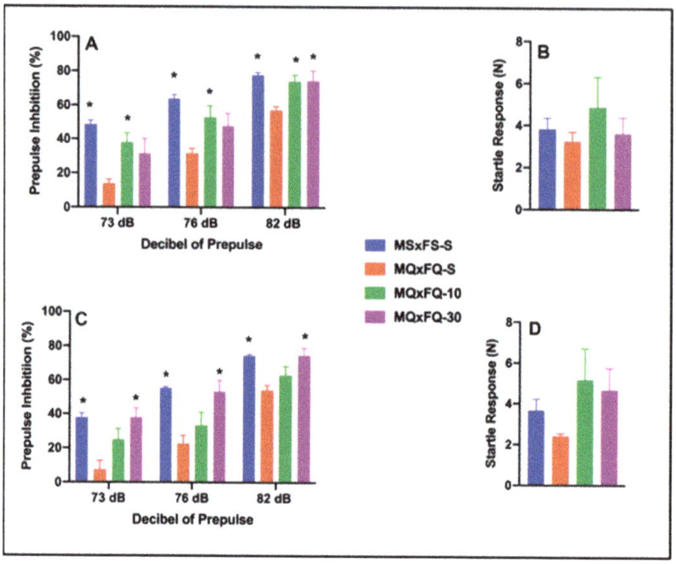

Figure 4. Prepulse inhibition (%) is presented as a function of MQxFQ groups in (**A**) (females) and (**C**) (males). In females, at 73 and 76 dB of prepulse, the MSxFS-S and MQxFQ-10 groups were equivalent and significantly greater than the MQxFQ-S group (indicated by *, $p < 0.05$). At 82 dB, the MSxFS, MQxFQ-10, and MQxFQ-30 groups were all equivalent and significantly greater than the MQxFQ-S group (indicated by *, $p < 0.05$). In males, the MSxFS-S and MQxFQ-30 groups were equivalent and demonstrated significantly greater levels of PPI than the MQxFQ-10 and MQxFQ-S groups at all decibel levels. Acoustic startle is presented as a function of MQxFQ groups in (**B**) (females) and (**D**) (males). There were no significant differences between groups.

Acoustic startle is presented in Figure 4B (females) and Figure 4D (males). There were no significant effects on startle.

3.5. CDPPB–MQxFS PPI and Acoustic Startle

Prepulse inhibition (%) is presented as a function of decibel of prepulse and group in Figure 5A (females) and Figure 5C (males). The number of animals per group for females was MSxFS-S = 9, MQxFS-S = 10, and 6 each in MQxFS-10 and MQxFS-30. For males, the number of animals was MSxFS-S = 10, MQxFS-S = 12, and 6 each in MQxFS-10 and MQxFS-30. A three-way repeated-measures ANOVA revealed a significant main effect of group ($F(3,50) = 23.50$, $p < 0.001$) and significant two-way interactions of sex × decibel of prepulse ($F(2,100) = 3.60$, $p < 0.031$) and group × decibel of prepulse ($F(6,100) = 3.43$, $p < 0.004$). In females, a two-way repeated-measures ANOVA revealed a significant main effect of group ($F(3,25) = 6.63$, $p < 0.001$) and a significant two-way interaction of group × decibel of prepulse ($F(6,50) = 3.89$, $p < 0.003$). In females, a two-way repeated-measures ANOVA revealed a significant main effect of group ($F(3,25) = 6.63$, $p < 0.001$) and a significant two-way interaction of group × decibel of prepulse ($F(6,50) = 3.90$, $p < 0.003$). Bonferroni post hoc analyses were conducted, and at 73 and 76 decibel levels of prepulse, MSxFS-S and MQxFS-30 were equivalent and both significantly greater than MQxFS-S and MQxFS-10. At 82 dB prepulse, MSxFS was significantly greater than MQxFS-10 but not different from the MQxFS-30 or MQxFS-S groups. For males, a two-way repeated-measures ANOVA revealed a significant main effect of group ($F(3,25) = 26.30$, $p < 0.001$). Across all decibel levels of prepulse, MSxFS-S was equivalent to MQxFS-30, and both were significantly greater than the MQxFS-10 and MQxFS-S groups.

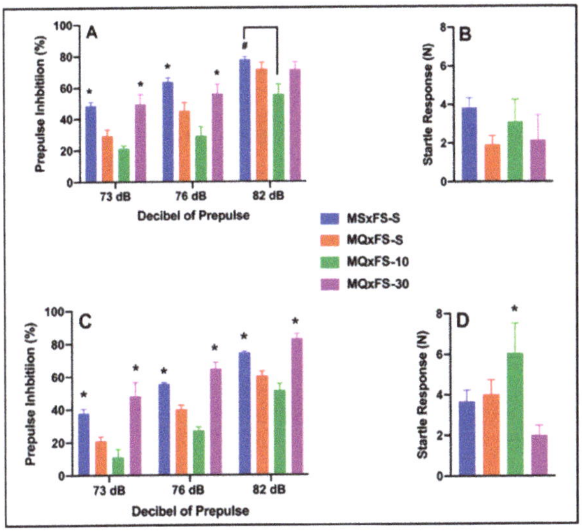

Figure 5. Prepulse inhibition (%) is presented as a function of MQxFS groups in (**A**) (females) and (**C**) (males). In females, the MSxFS-S and MQxFS-30 groups were equivalent at 73 and 76 dB of prepulse, and both demonstrated significantly greater PPI than the MQxFS-S and MQxFS-10 groups (indicated by *, $p < 0.05$). At 82 dB, only the MSxFS-S and MQxFS-10 groups were significantly different (indicated by #, $p < 0.05$). In males, the MSxFS-S and MQxFS-30 groups were equivalent at all dB levels of prepulse, and both demonstrated significantly greater PPI than the MQxFS-S and MQxFS-10 groups across all three dB levels of prepulse (indicated by *, $p < 0.05$). Acoustic startle is presented as a function of MQxFS groups in (**B**) (females) and (**D**) (males). There were no significant group differences in females, but in males, the MQxFS-10 group demonstrated a significant increase in acoustic startle relative to all other groups (indicated by *, $p < 0.05$).

Acoustic startle is presented in Figure 5B (females) and Figure 5D (males). Interestingly, with respect to startle response, a two-way ANOVA revealed both significant main effects of sex ($F(1,59) = 3.15$, $p < 0.047$) and group ($F(3,59) = 3.15$, $p < 0.033$). Males demonstrated a significantly higher startle response than females, and the MQxFS-10 males demonstrated a significantly higher startle response than the MQxFS-30 group.

3.6. CDPPB–MSxFQ PPI and Acoustic Startle

Prepulse inhibition (%) is presented as a function of decibel of prepulse and group in Figure 6A (females) and Figure 6C (males). The number of animals per group for females was nine in groups MSxFS-S and MSxFQ-S and eight in groups MSxFQ-10 and MSxFQ-30. For males, the number of animals was MSxFS-S = 10, MSxFQ-S = 9, and 6 each in MSxFQ-10 and MSxFQ-30. A three-way repeated-measures ANOVA revealed significant main effects of sex ($F(1,57) = 6.8$, $p < 0.012$) and group ($F(3,57) = 10.81$, $p < 0.001$) and an interaction of group × decibel of prepulse ($F(6,114) = 3.02$, $p < 0.009$). For females, a two-way ANOVA revealed a significant main effect of group ($F(3,33) = 5.7$, $p < 0.003$). Overall, MSxFS-S was equivalent to both the MSxFQ-10 and MSxFQ-30 groups, and MSxFS was significantly greater than the MSxFQ-S group. However, the MSxFQ-S group also did not significantly differ from the MSxFQ-10 or MSxFQ-30 groups. Since the group × decibel of prepulse interaction was not statistically significant, this effect did not affect overall performance across all three decibels of prepulse. In males, a two-way ANOVA revealed a significant main effect of group ($F(3,24) = 8.04$, $p < 0.001$) and a significant interaction of group × decibel of prepulse ($F(6,48) = 2.44$, $p < 0.039$). At 73 and 76 dB, MSxFS-S was equivalent to MSxFQ-30, and significantly greater than MSxFQ-10 and MSxFQ-S. At

82 dB, MSxFS-S was significantly greater than MSxFQ-10, but there were no other group differences, demonstrating that MSxFQ-S males did not demonstrate a significant deficit at 82 dB. Males in the MSxFQ group demonstrated a deficit when given the lower dose of CDPPB, which was a surprising result. However, the 30 mg/kg dose was effective. Acoustic startle is presented in Figure 6B (females) Figure 6D (males). There were no significant effects on startle.

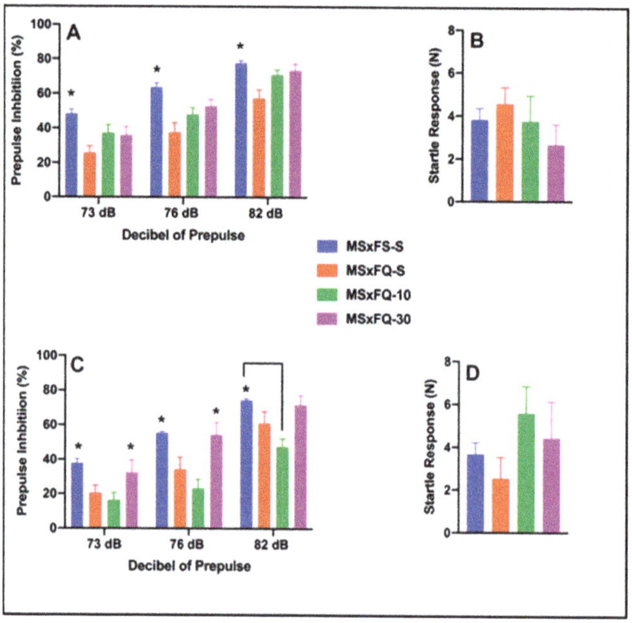

Figure 6. Prepulse inhibition (%) is presented as a function of MSxFQ groups in (**A**) (females) and (**C**) (males). In females, the MSxFS group was equivalent to both the MSxFQ-10 and MSxFQ-30 groups, and MSxFS was significantly greater than the MSxFQ-S group across all dB levels of prepulse (indicated by *, $p < 0.05$). In males, at 73 and 76 dB of prepulse, the MSxFS-S and MSxFS-30 groups were equivalent and significantly greater than the MSxFQ-S group in PPI (indicated by *, $p < 0.05$). At 82 dB of prepulse, only the MSxFS-S group was significantly greater than the MSxFQ-10 group (indicated by * and line, $p < 0.05$). Acoustic startle is presented as a function of MQxFQ groups in (**B**) (females) and (**D**) (males). There were no significant differences between groups.

3.7. A(2A) Agonist: CGS21680–CREB

3.7.1. Numbers of Animals for CREB Analyses

Across all CREB ELISA analyses, there were N values of four males and four females per group. The reason for the reduction in numbers compared to behavior was that all ELISAs were performed in 96-well plates, and there were not enough wells to represent all animals behaviorally tested in this study.

3.7.2. MQxFQ Group

CREB is presented as a function of group in Figure 7A. A two-way ANOVA revealed no significant main effect of sex. Therefore, we dropped sex as a factor, and a one-way ANOVA revealed a robust significant main effect of group ($F(3,35) = 22.42$, $p < 0.001$). The MQxFQ-S group was significantly higher than all other groups, demonstrating that both doses of CGS 21680 were effective in reducing the significant increase in CREB in the NAc.

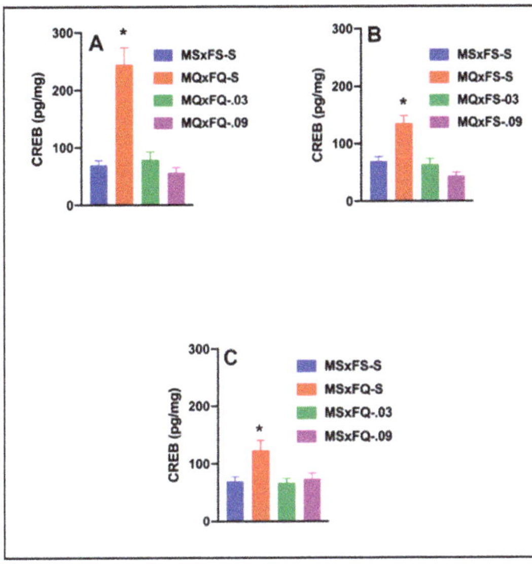

Figure 7. CREB is presented as a function of group for (**A**) MQxFQ, (**B**) MQxFS, and (**C**) MSxFQ in experiment 1. Across all three groups, the MQxFQ-S, MQxFS-S, and MSxFQ-S groups demonstrated significantly greater levels of CREB protein in the NAc than all other groups (indicated by *, $p < 0.05$).

3.7.3. MQxFS Group

CREB is presented as a function of group in Figure 7B. A two-way ANOVA revealed no significant main effect of sex. Therefore, we dropped sex as a factor, and a one-way ANOVA revealed a significant main effect of group ($F(3,37) = 16.97$, $p < 0.001$). The MQxFS-S group was significantly higher than all other groups, and there were no other significant differences between groups. As with the MQxFQ group, both doses of CGS 21680 were effective in reducing CREB in the NAc.

3.7.4. MSxFQ Group

CREB is presented as a function of group in Figure 7C. A two-way ANOVA revealed no significant main effect of sex. Therefore, we dropped sex as a factor, and a one-way ANOVA revealed a significant main effect of group ($F(3,33) = 5.09$, $p < 0.011$). The MSxFQ-S group was significantly higher for CREB in the NAc than MSxFS-S, but there were no other group differences.

3.8. mGlu5 PAM: CDPPB–CREB

3.8.1. MQxFQ Group

CREB is presented as a function of males in Figure 8A and females in Figure 8B. A two-way ANOVA revealed both a significant main effect of sex ($F(1,39) = 5.06$, $p < 0.032$) and a main effect of group ($F(3,39) = 16.19$, $p < 0.001$). Overall, females demonstrated a significantly higher level of CREB in the NAc than males, likely owing the lack of efficacy of CDPPB. Based on the significant main effect of sex, males and females were analyzed separately. In females, a one-way ANOVA revealed a significant main effect of group ($F(3,22) = 7.22$, $p < 0.001$). The MQxFQ-S group demonstrated significantly higher levels of CREB than the MQxFQ-10 and MSxFS-S groups but did not differ from the MQxFQ-10 group. There were no other significant differences between groups. In males, a one-way ANOVA revealed a significant main effect of group ($F(3,16) = 9.6$, $p < 0.001$). In males, the MQxFQ-S group demonstrated significantly higher CREB levels than all other groups, and there were no other significant differences between groups. This result suggests that CDPPB was not as effective in females as compared to males in reducing CREB in the NAc.

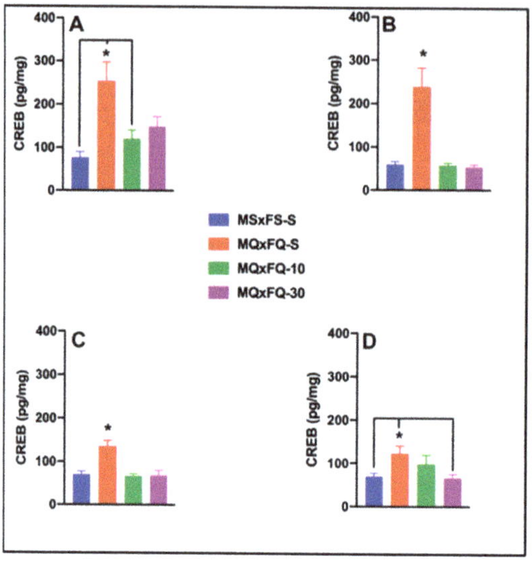

Figure 8. CREB is presented as a function of MQxFQ groups for females (**A**) and males (**B**). CREB is presented as a function of MQxFS groups in (**C**) and MSxFQ groups in (**D**). In female MQxFQ groups, the MQxFQ-S group demonstrated significantly greater CREB protein in the NAc than the MSxFS-S and MQxFQ-10 groups (indicated by *, $p < 0.05$ and line) but not significantly greater levels than the MQxFQ-30 group. In males, the MQxFQ-S group was significantly greater than all other groups (indicated by *, $p < 0.05$). As shown in (**C**), the MQxFS-S group demonstrated significantly greater levels of CREB than all other groups (indicated by *, $p < 0.05$). As shown in (**D**), MSxFQ-S was significantly greater than the MSxFS-S group and MSxFQ-10 group (indicated by * $p < 0.05$ and line) but was not significantly different from the MSxFQ-30 group.

3.8.2. MQxFS Group

CREB is presented as a function of group in Figure 8C. A two-way ANOVA revealed no significant main effect of sex. Therefore, we dropped sex as a factor, and a one-way ANOVA revealed a significant main effect of group ($F(3,35) = 10.96$, $p < 0.001$). Identical to the MQxFQ group, there was only a significant difference between the MSxFS-S and MQxFS-S groups, and the groups administered either dose of CDPPB did not significantly differ from controls (MSxFS-S) or the MQxFS-S group.

3.8.3. MSxFQ Group

CREB is presented as a function of group in Figure 8D. A two-way ANOVA revealed no significant main effect of sex. Therefore, we dropped sex as a factor, and a one-way ANOVA revealed a significant main effect of group ($F(3,35) = 3.94$, $p < 0.017$). In this case, the MSxFQ-S group was significantly higher than the MSxFS-S and MSxFQ-30 groups but not the MSxFQ-10 group. Thus, only the 30 mg/kg CDPPB dose was effective in this group in reducing CREB to control levels.

3.9. *F1-Generation Comparison of Offspring of NQ-Treated Rats in PPI and Acoustic Startle*

One final comparison was performed to compare saline-treated MSxFS, MQxFQ, MQxFS, and MSxFQ animals to determine if the MQxFQ group demonstrated more severe deficits in PPI than the other founder crosses. Prepulse inhibition (%) is presented as a function of decibel of prepulse and group for females in Figure 9A and for males in Figure 9C.

A three-way ANOVA with sex, group, and decibel of prepulse revealed a significant main effect of sex ($F(1,75) = 13.59$, $p < 0.001$) and founder cross ($F(3,75) = 28.55$, $p < 0.001$) and a significant two-way interaction of group × decibel of prepulse ($F(6,150) = 2.19$, $p < 0.047$). Overall, females demonstrated significantly improved performance as compared to males, as we revealed above. Based on the significant main effect of sex, we analyzed females and males separately. In females, a two-way ANOVA revealed a significant main effect of group ($F(3,36) = 10.58$, $p < 0.001$) and a significant two-way interaction of group × decibel of prepulse ($F(6,72) = 2.90$, $p < 0.014$). At 73 and 76 dB, the MQxFQ group was significantly less than all three other groups. However, at 82 dB, the MQxFQ and MSxFQ groups were equivalent but significantly less than the MQxFS group, as well as the MSxFS control group. In males, a three-way ANOVA revealed a significant main effect of group ($F(3,39) = 20.25$, $p < 0.001$), but the interaction of group × decibel of prepulse was not significant. Across all three decibels of prepulse, the MQxFQ founder cross demonstrated a significant deficit relative to the MQxFS founder cross (and MSxFS controls) but was equivalent to the MSxFQ group. In sum, within the MQxFQ group, females demonstrated a more robust PPI deficit than in males relative to the other crosses. However, in males, the MSxFQ cross was equivalent to MQxFQ in terms of the deficit in PPI.

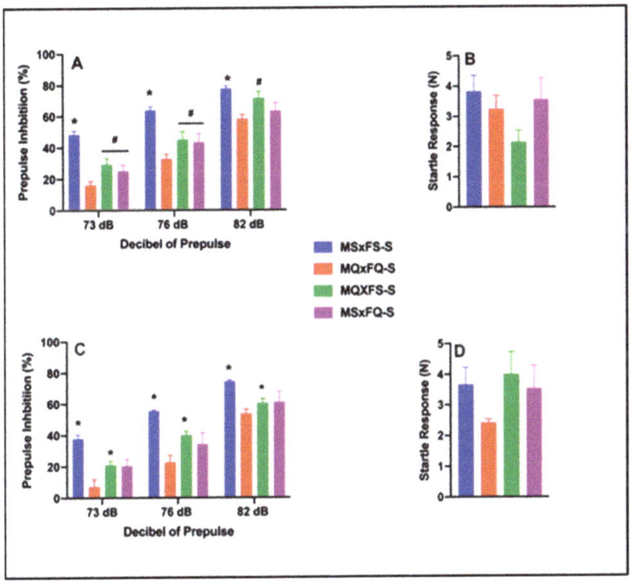

Figure 9. Prepulse inhibition (%) is presented as a function of all saline-treated founder crosses in females (**A**) and males (**C**). In females, the MSxFS-S group was significantly greater than all other groups across all dB values of prepulse (indicated by *, $p < 0.05$). The MQxFS-S group and MSxFQ-S group were significantly greater than the MQxFQ-S group at 73 and 76 dB of prepulse (indicated by #, $p < 0.05$). At 82 dB of prepulse, the MSxFS-S and MQxFS-S groups did not significantly differ but demonstrated greater levels of PPI than the MQxFQ-S and MSxFQ-S groups (indicated by *, $p < 0.05$). In males, the MSxFS-S group demonstrated significantly greater levels of PPI than all other groups across all dB values of prepulse (indicated by *, $p < 0.05$). Across all dB levels, the MQxFS-S group demonstrated significantly improved PPI relative to the MQxFQ-S group (indicated by *, $p < 0.05$). Acoustic startle is presented for females (**B**) and males (**D**). There were no significant differences between groups.

Results for acoustic startle are presented in Figure 9B for females and Figure 9D for males. There were no group differences in acoustic startle.

3.10. F1-Generation Comparison of Offspring of NQ-Treated Rats in NAc CREB

CREB is presented as a function of group in Figure 10. A two-way ANOVA did not reveal any significant main effect of sex, so this was dropped as a factor. The resulting one-way ANOVA revealed a significant main effect of group ($F(3,36) = 23.27$, $p < 0.001$). The MQxFQ-S group demonstrated a significantly higher level of CREB in the NAc than all other groups, and the MQxFS-S and MSxFQ-S groups also demonstrated a significantly higher level of CREB than MSxFS.

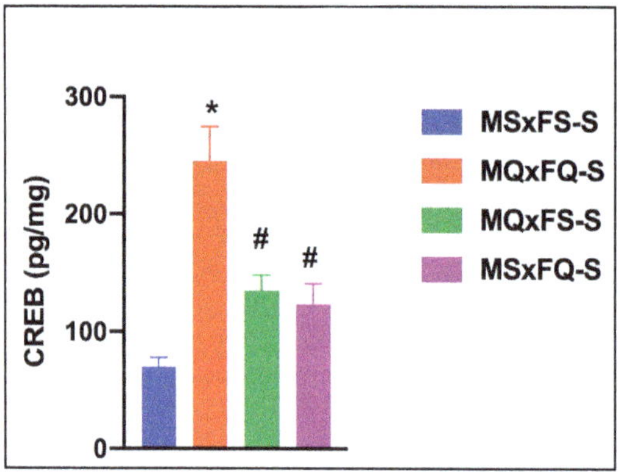

Figure 10. CREB is presented as a function of group for F1-generation vehicle-treated founder crosses. The MQxFQ-S group was significantly higher than all other groups (* indicates $p < 0.05$), and the MQxFS-S and MSxFQ-S groups were significantly higher than the MSxFS-S group (# indicates $p < 0.05$). This result supports the comparison of PPI across founder crosses, in that the MQxFQ-S group demonstrated the highest level of CREB protein in the NAc, as well as the most prominent deficits in PPI across the different founder crosses.

4. Discussion

This study demonstrates that stimulation of either the A(2A) or mGlu5 receptor results in an attenuation of sensorimotor gating deficits and a reduction in elevated CREB in the NAc in an established and validated heritable rodent model of psychosis. In experiment 1, administration of CGS21680 alleviated PPI deficits and significant increases in CREB in the NAc, and in experiment 2, administration of CDPPB was also showed efficacy in alleviating sensorimotor gating deficits in this model. This study is the first to show these findings in a heritable model of psychosis, with phenotypes that are consistent with SZ presenting in the offspring of NQ-treated animals. MQxFQ rats demonstrated the most robust deficits in PPI compared to all other founder crosses; aside from the MSxFQ rats, deficits were equivalent. This enhanced deficit is consistent with clinical genetic findings that indicate significantly increased risk of development of psychosis if an individual is the offspring of two affected parents [38]. In addition, the MQxFQ cross also demonstrated the most prominent increase in accumbal CREB, which may also be related to findings from RNAseq analysis in our previous work that reported that this group demonstrated a significant increase in the cortisol synthesis pathway in the dorsal striatum [21]. The findings here generally indicate that both treatment targets, the adenosine A(2A) receptor and the mGlu5 receptor, demonstrate potential for alleviating sensorimotor gating deficits in a model system of enhanced dopamine D2 receptor sensitivity. We previously reported that CDPPB was also effective in alleviating the enhanced behavioral response to nicotine in conditioned place preference expression and reinstatement, indicating that this target may have the potential to be effective in both alleviating sensorimotor gating deficits and dampening the rewarding impact of nicotine.

In experiments 1 and 2, among the MQxFQ group, females performed better in PPI overall compared to males, but administration of either dose of CGS21680 (0.03 mg/kg or 0.09 mg/kg) was successful in alleviating deficits in PPI to levels equal to MSxFS controls. This study is the first to demonstrate a sex difference in PPI within Sprague–Dawley rats. Previously, female Wistar rats have been shown to demonstrate more robust PPI deficits than males [39]; thus, this effect appears to be strain-dependent. In addition, the administration of either dose of CGS21680 reduced enhancements in CREB in MQxFQ rats relative to control animals. This is significant, as D_1 and D_2 receptor activation has been associated with CREB activation in the NAc [40]. This may indicate that stimulation of the A(2A) receptor results in a reduction in D_2 signaling by acting through striatal heteromers [41]. Sustained increases in accumbal CREB have also been associated with negative symptom presentation in psychosis [42], which may be related to the sensorimotor gating deficits observed here.

MQxFS rats demonstrated significant PPI deficits compared to the MSxFS control group, but these deficits were not as severe as deficits observed in MQxFQ rats. In experiment 1, both doses of CGS21680 attenuated PPI deficits in male subjects; however, the lower 0.03 mg/kg dose was not effective in females. This finding was mirrored in the MSxFQ epigenetic cross, where both doses of CGS21680 successfully attenuated deficits in males but no differences were found within drug groups among females. In the clinical population, it has been reported that females present 40% lower A(2A) receptor expression at the mRNA level compared to males [43]. This may explain the reduced behavioral response to CGS21680 observed in females in our MQxFS and MSxFQ epigenetic crosses. In Sprague–Dawley rats, no significant sex differences have been found regarding binding affinity and selectivity of the A(2A) receptor [44]. However, this could indicate that the differential expression of these receptors is primarily responsible for the observed sex differences in PPI and not the affinity of the receptor itself. Interestingly, we found that the startle response was significantly reduced in MQxFS rats by the lower 0.03 mg/kg dose of CGS21680, but no differences in startle were observed among other epigenetic crosses.

Among MQxFS and MSxFQ subjects, 0.03 mg/kg and 0.09 mg/kg CGS21680 attenuated significant increases in CREB observed in saline-treated rats. No sex differences were observed in total CREB in experiment 1, a finding previously shown in Sprague–Dawley rats [45]. These findings suggest a significant role of the A(2A) receptor in psychosis pathology and sensorimotor gating deficits. A knockout of the A(2A) receptor has been previously shown to induce PPI deficits [46]. Stimulation of the A(2A) receptor prior to testing restored deficits in adenosine modulation and sensorimotor gating previously associated with negative and cognitive symptoms of SZ [47]. Additionally, the consistent restoration of CREB levels to those of controls in the NAc across sex and epigenetic crosses suggests that CREB-linked emotional reactivity may be heavily involved in sensorimotor gating responses in psychosis. Future studies will seek to analyze changes in A(2A)-D_2 heteromers in the striatum in the F1 generation because past work in post-mortem SZ brain tissue has shown reductions in D2-A(2A) heteromers in the caudate nucleus [48]. The CREB response to restraint stress and nicotine reinstatement will additionally be explored to determine the role of CREB in drug-induced plasticity and emotional responses to stress.

In experiment 2, the mGlu5 PAM CDPPB attenuated PPI deficits and normalized enhanced CREB in the NAc but with less overall efficacy than CGS21680. Sex differences were observed within all epigenetic crosses in experiment 2. As in experiment 1, MQxFQ females performed better than males. Interestingly, the 10 mg/kg dose of CDPPB was more effective in female MQxFQs, whereas the 30 mg/kg dose was more effective in males. Clinically, females have demonstrated increased levels of glutamate in the striatum compared to males [49]. It has been shown mGlu5 receptors, in particular, are expressed at higher levels in females [50]. This may indicate that females are more sensitive to glutamate and explain why the lower dose of CDPPB was more effective in PPI. The 30 mg/kg dose of CDPPB may not be as efficacious in females because of the increased number of mGlu5 receptors in females, but this has yet to be determined. In males, as expected, the 10 and 30 mg/kg doses of CDPPB were effective in reducing enhanced levels of CREB in the NAc. Females demonstrated higher levels of total CREB, and CDPPB was not as effective. This enhanced baseline increase in females may be responsible for the lack of effectiveness of CDPPB in reducing CREB to control levels. Female rats have previously demonstrated enhanced excitatory signaling in the NAc compared to males [51], which may be related to this effect. Increased signaling in the D_2-rich mesolimbic pathway may also play a role in generating increases in CREB, as in vitro quinpirole treatment has been shown to enhance the expression and activation of CREB [52].

A(2A) agonist CGS21680 demonstrated greater efficacy than CDPPB in reducing accumbal CREB across the founder crosses. The precise mechanism of this effect is unknown. However, it is important to note that A(2A) agonist CGS 21680 has a dopamine D2 receptor antagonistic-like action, in that it indirectly decreases dopamine D2 signaling. Past work has shown that dopamine D2 antagonist sulpiride slows increases in phosphorylated CREB produced by D1 agonist SKF 38393 in the caudate nucleus [53]. Interestingly, antipsychotic clozapine, given alone, was shown to decrease phosphorylation of CREB in the mouse dorsal striatum, which may be related to its ability to inhibit ERK1/2 [54] and reduce intracellular calcium levels. Using in vitro preparations, Fuxe and colleagues reported that CGS 21680 was effective at reducing increases in intracellular calcium induced by quinpirole [55]. On the other hand, CDPPB, as a positive allosteric modulator of mGlu5, likely works through a different mechanism. A 10 mg/kg dose of CDPPB has been shown to increase the phosphorylation of CREB in the prefrontal cortex and hippocampus in past work, but a 30 mg/kg dose of CDPPB was not shown to have any effect [37]. This is the first study to analyze the effect of CDPPB on a system in which CREB is elevated, as well as the effects of CDPPB specifically on CREB in the NAc.

The MQxFS epigenetic cross demonstrated similar results to MQxFQ rats in PPI, although the CREB levels in the NAc were reduced compared to the MQxFQ group. Sex differences were noted in both crosses. Whereas the 10 mg/kg dose of CDPPB was generally ineffective across the male and female MQxFS groups, the 30 mg/kg dose was far more effective in male animals. Interestingly, males demonstrated an enhanced startle response that was significantly reduced by the 30 mg/kg dose of CDPPB. Male Long–Evans rats have previously demonstrated enhancements in acoustic startle response [37]. The dose dependency of CDPPB has been previously explored in novel object recognition memory, with lower doses displaying greater efficacy [37]. Our findings demonstrate that the dose dependency of CDPPB is not only affected by the behavioral task but also by sex and founder cross. CREB levels in the NAc of the MQxFS group support our original hypothesis, as both 10 and 30 mg/kg of CDPPB reduced CREB to control levels. Both doses were successful in attenuating aberrant CREB levels, likely due to the D2-genotype/phenotype contributions coming from only one founder. Clinically, having only one parent diagnosed with psychosis, as opposed to two, decreases the chance of illness development by over 20% [56].

MSxFQ rats, likewise, demonstrated sex differences in PPI, with neither dose of CDPPB alleviating PPI deficits observed in saline-treated animals. PPI deficits in MSxFQ males were rescued by the 30 mg/kg dose of CDPPB. Interestingly, MSxFQ males given saline did not demonstrate PPI deficits at 82 dB. Our laboratory has previously hypothesized that the male parent contributes more to the psychosis phenotype via epigenetics, as MSxFQ rats consistently perform better than their MQxFS counterparts. One such instance of increased risk due to paternal genetic contributions arises when the male is significantly older. Specific mutant clone lines expand as men age, which can contribute to changes in neurodevelopment, which increases risk of neuropsychiatric disorders [57]. MSxFQ rats given 30 mg/kg CDPPB demonstrated significantly reduced CREB in the NAc; however, the 10 mg/kg dose failed to lower CREB. In summary, mGlu5 PAM CDPPB demonstrated reduced efficacy compared to CGS21680 across all epigenetic crosses in both PPI and levels of CREB. Adenosine is heavily involved in PPI, primarily acting through A(2A) receptors in the NAc [58], suggesting that CGS21680 may be beneficial in alleviating negative cognitive symptoms of SZ. Our laboratory has previously demonstrated the effectiveness of CDPPB in PPI as well but in F0-generation rats that were neonatally treated with quinpirole [13]. Future studies will aim to evaluate the genetic contribution of the male spermatozoa in our epigenetic model of psychosis. Additionally, behavioral changes and responses to CDPPB will be analyzed in varying estrous cycles to determine if this played a role in the numerous sex differences observed in experiment 1.

While our results suggest a positive outlook on the stimulation of the adenosine and glutamate systems in treating psychosis, clinical shortcomings within these areas have been previously documented [59,60]. However, great interest and success in targeting the mGlu5 and A(2A) receptors persists, as the etiology of psychosis remains poorly understood and current antipsychotics continue to induce intolerable side effects. It has recently been shown that CDPPB is effective in restoring cognitive function in an MK-801 model of SZ via reduction of extracellular glutamate in the medial prefrontal cortex [61]. Additionally, CGS21680 has been shown to alleviate antipsychotic-induced tardive dyskinesia in mice [62]. This finding reveals alternative roles for A(2A) stimulation in the management of antipsychotic side effects. The complexity and vast symptomology of psychosis contribute to the difficulty of engineering an effective treatment. With several neurotransmitters and neuromodulators involved in the presentation of symptoms, it is valuable to continue targeting mGlu5 and A(2A) receptors as we further elucidate aspects of psychosis that can contribute to improved patient outcomes.

In conclusion, this study revealed that CGS21680 and CDPPB are effective in rescuing PPI deficits in a transgenerational model of psychosis. However, CGS21680 demonstrated enhanced efficacy over CDPPB; in many cases, only one dose of the drug was successful in alleviating sensorimotor gating deficits. Both drugs additionally reduced CREB in our model of psychosis, with CGS21680 again demonstrating enhanced effectiveness. Several sex differences across groups arose within experiments 1 and 2, which we hypothesize were due to differences in sensitivity to mGlu5 and A(2A) receptors across males and females. These findings have strong implications for both the adenosine and glutamate systems in the treatment of psychosis and raise interest in further analyzing striatal heteromers and their role in psychosis. In addition, an important finding here is the level of severity, which differs across these founder crosses, with the most severe deficit in PPI and the most prominent effect on CREB being present in the MQxFQ cross. This is consistent with clinical work, in that levels of severity do appear to vary across individuals diagnosed with psychosis [63,64], providing further validity for this model. Overall, future work will focus on further modulating these systems to elucidate precise neurocircuitry; reduce supersensitized D_2 signaling; and, in turn, improve health outcomes for this vulnerable population.

Author Contributions: Conceptualization, R.W.B.; methodology, R.W.B. and J.T.G.; formal analysis, R.W.B.; investigation, A.M.C., L.D.P., C.D.A. and L.J.W.; resources, R.W.B.; data curation, A.M.C., L.D.P. and L.J.W.; writing—original draft preparation, A.M.C.; writing—review and editing, R.W.B., L.D.P., C.D.A. and L.J.W.; visualization, R.W.B., L.D.P., A.M.C., C.D.A. and L.J.W.; supervision, R.W.B. and J.T.G.; project administration, A.M.C., L.D.P. and R.W.B.; funding acquisition, R.W.B. All authors have read and agreed to the published version of the manuscript.

Funding: This research was funded by NIH grant R15DA046926 awarded to RWB and VA grant 1I01 BX005367-01A2 to JTG.

Institutional Review Board Statement: The animal study protocol was approved by the Institutional Animal Care and Use Committee (IACUC) of East Tennessee State University (protocol code P220102-VA-119; approved 8 February 2022) for studies involving animals.

Informed Consent Statement: Not applicable.

Data Availability Statement: The data that support the findings of the current study are available from the corresponding author upon reasonable request.

Conflicts of Interest: The authors declare no conflicts of interest.

References

1. Dunwiddie, T.V.; Masino, S.A. The Role and Regulation of Adenosine in the Central Nervous System. *Annu. Rev. Neurosci.* **2001**, *24*, 31–55. [CrossRef] [PubMed]
2. Cunha, R.A. How Does Adenosine Control Neuronal Dysfunction and Neurodegeneration? *J. Neurochem.* **2016**, *139*, 1019–1055. [CrossRef] [PubMed]
3. Shen, H.-Y.; Chen, J.-F. Adenosine A2A Receptors in Psychopharmacology: Modulators of Behavior, Mood and Cognition. *Curr. Neuropharmacol.* **2009**, *7*, 195. [CrossRef] [PubMed]
4. Brown, R.W.; Bhide, P.G.; Gill, W.D.; Peeters, L.D. The Adenosine A(2A) Receptor Agonist CGS 21680 Alleviates Auditory Sensorimotor Gating Deficits and Increases in Accumbal CREB in Rats Neonatally Treated with Quinpirole. *Psychopharmacology* **2020**, *237*, 3519. [CrossRef]
5. Mena, A.; Ruiz-Salas, J.C.; Puentes, A.; Dorado, I.; Ruiz-Veguilla, M.; De la Casa, L.G. Reduced Prepulse Inhibition as a Biomarker of Schizophrenia. *Front. Behav. Neurosci.* **2016**, *10*, 202. [CrossRef] [PubMed]
6. Brisch, R.; Saniotis, A.; Wolf, R.; Bielau, H.; Bernstein, H.-G.; Steiner, J.; Bogerts, B.; Braun, K.; Jankowski, Z.; Kumaratilake, J.; et al. The Role of Dopamine in Schizophrenia from a Neurobiological and Evolutionary Perspective: Old Fashioned, but Still in Vogue. *Front. Psychiatry* **2014**, *5*, 47. [CrossRef]
7. Kesby, J.P.; Eyles, D.W.; McGrath, J.J.; Scott, J.G. Dopamine, Psychosis and Schizophrenia: The Widening Gap between Basic and Clinical Neuroscience. *Transl. Psychiatry* **2018**, *8*, 30. [CrossRef] [PubMed]
8. Singer, P.; Yee, B.K. The Adenosine Hypothesis of Schizophrenia into Its Third Decade: From Neurochemical Imbalance to Early Life Etiological Risks. *Front. Cell. Neurosci.* **2023**, *17*, 1120532. [CrossRef] [PubMed]
9. Boison, D.; Singer, P.; Shen, H.-Y.; Feldon, J.; Yee, B.K. Adenosine Hypothesis of Schizophrenia –Opportunities for Pharmacotherapy. *Neuropharmacology* **2011**, *62*, 1527. [CrossRef] [PubMed]
10. Institute of Medicine, Board on Health Sciences Policy, Forum on Neuroscience and Nervous System Disorders. Overview of the Glutamatergic System. In *Glutamate-Related Biomarkers in Drug Development for Disorders of the Nervous System: Workshop Summary*; National Academies Press (US): Cambridge, MA, USA, 2011.
11. Budgett, R.F.; Bakker, G.; Sergeev, E.; Bennett, K.A.; Bradley, S.J. Targeting the Type 5 Metabotropic Glutamate Receptor: A Potential Therapeutic Strategy for Neurodegenerative Diseases? *Front. Pharmacol.* **2022**, *13*, 893422. [CrossRef] [PubMed]
12. Matosin, N.; Newell, K.A. Metabotropic Glutamate Receptor 5 in the Pathology and Treatment of Schizophrenia. *Neurosci. Biobehav. Rev.* **2013**, *37*, 256–268. [CrossRef]
13. Brown, R.W.; Varnum, C.G.; Wills, L.J.; Peeters, L.D.; Gass, J.T. Modulation of mGlu5 Improves Sensorimotor Gating Deficits in Rats Neonatally Treated with Quinpirole through Changes in Dopamine D2 Signaling. *Pharmacol. Biochem. Behav.* **2021**, *211*, 173292. [CrossRef] [PubMed]
14. Mei, Y.-Y.; Wu, D.C.; Zhou, N. Astrocytic Regulation of Glutamate Transmission in Schizophrenia. *Front. Psychiatry* **2018**, *9*. [CrossRef] [PubMed]
15. Wang, H.-Y.; MacDonald, M.L.; Borgmann-Winter, K.E.; Banerjee, A.; Sleiman, P.; Tom, A.; Khan, A.; Lee, K.-C.; Roussos, P.; Siegel, S.J.; et al. mGluR5 Hypofunction Is Integral to Glutamatergic Dysregulation in Schizophrenia. *Mol. Psychiatry* **2018**, *25*, 750. [CrossRef] [PubMed]
16. Janus, A.; Lustyk, K.; Pytka, K. MK-801 and Cognitive Functions: Investigating the Behavioral Effects of a Non-Competitive NMDA Receptor Antagonist. *Psychopharmacology* **2023**, *240*, 2435–2457. [CrossRef]
17. Azargoonjahromi, A. Current Findings and Potential Mechanisms of KarXT (Xanomeline-Trospium) in Schizophrenia Treatment. *Clin. Drug Investig.* **2024**, *44*, 471–493. [CrossRef]
18. Ferré, S.; Karcz-Kubicha, M.; Hope, B.T.; Popoli, P.; Burgueño, J.; Gutiérrez, M.A.; Casadó, V.; Fuxe, K.; Goldberg, S.R.; Lluis, C.; et al. Synergistic Interaction between Adenosine A2A and Glutamate mGlu5 Receptors: Implications for Striatal Neuronal Function. *Proc. Natl. Acad. Sci. USA* **2002**, *99*, 11940–11945. [CrossRef] [PubMed]

19. Beggiato, S.; Tomasini, M.C.; Borelli, A.C.; Borroto-Escuela, D.O.; Fuxe, K.; Antonelli, T.; Tanganelli, S.; Ferraro, L. Functional Role of Striatal A2A, D2, and mGlu5 Receptor Interactions in Regulating Striatopallidal GABA Neuronal Transmission. *J. Neurochem.* **2016**, *138*, 254–264. [CrossRef]
20. Kostrzewa, R.M.; Wydra, K.; Filip, M.; Crawford, C.A.; McDougall, S.A.; Brown, R.W.; Borroto-Escuela, D.O.; Fuxe, K.; Gainetdinov, R.R. Dopamine D2 Receptor Supersensitivity as a Spectrum of Neurotoxicity and Status in Psychiatric Disorders. *J. Pharmacol. Exp. Ther.* **2018**, *366*, 519–526. [CrossRef] [PubMed]
21. Gill, W.D.; Burgess, K.C.; Vied, C.; Brown, R.W. Transgenerational Evidence of Increases in Dopamine D2 Receptor Sensitivity in Rodents: Impact on Sensorimotor Gating, the Behavioral Response to Nicotine and BDNF. *J. Psychopharmacol.* **2021**, *35*, 1188–1203. [CrossRef]
22. Brown, R.W.; Maple, A.M.; Perna, M.K.; Sheppard, A.B.; Cope, Z.A.; Kostrzewa, R.M. Schizophrenia and Substance Abuse Comorbidity: Nicotine Addiction and the Neonatal Quinpirole Model. *Dev. Neurosci.* **2012**, *34*, 140–151. [CrossRef]
23. Kostrzewa, R.M.; Nowak, P.; Brus, R.; Brown, R.W. Perinatal Treatments with the Dopamine D2-Receptor Agonist Quinpirole Produces Permanent D2-Receptor Supersensitization: A Model of Schizophrenia. *Neurochem. Res.* **2016**, *41*, 183–192. [CrossRef] [PubMed]
24. Ding, J.B.; Hu, K. Cigarette Smoking and Schizophrenia: Etiology, Clinical, Pharmacological, and Treatment Implications. *Schizophr. Res. Treat.* **2021**, *2021*, 7698030. [CrossRef] [PubMed]
25. Peeters, L.D.; Wills, L.J.; Cuozzo, A.M.; Ahmed, C.D.; Massey, S.R.; Chen, W.; Chen, Z.; Wang, C.; Gass, J.T.; Brown, R.W. Effects of Positive mGlu5 Modulation on D2 Signaling and Nicotine-Conditioned Place Preference: Mechanisms of Epigenetic Inheritance in a Transgenerational Model of Drug Abuse Vulnerability in Psychosis. *J. Psychopharmacol.* **2024**. [CrossRef]
26. Halberstadt, A.L. Recent Advances in the Neuropsychopharmacology of Serotonergic Hallucinogens. *Behav. Brain Res.* **2015**, *277*, 99–120. [CrossRef] [PubMed]
27. Wan, F.J.; Taaid, N.; Swerdlow, N.R. Do D1/D2 Interactions Regulate Prepulse Inhibition in Rats? *Neuropsychopharmacology* **1996**, *14*, 265–274. [CrossRef]
28. Braff, D.L. Prepulse Inhibition of the Startle Reflex: A Window on the Brain in Schizophrenia. *Curr. Top. Behav. Neurosci.* **2010**, *4*, 349–371. [CrossRef] [PubMed]
29. Powell, S.B.; Zhou, X.; Geyer, M.A. Prepulse Inhibition and Genetic Mouse Models of Schizophrenia. *Behav. Brain Res.* **2009**, *204*, 282–294. [CrossRef]
30. Yamashima, T. 'PUFA–GPR40–CREB Signaling' Hypothesis for the Adult Primate Neurogenesis. *Prog. Lipid Res.* **2012**, *51*, 221–231. [CrossRef] [PubMed]
31. Ren, X.; Rizavi, H.S.; Khan, M.A.; Bhaumik, R.; Dwivedi, Y.; Pandey, G.N. Alteration of Cyclic-AMP Response Element Binding Protein in the Postmortem Brain of Subjects with Bipolar Disorder and Schizophrenia. *J. Affect. Disord.* **2014**, *152–154*, 326–333. [CrossRef]
32. Barrot, M.; Olivier, J.D.A.; Perrotti, L.I.; DiLeone, R.J.; Berton, O.; Eisch, A.J.; Impey, S.; Storm, D.R.; Neve, R.L.; Yin, J.C.; et al. CREB Activity in the Nucleus Accumbens Shell Controls Gating of Behavioral Responses to Emotional Stimuli. *Proc. Natl. Acad. Sci. USA* **2002**, *99*, 11435–11440. [CrossRef] [PubMed]
33. Bahlinger, K.; Lincoln, T.M.; Clamor, A. Do Deficits in Subjective Stress Recovery Predict Subsequent Stress Sensitivity and Symptoms in Schizophrenia Spectrum Disorders? *Schizophr. Res.* **2024**, *264*, 170–177. [CrossRef]
34. Li, L.; Du, Y.; Li, N.; Wu, X.; Wu, Y. Top–down Modulation of Prepulse Inhibition of the Startle Reflex in Humans and Rats. *Neurosci. Biobehav. Rev.* **2009**, *33*, 1157–1167. [CrossRef]
35. Kostrzewa, R.M. Dopamine receptor supersensitivity. *Neurosci. Biobehav. Rev.* **1995**, *19*, 1–17. [CrossRef] [PubMed]
36. Maple, A.M.; Smith, K.J.; Perna, M.K.; Brown, R.W. Neonatal Quinpirole Treatment Produces Prepulse Inhibition Deficits in Adult Male and Female Rats. *Pharmacol. Biochem. Behav.* **2015**, *137*, 93–100. [CrossRef] [PubMed]
37. Uslaner, J.M.; Parmentier-Batteur, S.; Flick, R.B.; Surles, N.O.; Lam, J.S.H.; McNaughton, C.H.; Jacobson, M.A.; Hutson, P.H. Dose-Dependent Effect of CDPPB, the mGluR5 Positive Allosteric Modulator, on Recognition Memory Is Associated with GluR1 and CREB Phosphorylation in the Prefrontal Cortex and Hippocampus. *Neuropharmacology* **2009**, *57*, 531–538. [CrossRef]
38. Kahn, R.S.; Sommer, I.E.; Murray, R.M.; Meyer-Lindenberg, A.; Weinberger, D.R.; Cannon, T.D.; O'Donovan, M.; Correll, C.U.; Kane, J.M.; van Os, J.; et al. Schizophrenia. *Nat. Rev. Dis. Primers* **2015**, *1*, 15067. [CrossRef] [PubMed]
39. Monte, A.S.; Mello, B.S.F.; Borella, V.C.M.; da Silva Araujo, T.D.S.; Da Silva, F.E.R.; De Sousa, F.C.F.; De Oliveira, A.C.P.; Gama, C.S.; Seeman, M.V.; Vasconcelos, S.M.M.; et al. Two-Hit Model of Schizophrenia Induced by Neonatal Immune Activation and Peripubertal Stress in Rats: Study of Sex Differences and Brain Oxidative Alterations. *Behav. Brain Res.* **2017**, *331*, 30–37. [CrossRef] [PubMed]
40. Blendy, J.A. The Role of CREB in Depression and Antidepressant Treatment. *Biol. Psychiatry* **2006**, *59*, 1144–1150. [CrossRef] [PubMed]
41. Ferre, S.; von Euler, G.; Johansson, B.; Fredholm, B.B.; Fuxe, K. Stimulation of High-Affinity Adenosine A2 Receptors Decreases the Affinity of Dopamine D2 Receptors in Rat Striatal Membranes. *Proc. Natl. Acad. Sci. USA* **1991**, *88*, 7238–7241. [CrossRef]

42. Muschamp, J.W.; Veer, A.V.; Parsegian, A.; Gallo, M.S.; Chen, M.; Neve, R.L.; Meloni, E.G.; Carlezon, W.A. Activation of CREB in the Nucleus Accumbens Shell Produces Anhedonia and Resistance to Extinction of Fear in Rats. *J. Neurosci.* **2011**, *31*, 3095–3103. [CrossRef] [PubMed]
43. Lassen, M.L.; Byrne, C.; Sheykhzade, M.; Wissenberg, M.; Hurry, P.K.; Schmedes, A.V.; Kjaer, A.; Hasbak, P. Sex Differences and Caffeine Impact in Adenosine-Induced Hyperemia. *J. Nucl. Med.* **2022**, *63*, 431–437. [CrossRef] [PubMed]
44. Janse van Rensburg, H.D.; Terre'Blanche, G.; Van der Walt, M.M. On the Basis of Sex: Male vs. Female Rat Adenosine A1/A2A Receptor Affinity. *BMC Res. Notes* **2023**, *16*, 165. [CrossRef]
45. Auger, A.P.; Hexter, D.P.; McCarthy, M.M. Sex Difference in the Phosphorylation of cAMP Response Element Binding Protein (CREB) in Neonatal Rat Brain. *Brain Res.* **2001**, *890*, 110–117. [CrossRef]
46. Taura, J.; Valle-León, M.; Sahlholm, K.; Watanabe, M.; Van Craenenbroeck, K.; Fernández-Dueñas, V.; Ferré, S.; Ciruela, F. Behavioral Control by Striatal Adenosine A2A-Dopamine D2 Receptor Heteromers. *Genes Brain Behav.* **2018**, *17*, e12432. [CrossRef] [PubMed]
47. Lin, Y.; Roy, K.; Ioka, S.; Otani, R.; Amezawa, M.; Ishikawa, Y.; Cherasse, Y.; Kaushik, M.K.; Klewe-Nebenius, D.; Zhou, L.; et al. Positive Allosteric Adenosine A2A Receptor Modulation Suppresses Insomnia Associated with Mania- and Schizophrenia-like Behaviors in Mice. *Front. Pharmacol.* **2023**, *14*, 1138666. [CrossRef]
48. Valle-León, M.; Callado, L.F.; Aso, E.; Cajiao-Manrique, M.M.; Sahlholm, K.; López-Cano, M.; Soler, C.; Altafaj, X.; Watanabe, M.; Ferré, S.; et al. Decreased striatal adenosine A2A-dopamine D2 receptor heteromerization in schizophrenia. *Neuropsychopharmacology* **2021**, *46*, 665–672. [CrossRef] [PubMed]
49. Wickens, M.M.; Bangasser, D.A.; Briand, L.A. Sex Differences in Psychiatric Disease: A Focus on the Glutamate System. *Front. Mol. Neurosci.* **2018**, *11*, 197. [CrossRef] [PubMed]
50. Wang, Y.; Ma, Y.; Hu, J.; Cheng, W.; Jiang, H.; Zhang, X.; Li, M.; Ren, J.; Li, X. Prenatal Chronic Mild Stress Induces Depression-like Behavior and Sex-Specific Changes in Regional Glutamate Receptor Expression Patterns in Adult Rats. *Neuroscience* **2015**, *301*, 363–374. [CrossRef]
51. Knouse, M.C.; Deutschmann, A.U.; Nenov, M.N.; Wimmer, M.E.; Briand, L.A. Sex Differences in Pre- and Post-Synaptic Glutamate Signaling in the Nucleus Accumbens Core. *Biol. Sex Differ.* **2023**, *14*, 52. [CrossRef] [PubMed]
52. Yan, Z.; Feng, J.; Fienberg, A.A.; Greengard, P. D_2 Dopamine Receptors Induce Mitogen-Activated Protein Kinase and cAMP Response Element-Binding Protein Phosphorylation in Neurons. *Proc. Natl. Acad. Sci. USA* **1999**, *96*, 11607–11612. [CrossRef]
53. Nijholt, I.; Blank, T.; Ahi, J.; Spiess, J. In Vivo CREB Phosphorylation Mediated by Dopamine and NMDA Receptor Activation in Mouse Hippocampus and Caudate Nucleus. *Gene Expr. Patterns* **2002**, *1*, 101–106. [CrossRef] [PubMed]
54. Pozzi, L.; Håkansson, K.; Usiello, A.; Borgkvist, A.; Lindskog, M.; Greengard, P.; Fisone, G. Opposite Regulation by Typical and Atypical Anti-Psychotics of ERK1/2, CREB and Elk-1 Phosphorylation in Mouse Dorsal Striatum. *J. Neurochem.* **2003**, *86*, 451–459. [CrossRef]
55. Yang, S.N.; Dasgupta, S.; Lledo, P.M.; Vincent, J.D.; Fuxe, K. Reduction of Dopamine D2 Receptor Transduction by Activation of Adenosine A2a Receptors in Stably A2a/D2 (Long-Form) Receptor Co-Transfected Mouse Fibroblast Cell Lines: Studies on Intracellular Calcium Levels. *Neuroscience* **1995**, *68*, 729–736. [CrossRef]
56. Lo, L.E.; Kaur, R.; Meiser, B.G.; Green, M.J. Risk of Schizophrenia in Relatives of Individuals Affected by Schizophrenia: A Meta-Analysis. *Psychiatry Res.* **2020**, *286*, 112852. [CrossRef]
57. Malaspina, D.; Gilman, C.; Kranz, T.M. Paternal Age and Mental Health of Offspring. *Fertil. Steril.* **2015**, *103*, 1392–1396. [CrossRef] [PubMed]
58. Hauber, W.; Koch, M. Adenosine A2a Receptors in the Nucleus Accumbens Modulate Prepulse Inhibition of the Startle Response. *Neuroreport* **1997**, *8*, 1515–1518. [CrossRef]
59. Bugarski-Kirola, D.; Blaettler, T.; Arango, C.; Fleischhacker, W.W.; Garibaldi, G.; Wang, A.; Dixon, M.; Bressan, R.A.; Nasrallah, H.; Lawrie, S.; et al. Bitopertin in Negative Symptoms of Schizophrenia—Results From the Phase III FlashLyte and DayLyte Studies. *Biol. Psychiatry* **2017**, *82*, 8–16. [CrossRef]
60. Weiser, M.; Gershon, A.A.; Rubinstein, K.; Petcu, C.; Ladea, M.; Sima, D.; Podea, D.; Keefe, R.S.; Davis, J.M. A randomized controlled trial of allopurinol vs. placebo added on to antipsychotics in patients with schizophrenia or schizoaffective disorder. *Schizophr. Res.* **2012**, *138*, 35–38. [CrossRef] [PubMed]
61. LaCrosse, A.L.; May, C.E.; Griffin, W.C.; Olive, M.F. mGluR5 positive allosteric modulation prevents MK-801 induced increases in extracellular glutamate in the rat medial prefrontal cortex. *Neuroscience* **2024**, *555*, 83–91. [CrossRef] [PubMed]
62. Nagaoka, K.; Asaoka, N.; Nagayasu, K.; Shirakawa, H.; Kaneko, S. Enhancement of adenosine A2A signaling improves dopamine D2 receptor antagonist-induced dyskinesia via β-arrestin signaling. *Front. Neurosci.* **2023**, *16*, 1082375. [CrossRef] [PubMed]

63. Yamada, Y.; Matsumoto, M.; Iijima, K.; Sumiyoshi, T. Specificity and Continuity of Schizophrenia and Bipolar Disorder: Relation to Biomarkers. *Curr. Pharm. Des.* **2020**, *26*, 191–200. [CrossRef] [PubMed]
64. Stentebjerg-Olesen, M.; Pagsberg, A.K.; Fink-Jensen, A.; Correll, C.U.; Jeppesen, P. Clinical Characteristics and Predictors of Outcome of Schizophrenia-Spectrum Psychosis in Children and Adolescents: A Systematic Review. *J. Child. Adolesc. Psychopharmacol.* **2016**, *26*, 410–427. [CrossRef] [PubMed]

cells MDPI

Article

A Novel ALDH2 Inhibitor for the Treatment of Alcohol Use Disorder: Preclinical Findings

Randall D. Marshall *, Andrew Fowlie and Adam Sabouni

Sophrosyne Pharmaceuticals Inc., 540 W. Madison Street, Suite 2500, Chicago, IL 60661, USA; afowlie@sophrosynepharma.com (A.F.)
* Correspondence: rmarshallmd@sophrosynepharma.com

Highlights

What are the main findings?

- SOPH-110S is a novel ALDH2 inhibitor intended to be developed as a long-acting injectable for the treatment of alcohol use disorder
- Preclinical studies that supported an approved IND in the US demonstrated high potency, acceptable safety, no off-target effects in a panel of 84 common targets, no CYP interactions, and positive proof-of-mechanism data in rats.

What are the implications of the main findings?

- SOPH-110S has preclinical promise supporting clinical evaluation.
- SOPH-110S data, to date, support its potential as a safer, better-tolerated, long-acting injectable therapy for AUD that is superior to oral disulfiram.

Abstract

Background: Alcohol use disorder is a common condition with high morbidity and mortality and no highly effective treatments. Achieving and maintaining abstinence is necessary or desired for many persons with AUD, but is difficult due to the nature of the condition. Pharmacologic inhibition of the enzyme ALDH2, which increases levels of the substrate acetaldehyde when alcohol is imbibed, can serve as a powerful enforcer of efforts to remain abstinent. Disulfiram is an approved ALDH2 inhibitor via its active metabolite DETC-MeSO, but has many limitations, including numerous adverse effects, hepatotoxicity, oral administration, and unpredictable mechanistic activity. Methods: SOPH-110S, an analog of DETC-MeSO, was evaluated in a series of experiments to assess mechanism, pharmacokinetics in male beagle dogs, cardiovascular safety in telemeterized male beagle dogs, selectivity, off-target activity, CYP inhibition, and proof of mechanism in a rat model that included dosing and alcohol challenge followed by analysis of liver ALDH2 inhibition. Results: SOPH-110S showed high potency with a comparable IC_{50} vs. positive controls and no physiologically relevant off-target binding in an 84-target panel. It did not inhibit or induce any major CYP enzymes or meaningfully inhibit the hERG channel. After 10 days' dosing in rats, followed by administration of alcohol, SOPH-110S was a highly potent, dose-dependent inhibitor of ALDH2, comparable to DETC-MeSO. No cardiovascular safety concerns were found at multiples above expected clinical doses. Conclusions: The preclinical data support further clinical study of SOPH-110S as a potential ALDH2 inhibitor treatment for AUD. The FDA approved the IND to conduct a first-in-man phase 1 study in September 2025.

Cells **2026**, *15*, 123 https://doi.org/10.3390/cells15020123

Keywords: sophrosyne; SOPH-110S; ALDH2 inhibitor; disulfiram; alcohol use disorder; AUD; DETC-MeSO; alcohol; ethanol

1. Introduction

Alcohol use disorder (AUD) is a chronic disorder characterized by an impaired ability to stop or control alcohol use despite adverse social, occupational, or health consequences, with an estimated incidence of 3.7% of adults globally. AUD was responsible for an estimated 2.6 million deaths globally in 2019, and for 13% of all deaths among persons 20–39 years of age [1].

It is the most common substance use disorder in the US, with an estimated prevalence in 2022 of 28.1 million adults (10.9% in the 18-year and older age group) and 757,000 adolescents (2.9% in the youth ages 12 to 17 age group) [2]. In US public facilities alone, which only represent about half of all treatment facilities (17,561/28,900), 440,540 individuals received inpatient treatment for AUD in 2022 [2,3]. Thus, it is estimated that nearly 900,000 patients per year receive inpatient treatment for AUD in the US. Based on clinical practice, it is expected that the majority are discharged with the explicit goal of long-term abstinence.

The relative risk for mortality increases across the spectrum of daily alcohol intake, beginning from low-volume drinking to heavy drinking [4]. Chronic heavy drinking is also associated with increased risk for cancers, pancreatitis, and the spectrum of alcohol-related liver disease, which includes steatosis, fibrosis, alcohol-associated hepatitis, and cirrhosis [5]. It is estimated that the prevalence of alcohol-associated liver disease is 3.5–4.8% in the general population, and 51% of all patients with AUD [5,6].

AUD has a strong biological component, associated with high heritability [7]: 25–30% of persons with an alcoholic parent also develop AUD. The specific mechanisms of heritability are largely unknown and under active investigation, involve hundreds of variants across the genome, and may involve gene influences on systems such as neurotransmission, reward-related learning and memory, and the stress response [8,9]. Persons with AUD have a lower level of clinical response to alcohol and a reduced size of brain neuroanatomy associated with executive functions (dorsolateral prefrontal cortex), risk taking, and decision making (insular cortex) [8]. In fact, negative associations between alcohol intake and brain macrostructure and microstructure are already apparent in individuals consuming an average of only one to two daily units and become stronger as alcohol intake increases [10]. Other neurobiological findings in AUD patients are changes in gene expression over time [11], increased brain activity in the PFC and the anterior thalamus vs. social drinkers [12], higher craving in response to alcohol cues with increased PFC and limbic activity [12], hyper-responsiveness to alcohol-related stimuli [13], and altered striatal dopamine release long-term [14]. With abstinence, some brain volume changes are reversible [15].

A subpopulation, primarily of East Asian descent, demonstrates a flushing reaction similar to the disulfiram-alcohol reaction due to a genetic mutation of chromosome 12q24, which renders ALDH2 nonfunctional in catalyzing acetaldehyde. In vivo measurement of ALDH2 activity in the liver of heterozygotes was found to be 17% of that of homozygotes [16]. No catalytic activity is found in homozygotes.

Patients with AUD are also one of the most underserved populations with psychiatric disorders. In the US, only 7.3% of persons with AUD received treatment of any kind, and only 1.6% were prescribed a medication to treat AUD [17]. After attempting abstinence, approximately 70–80% of patients relapse within a year of treatment [18].

https://doi.org/10.3390/cells15020123

Despite the considerable unmet medical need, there are only four approved drug therapies for AUD with three distinct mechanisms, approved decades ago, all of which have major limitations in efficacy and/or safety: acamprosate [19], naltrexone oral/naltrexone ER (injectable) [19,20], and disulfiram [21,22]. Importantly, acamprosate and naltrexone do not rigorously support abstinence, as both can be taken even as the patient continues to drink alcohol.

For many individuals, abstinence from alcohol is the optimal goal. Disulfiram is an oral treatment for AUD discovered to have an aversive effect when alcohol is imbibed in the 1940s, called the disulfiram–alcohol reaction (DAR) in the literature. Alcohol is metabolized by alcohol dehydrogenase to acetaldehyde, which is then metabolized by aldehyde dehydrogenase to acetic acid and eventually to carbon dioxide and water. Disulfiram is metabolized by CYP-P450 enzymes to many metabolites, including its primary active metabolite, S-methyl N,N-diethylthiolcarbamate (DETC-MeSO), which irreversibly binds and inhibits the ALDH2 enzyme [23,24]. In contrast with disulfiram, which is dosed orally as 250 mg/day, the active metabolite DETC-MeSO has very high potency.

When persons taking disulfiram drink alcohol, the build-up of the toxic substrate acetaldehyde generates a range of adverse effects that can include general discomfort, flushing, drowsiness, headache, blurry vision, sweating, nausea or vomiting, shortness of breath, irregular heartbeat, weakness or fainting, and confusion. The reaction may last for a few hours.

Disulfiram can be effective when used as prescribed [21], but it is an oral drug with documented high nonadherence in clinical trials [22]. Its clinical use is further constrained by hepatotoxicity in a hepatically compromised population, oral dosing requirements, multiple drug–drug interactions, and poor adherence [22]. Adverse effects, including hepatitis, neuropathy [25,26], and rare fatalities, have further limited adoption.

SOPH-110S is a novel ALDH2 inhibitor analog of DETC-MeSO under development for AUD as an injectable formulation. To support its clinical safety and efficacy testing, SOPH-110S was evaluated in a series of nonclinical regulatory-compliant experiments, as appropriate, to assess the mechanism, pharmacokinetics, safety, selectivity, off-target activity, CYP inhibition, and proof-of-concept efficacy in a rat model that included dosing and alcohol challenge followed by analysis of liver ALDH2 inhibition.

2. Materials and Methods

SOPH-110S is the S enantiomer of a thiocarbamate-S-oxide containing a stereogenic sulfur center (Figure 1). Sulfoxides are conformationally stable at room temperature and, therefore, can be separated into pure enantiomers [25]. The initial racemic synthesis of SOPH-110 was performed at Raybow USA, Brevard, NC, USA, and the chiral separation to produce SOPH-110S was performed under the Raybow GMP protocol (Lotus Separations LLC, Princeton, NJ, USA) resulting in >98% chemical purity and >99% chiral purity.

2.1. In Vitro Study Evaluating ALDH2 Inhibition

An in vitro study was performed to evaluate the effects of various compounds and SOPH-110 in both its racemate and S-enantiomer forms on ALDH2 enzyme (Confluence Discovery Technologies, St Louis, MO, USA). Comparison was made with the positive controls known to inhibit ALDH2, i.e., disulfiram and its active metabolite S-methyl-N, N-diethylthiocarbamate sulfoxide (DETC-MeSO).

The assay was conducted close to the optimum pH of the activity peak for ALDH2 (pH 9.0) and at the more physiological pH 6.5.

https://doi.org/10.3390/cells15020123

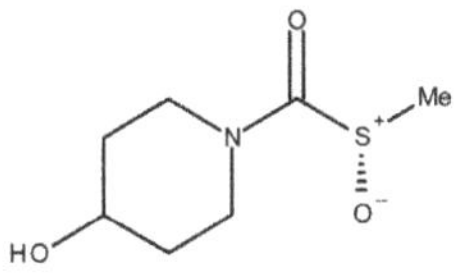

Formula Weight : 191.248(8)
Formula : $C_7H_{13}NO_3S$
Composition : C 43.96% H 6.85% N 7.32% O 25.10% S 16.77%
clogP=-2.61+/- 0.64

Figure 1. Structure and composition of SOPH-110S. The chirality of SOPH-110S was determined by vibrational circular dichroism (VCD) and confirmed by X-ray structure determination. We have a qualified chiral HPLC method to accurately measure any changes in the %ee. We have stability for up to 12 months. Five degrees below 0 centigrade and below is best for storage.

The compounds, reconstituted in a 0.2% final concentration of DMSO, were added to 96-well plates. An ALDH2 enzyme solution was prepared from a 1 mg/mL stock of active recombinant human ALDH2 protein (rhALDH2; Abcam Inc., Waltham, MA, USA; catalog number: ab87415) to a concentration of 30 ng per 25 µL in base buffer (pH 9.0) (plate A) or a 50 mM sodium phosphate buffer (pH 6.5) (plate B) and added to respective wells. The plates were incubated with gentle shaking at room temperature (RT) for two hours. At the end of the two-hour preincubation, an assay mix consisting of acetaldehyde and NAD+ substrates in base buffer was added to catalyze the reaction for conversion of NAD+ to NADH by ALDH2. The plates were incubated for 30 min for enzymatic conversion, and NADH product formation was quantitated by adding the NAD(P)H-Glo detection reagent (Promega Corporation, Madison, WI, USA; Catalog Number: G9062) to each well. The enzyme reductase used NADH to reduce a pro-luciferin substrate to form luciferin. Luciferin was then quantitated using Ultra-Glo Recombinant Luciferase, and the light signal produced (relative light units: RLU) was proportional to the amount of NADH in the sample. Luminescence readings were obtained using an Analyst HT plate reader (Molecular Devices Corporation, Sunnyvale, CA, USA). The luminescence values were converted to pmol NADH product per well using the standard curves. ALDH2 activity was expressed as the specific activity (SA) given the time of reaction (30 min) and the amount of enzyme (30 ng per well). The activity of the tested compound (i.e., SOPH-110S, DETC-MeSO, or disulfiram) was expressed as nmol of NADH formed per min based on a standard curve with NADH. The measurement was conducted using the NADH-Glo assay. The inhibition of the enzyme by each test compound was determined as a percentage of the negative control, which consisted of 30 ng of rhALDH2 without inhibitors (test articles) as 100%.

The CriterionHost software version 2 (LJL Biosystems, Sunnyvale, CA, USA), Excel® 2010 (Microsoft®, Redmond, WA, USA), and GraFit® version 5 or 7 (Erithacus, East Grinstead, West Sussex, UK) were used for data analysis. Graphs were analyzed using a four-parameter logistic model, and IC_{50}s were determined after iterating to the best least-squares fit. The experiment used two (n = 2) wells per condition.

2.2. *Off-Target Binding Assay*

SOPH-110S (10 µM) was assessed in a panel of 84 binding targets (including receptors, ion channels, transporters, kinases, and uptake assays). In each experiment, the respective reference compound was tested concurrently with SOPH-110S, and the data were compared with historical values determined at Eurofins (Celle-Lévescault, France).

https://doi.org/10.3390/cells15020123

Compound binding was calculated as the percent inhibition of the binding of a radioactively labeled ligand specific for each target. The compound enzyme inhibition effect was calculated as the percent inhibition of control enzyme activity. Each assay was performed in duplicate, and the mean of the two duplicates was calculated.

2.3. Inhibition of CYP Enzymes in Human Liver Microsomes

An in vitro study was conducted to determine the potential of SOPH-110S to cause direct or time-dependent inhibition (TDI) of the major drug-metabolizing liver cytochrome P450 enzymes (CYP1A2, 2B6, 2C8, 2C9, 2C19, 2D6, and 3A) using pooled human liver microsomes (HLMs) and standard probe substrates (Frontage Laboratories Inc., Exton, PA, USA). Direct inhibition was assessed at a range of SOPH-110S concentrations (0.05 to 100 μM) by monitoring CYP-specific metabolite formation and calculating the concentration that caused a 50% reduction in the metabolite formation rate (IC_{50}). The potential of SOPH-110S to be a TDI of CYP enzymes was also assessed using the "IC_{50} shift" method, in which the IC_{50} was determined following 30 min pre-incubations of SOPH-110S (0.05 to 100 μM) in pooled HLM with and without NADPH.

2.4. Induction of CYP Enzymes in Plateable Human Hepatocytes

An in vitro study was conducted to determine the potential of SOPH-110S to induce CYP1A2, CYP2B6, and CYP3A4 enzymes (Frontage Laboratories Inc., Monmouth Junction, NJ, USA). This study was conducted using primary human hepatocytes from three donors (BEI, BRX, and ZNE) Omeprazole, phenobarbital, and rifampin were used as the positive controls for CYP1A2, CYP2B6, and CYP3A4, respectively. Flumazenil was used as the negative control (no induction). The mRNA levels were determined by real-time qPCR. The baseline values of mRNA levels were obtained from the hepatocytes treated with a vehicle under the same conditions.

2.5. hERG Current Amplitude Study

A regulatory-compliant study was conducted to examine the effects of SOPH-110S on the hERG current amplitude recorded under physiological temperatures at a nominal concentration of 50 μM (Nova Research Laboratories, New Orleans, LA, USA). Dofetilide (nominal concentrations of 0.5, 10, and 20 nM) was used as a positive control.

The hERG current was recorded from stably expressing cell lines HEK/hERG, which originated from the human kidney (Nova Research Laboratories, New Orleans, LA, USA). Cells were stored in liquid nitrogen, and a new vial was thawed on the experimental day. Currents were measured using the whole-cell variant of the patch clamp method. Glass pipettes were pulled from borosilicate glass by a horizontal puller (Sutter Instruments, Novato, CA, USA). The internal and external recording solutions were composed of NaCl (137 mM), KCl (4.0 mM), $MgCl_2$ (1.0 mM), $CaCl_2$ (1.8 mM), HEPES (10.0 mM), and dextrose (11.0 mM), adjusted to a pH of 7.4 with NaOH. The bath temperature was measured by a thermistor placed near the cell under study and was maintained by a thermoelectric device. All experiments were conducted at 37 ± 1 °C. For SOPH-110S-exposed cells, n = 5 cells per concentration were tested. For the positive control, n = 3, and for the vehicle control, n = 3 cells each were tested.

An Axopatch 1-D amplifier (Axon Instruments/Molecular Devices San Jose, CA, USA) was used for whole-cell voltage clamping. Creation of voltage clamp pulses and data acquisition were controlled by a computer running the pClamp software version 10.7 (Molecular Devices San Jose, CA, USA).

https://doi.org/10.3390/cells15020123

2.6. Cardiovascular and Body Temperature Effects in Dogs

A regulatory-compliant study was conducted to evaluate cardiovascular effects and body temperature in dogs (Attentive Sciences, Overland Park, KS, USA). Four drug-naïve male beagle dogs, 20 to 23 months old, animals previously implanted with telemetry transmitters (Data Sciences International, St. Paul, MN, USA) for cardiovascular monitoring, were selected from the telemetry colony for use in the study based on a health assessment and confirmation of signal. The dogs received the vehicle or SOPH-110S (0.3, 1, or 3 mg/kg) via subcutaneous injection in a Latin square-dosing regimen, such that each animal received all treatments. There was a minimum off-treatment period of seven days between dose administration. The number of animals used in this study was considered the minimum required to achieve the study objectives, based on regulatory requirements, statistical power, and/or availability of historical data, and was within IACUC-approved guidelines. Only male animals were used because no sex differences were anticipated.

Computer hardware and software supplied by Data Sciences International (DSI, St Paul, MN, USA) were used for radio-telemetry data acquisition. The PhysioTel Digital system with M11R implants was used. These implants, connected to the Dataquest™ OpenART™ Acquisition Interface, provide direct digital signals to the DSI PONEMAH software (version 5.50). ECG and ABP waveforms were sampled at a minimum of 500 Hz. Temperature data was sampled at a minimum of 10 Hz. Data acquired continuously were logged every 60 s.

On each day of dosing, radio-telemetry recording began at ~2 h prior to dosing and continued to ~24 h post-dose.

ECG and blood pressure signals may at times be uninterpretable due to interference caused by an animal's activity or body positioning. These waveforms were excluded, but all other ECG waveforms that were recognized by the software were included. Qualitative and quantitative assessment of ECG waveforms was performed by an experienced cardiovascular data analyst.

The data captured included heart rate (HR), systolic arterial pressure (SBP), diastolic arterial pressure (DBP), mean arterial pressure (MAP), body temperature (BT), PR interval, RR interval, QRS duration, QT interval, QtcI (individual) interval, and QTcVW (Van de Water's) interval using the Ponemah Acquisition System.

The following parameters were analyzed post-acquisition on a validated system with EMKA ecgAUTO version 3.5.9 (emka Technologies, Stirling, VA, USA) using a pattern recognition approach: heart rate (HR), systolic arterial pressure (SBP), diastolic arterial pressure (DBP), mean arterial pressure (MAP), body temperature (BT), PR interval, RR interval, QRS interval, QT interval, QTcI (individual), and QTcVW [27]. QTcI interval values were calculated using an individual animal correction factor (IACF). For each animal, the IACF value was calculated from data collected during the vehicle dose day for that animal and applied to each dose level for the same animal. The IACF was generated by fitting a linear regression line to the 1 min means of the QT interval versus heart rate data. QTcI values were calculated using the formula $QTc = QT - IACF \times (HR - 75)$ [28].

The data for each required time point was analyzed and assessed on completion of data acquisition, and was the average of 60 min data bins.

Statistical analysis was performed in 5 phases:

Phase I: Four time intervals of 60 min each for hours 1–4 post-dose.
Phase II: Four time intervals of 60 min each for hours 5–8 post-dose.
Phase III: Six time intervals of 60 min each for hours 9–14 post-dose.
Phase IV: Six time intervals of 60 min each for hours 15–20 post-dose.
Phase V: Four time intervals of 60 min each for hours 21–24 post-dose.

https://doi.org/10.3390/cells15020123

Each parameter was analyzed in the analysis phase with a repeated-measure analysis of covariance (RANCOVA). The factors in the model included baseline (BASE) as a covariate, dosing day (PERIOD), ANIMAL, treatment (TRT), time after dose (TIME), and the interaction of the treatment group and time after dose (TRT*TIME). The SAS® (version 9.4) procedure PROC MIXED was used for analysis, and the covariance structure across time was selected by comparing the corrected Akaike's Information Criterion (AICC) from the following covariance structures: compound symmetric, autoregressive order 1, compound symmetric heterogeneous, and autoregressive heterogeneous order 1. The covariate was defined for each subject as the time-matched baseline observation from the 2 h baseline immediately preceding each dosing day.

Dunnett's test was used to compare the treated groups with the vehicle control. If the TRT*TIME interaction was statistically significant, Dunnett's test was conducted individually for each time interval. Otherwise, Dunnett's test was conducted overall across the pooled time intervals only. All tests were conducted at the 0.05 significance level.

2.7. Pharmacokinetics of SOPH-110S Following Intravenous (IV), Subcutaneous (SQ), and Intramuscular (IM) Administration to Beagle Dogs

The PK parameters of SOPH-110S and its primary metabolite SOPH-120 were determined following a single IV (3 mg/kg), SQ (0.3, 1, and 3 mg/kg), or IM (3 mg/kg) administration to three healthy, adult male Beagle dogs (Attentive Sciences, Overland Park, KS, USA). The same animals (3 dogs) were used to assess PK following SC at 1 and 3 mg/kg, and 3 mg/kg IM with a dosing of at least 7 days between doses. A different set of animals (a total of 6 dogs, 3 dogs for each group) were dosed SQ at 0.3 mg/kg and 3 mg/kg IV.

IV injections were administered into a cephalic vein. SQ injections were administered to the intrascapular region using an appropriately sized needle and syringe. IM injections were delivered into the caudal thigh of a hind limb via an appropriately sized needle and syringe.

Blood samples were collected from the jugular vein at 0.083, 0.25, 0.5, 1, 2, 4, 8, and 24 hr post-dosing (IV) or 0.25, 0.5, 1, 2, 4, 8, 24. and 48 h post-dose (SQ and IM). Blood samples were kept on ice until centrifugation at settings of 2000× g, 10 min, and 5 °C within 1 h of collection. Plasma was directly transferred to tubes and stored at −70 °C ± 10 °C until analysis using a qualified LC/MS/MS analytical method. The limit of quantitation (LLOQ) was 5 ng/mL for SOPH-110S and 10 ng/mL for SOPH-120.

Individual animal plasma SOPH-110S and SOPH-120 concentration versus time data were imported into Phoenix® WinNonlin® version 8.3.4.295 (Certara; Princeton, NJ, USA) for analysis. The concentration versus time data were analyzed using a noncompartmental extravascular administration model with uniform weighting. Nominal sample collection times were used in estimating the TK parameters. Concentrations that were below the LLOQ were treated as missing for the TK analyses.

The following parameters were estimated whenever possible:

- C_{max}: The maximum observed concentration;
- T_{max}: The time to reach the peak concentration;
- T_{last}: The timepoint with the last measurable concentration;
- $AUC_{0\text{-}t}$: The area under the concentration–time curve from time zero to the timepoint with the last measurable concentration, estimated by the linear up/log down trapezoidal method;
- $t_{1/2}$: The terminal half-life.

All parameters were reported to 3 significant figures. $T_{1/2}$ was only reported if the goodness-of-fit parameter, R2, was greater than or equal to 0.9 and the percent of AUC_{inf}

https://doi.org/10.3390/cells15020123

extrapolated to time infinity ($AUC_{\%extrap}$) was less than or equal to 20%. The data are reported as the means of the calculated values for individual animals.

2.8. In Vivo Study of Liver ALDH2 Inhibition in the Presence of Alcohol in Rats

The effects of the SOPH-110S and R enantiomers on liver ALDH2 activity and ALDH2 protein levels following ethanol challenge were evaluated in Sprague-Dawley (SD) rats at Eurofins (New Taipei City, Taiwan). This study was conducted on 8–9-week-old male rats, and various treatments were evaluated, including a negative control (PBS at pH 5.5), a positive control (DETC-MeSO racemate at 6.5 mg/kg), and the two stereoisomeric forms of SOPH-110 (SOPH-110S and SOPH-110R) at 1, 4, and 8 mg/kg. All test materials were injected intravenously (IV) in a volume of 1 mL/rat once daily (QD) from day 1 to day 10 for a total of 10 days (n = 6 rats per group). As another positive control, oral disulfiram at 6.5 mg/kg was also given QD during the same study period. Animals then received 20% ethanol at 1 mL/kg (IP) over 8 h following the 10th dose of the test article. Thirty minutes following ethanol challenge, the animals were sacrificed, and the whole liver was harvested, rinsed in cold saline, and snap-frozen to measure ALDH2 activity (Mitochondrial Aldehyde Dehydrogenase (ALDH2) Activity Assay Kit (abcam ab115348)) and ALDH2 protein (using rat aldehyde dehydrogenase 2 family (mitochondrial) ELISA Kit MBS700110; MyBioSource, San Diego, CA, USA). The mean and SEM were calculated. One-way ANOVA and Dunnett's test were used to determine the statistical significance between the vehicle control and the treated groups. Differences were considered significant at $p < 0.05$ vs. the negative control.

3. Results

3.1. In Vitro Study Evaluating ALDH2 Inhibition: Results

SOPH-110S, DETC-MeSO S-enantiomer, and disulfiram had comparable IC_{50}s to each other (1.46 µM, 1.94 µM, and 1.80 µM, respectively) and to the literature (1.45 µM for disulfiram) [16] under standard assay conditions with pH 9.0 (Table 1 and Figure 2a).

Table 1. IC_{50} values for ALDH2 inhibition by SOPH-110S, DETC-MeSO racemate, DETC-MeSO S enantiomer, and disulfiram at pH values of 9.0 and 6.5.

pH	Compound	IC_{50} (µm)	Standard Error (SE)	Slope	Max.	Min.
9.0	SOPH-110S	1.46	0.27	2.50	102	*
	DETC-MeSO Racemate	3.50	0.59	1.41	114	*
	DETC-MeSO S-enantiomer	1.94	0.50	1.12	95.1	*
	Disulfiram	1.80	0.44	0.92	**	*
6.5	SOPH-110S	2.34	0.796	1.56	79.4	*
	DETC-MeSO Racemate	4.10	0.630	1.28	56.4	*
	DETC-MeSO S-enantiomer	0.657	2.31	0.462	106	*
	Disulfiram	0.463	0.151	1.49	71.7	*

* Minimum set to 0. ** 0 to 100% fit.

The activities were much lower and less reliable using the two-hour preincubation at pH 6.5, followed by an in vitro dehydrogenase assay at pH 7.85 (Table 1). The activity was less than 3% of that in the standard assay (pH 9.0). This was not unexpected based on the known pH activity profile of ALDH2, which peaks at a non-physiological alkalinity near pH 9.0, though it is uncertain whether the loss was due to lower catalytic activity or to protein denaturation. The IC_{50} values, though less accurate, tended to still have comparable relative potencies (Figure 2b), suggesting little relative difference in compound stabilities at a lower pH. The IC_{50}s were 2.34 μM, 4.10 μM, 0.657 μM, and 0.493 μM, respectively. The IC_{50}s for the DETC-MeSO-S-enantiomer and disulfiram were considerably lower than at pH 9.0.

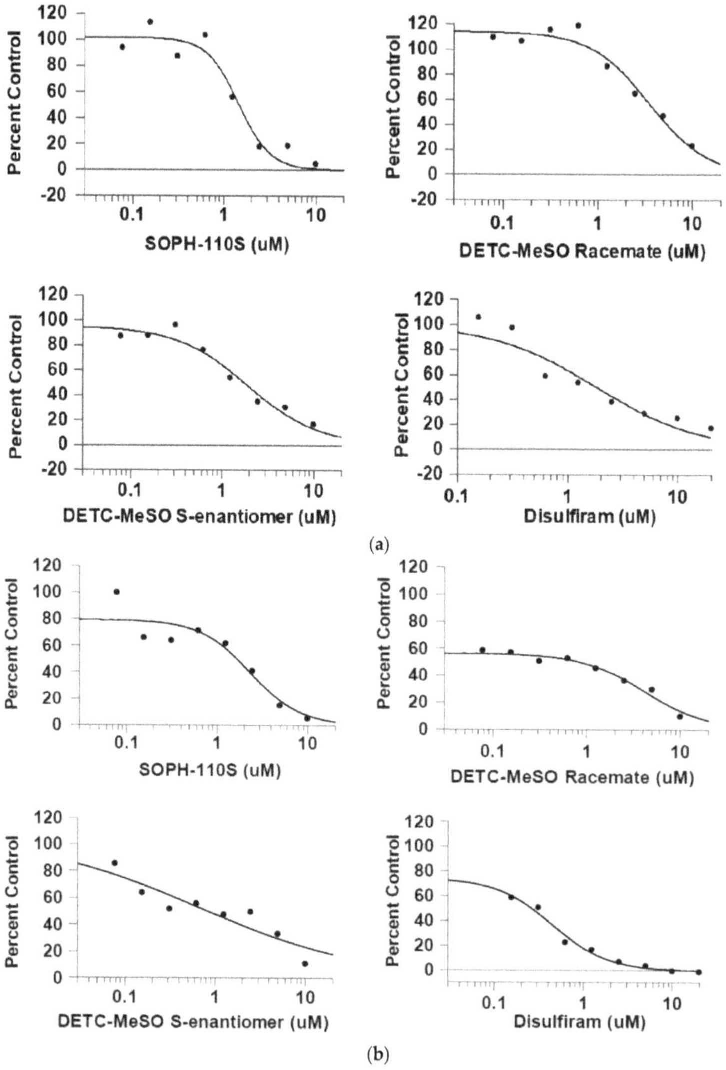

Figure 2. (**a**) Dose-response curves for ALDH2 inhibition by SOPH-110S, DETC-MeSO racemate, DETC-MeSO S enantiomer, and disulfiram at pH 9.0. (**b**) Dose–response curves for ALDH2 inhibition by SOPH-110S, DETC-MeSO racemate, DETC-MeSO S enantiomer, and disulfiram at pH 6.5.

The IC_{50} values for the compounds SOPH-110S, DETC-MeSO racemate, DETC-MeSO S-enantiomer, and disulfiram at pH 6.5 were considerably higher than at pH 9.0 (Table 1).

3.2. Off-Target Binding Assay: Results

Results showing inhibition or stimulation higher than 50% are considered to represent significant effects of the test compound, requiring further investigation (determination of

https://doi.org/10.3390/cells15020123

IC_{50} or EC_{50} values from concentration–response curves). Such effects were not observed at any of the receptor targets studied (Table 2).

Table 2. Mean inhibition by SOPH-110S (10 μM) in in vitro off-target assays.

Target	% Inhibition	Target	% Inhibition	Target	% Inhibition
A_1	−1.8	CXCR2 (IL-8B)	32.6	GR	0.3
A_{2A}	−0.3	CCR1	0.2	Estrogen ER alpha	9.5
α_{1A}	3.0	H_1	−4.9	PR	8.7
α_{1B}	−3.8	H_2	−5.6	AR	−3.3
α_{1D}	6.8	$CysLT_1$ (LTD_4)	−7.0	V_{1a}	−1.9
$\alpha 2_A$	−3.5	MC_1	8.0	Ca^{2+} channel (L, dihydropyridine site)	−6.0
$\alpha 2_B$	−4.0	MC_4	3.8	Ca^{2+} channel (L, diltiazem site)	14.3
β_1	3.6	MAO-A	9.5	Ca^{2+} channel (L, verapamil site)	4.5
β_2	−5.1	M_1	2.0	Ca^{2+} channel (N)	−3.7
AT_1	−13.6	M2	−5.7	Potassium channel hERG	16.9
BZD	−13.7	M3	4.8	Kv channel	0.2
Cl- channel (GABA-gated)	26.4	M4	15.1	Na^+ channel (site 2)	−2.9
B_2	−2.6	NK1	−19.9	Adenosine transporter	0.6
CB_1	−6.7	Y1	−6.8	Norepinephrine transporter	0.6
CB_2	4.6	N neuronal $\alpha 4\beta 2$	−7.6	Dopamine transporter	12.2
CCK_1	22.8	N muscle-type	−0.6	GABA transporter	2.71
CCK_2	−13.3	δ (DOP)	4.5	5-HT transporter	−11.1
D_1	−5.5	Kappa (KOP	6.9	COX1	11.8
D_{2S}	−3.2	μ (MOP)	2.1	COX2	−11.3
D_{2L}	−4.4	PPARγ	−2.1	PDE3A	8.6
ET_A	0.0	PCP	−2.5	PDE4D2	−2.6
$GABA_{A1}$	−7.3	RARα	−1.6	ACE	−6.4
$GABA_{B(1b)}$	−7.4	$5\text{-}HT_{1A}$	3.4	Cathepsin G	1.3
AMPA	−6.3	$5\text{-}HT_{1B}$	23.8	IRK	25.2
Kainate	−4.8	$5\text{-}HT_{2A}$	−8.3	Lck kinase	−20.7
NMDA	−5.8	$5\text{-}HT_{2B}$	−25.5	PKCa	4.0
Glycine (strychnine-sensitive)	−3.9	$5\text{-}HT_{2C}$	−1.0	acetylcholinesterase	5.0
Glycine (strychnine-insensitive)	−8.1	$5\text{-}HT_3$	5.6		
ATPase	0.6	MAO-b	−5.6		

Results showing an inhibition (or stimulation) between 25% and 50% are indicative of weak to moderate and are within a range where more inter-experimental variability can occur. Three targets were observed with inhibition in the 25% to 50% range (CXCR2 at 32.6%, GABA-gated Cl^- channel at 26.4%, and IRK at 25.2%). The significance of this, if any, will be determined in future nonclinical studies.

3.3. Inhibition of CYP Enzymes in Human Liver Microsomes: Results

SOPH-110S did not inhibit the activity of CYP1A2, CYP2B6, CYP2C8, CYP2C9, CYP2C19, CYP2D6, and CYP3A enzymes (IC_{50} > 100 μM) (Table 3). Direct positive control inhibitors inhibited CYP activity by 77.2–95% compared with solvent control.

https://doi.org/10.3390/cells15020123

Table 3. Summary of the direct and time-dependent inhibition of CYP enzymes by SOPH-110S.

CYP	Substrate	Direct Inhibition IC_{50} (μM)	Time-Dependent Inhibition IC_{50} (μM)		
			No NADPH Pre-Incubation [1]	With NADPH Pre-Incubation [2]	Fold Δ in IC_{50}
1A2	Phenacetin	>100	>100	>100	None
2B6	Bupropion	>100	>100	>100	None
2C8	Amodiaquine	>100	>100	>100	None
2C9	Diclofenac	>100	>100	>100	None
2C19	S-mephenytoin	>100	>100	>100	None
2D6	Dextromethorphan	>100	>100	>100	None
3A	Midazolam	>100	>100	>100	None
3A	Testosterone	>100	>100	>100	None

[1] Microsomes, buffer, and SOPH-110S pre-incubated for 30 min prior to initiation of reaction with NADPH and probe substrate. [2] Microsomes, buffer, and SOPH-110S pre-incubated for 30 min with NADPH prior to initiation of reaction with probe substrate. The fold change observed does not meet the threshold of significance for a TDI effect (typically when an IC_{50} shift $\geq$ 2.0).

Since SOPH-110S is not a time-dependent inhibitor of the common hepatic drug-metabolizing P450 enzymes, it is not expected to be associated with liver toxicity like disulfiram. Urinary excretion is considered the major excretion pathway of SOPH-110S.

3.4. Induction of CYP Enzymes in Plateable Human Hepatocytes: Results

In human hepatocytes from three donors, SOPH-110S at concentrations of up to 100 μM did not produce induction (<2-fold) in CYP1A2 mRNA levels at all concentrations up to 100 μM. The percentage of fold change relative to the positive control (omeprazole) was less than 1% in all preparations at all concentration levels. Under the same experimental conditions, omeprazole (25 μM) produced the expected increase in CYP1A2 mRNA levels (63–332-fold) compared with the vehicle control in all three preparations. The negative control (flumazenil (20 μM)) used in this study produced no change in mRNA levels (1.0–1.5-fold) compared with the vehicle control. Therefore, SOPH-110S is not an inducer of CYP1A2 in human hepatocytes.

Treatment with SOPH-110S did not produce induction (<2-fold) in CYP2B6 mRNA levels at all tested concentrations up to 100 μM in all preparations. The percentage of fold change relative to the positive control (phenobarbital (2 mM)) was less than 3% in all preparations at all concentrations. Under the same experimental conditions, the positive control, phenobarbital, produced the expected increases in CYP2B6 mRNA levels (5–14-fold) compared with the vehicle control in all three preparations. The negative control, flumazenil, produced no change in CYP2B6 mRNA levels (0.8–1-fold) compared with the vehicle control. These results suggest SOPH-110S is not an inducer of CYP2B6 in primary human hepatocytes.

Treatment with SOPH-110S did not produce induction (<2-fold) in CYP3A4 mRNA levels at all concentrations up to 100 μM in all preparations. The percentage of positive control (rifampin) for the fold-change of the CYP3A4 mRNA was less than 0% at all concentration levels in all subjects. Therefore, SOPH-110S is not an inducer of CYP3A4 as well as in primary human hepatocytes.

3.5. hERG Current Amplitude Study: Results

SOPH-110S had no effect on hERG current, with the highest concentration tested (50 μM) being associated with a 1.6 ± 1.0% reduction. An IC_{50} value could not be calculated. The response to the positive control, dofetilide (20.9% inhibition at 0.5 nM; 94.0% inhibition at 10 nM; and complete block at 20 nM), and to the vehicle–time control demonstrated proper system performance.

https://doi.org/10.3390/cells15020123

SOPH-110S does not bind to human protein. Allometric scaling was used to predict human exposure SOPH-110S following subcutaneous dosing. Clinically relevant doses are predicted to result in a safety margin of greater than 50-fold for potential hERG interactions and QT/QTc interval prolongation and proarrhythmic potential. However, clinical studies are needed to confirm these results.

3.6. Cardiovascular and Body Temperature Effects in Dogs: Results

No biologically relevant effects on measured arterial pressure, ECG, or body temperature parameters were noted.

3.7. Pharmacokinetics of SOPH-110S Following Intravenous (IV), Subcutaneous (SQ), and Intramuscular (IM) Administration to Beagle Dogs: Results

3.7.1. SOPH-110S

Following a single SC or IM administration of SOPH-110S, the parent compound (SOPH-110s) was quantifiable for 1 hr post-dose. Following a single IV administration of SOPH-110S, the parent compound (SOPH-110s) was quantifiable for 0.5 hr post-dose. As summarized in Table 4, the T_{max} of SOPH-110S was observed at 0.333 hr following single SC administration of 0.3, 1, or 3 mg/kg of SOPH-110S. Following single IV and IM administration of 3 mg/kg of SOPH-110S, the T_{max} was observed at 0.0833 and 0.250 hr, respectively. Following SC dosing, exposure of SOPH-110S based on C_{max} increased in a less than dose-proportional manner. As the dose increased in a ratio of 1:3.33:10, the C_{max} of SOPH-110S increased in a ratio of 1:1.58:8.78. The $AUC_{0\text{-}t}$ increased in a dose-proportional manner in a ratio of 1:2.07:8.23. The terminal phase of the TK profiles was insufficiently characterized to estimate the $T_{1/2}$ of SOPH-110s in all groups.

Table 4. Mean and SD PK parameters of SOPH-110S following a single subcutaneous (SC; 0.3, 1, or 3 mg/kg), intramuscular (IM; 3.0 mg/kg), or intravenous (IV, 3 mg/kg) dose of SOPH-110S to male Beagle dogs.

Route	Dose (mg/kg)	Animal ID	T_{max} (h)	C_{max} (ng/mL)	T_{last} (h)	$AUC_{0\text{-}t}$ (h.ng/mL)	$t_{1/2}$ (hr)
SC	0.3	**Mean**	0.333	32.2	0.833	16.4	NE
		SD	0.144	8.51	0.289	4.19	N/A
SC	1.0	**Mean**	0.333	50.9	1.00	33.9	NE
		SD	0.144	8.23	0.00	3.05	N/A
SC	3.0	**Mean**	0.333	283	1.00	135	NE
		SD	0.144	108	0.00	30.8	N/A
IM	3.0	**Mean**	0.250	566	0.833	170	NE
		SD	0.00	282	0.289	111	N/A
IV	3.0	**Mean**	0.0833	768	0.50	98.8	NE
		SD	0.000	124	0.00	15.2	N/A

NE: not estimated (insufficient characterization of the terminal phase to estimate this parameter); N/A: not applicable ($N \leq 2$).

3.7.2. SOPH-120 (Primary Metabolite)

Following a single IV, SQ, or IM administration of SOPH-110S, SOPH-120 was quantifiable for 4 hr post-dose. Following a single IV, SC, or IM administration of SOPH-110S, the metabolite (SOPH-120) was quantifiable for 4 hr post-dose. As summarized in Table 5, the T_{max} of SOPH-120 was observed at 0.833, 0.667, and 0.799 hr following single SC administration of 0.3, 1, and 3 mg/kg of SOPH-110S, respectively. Following single IV and IM administration of 3 mg/kg of SOPH-120, the T_{max} was observed at 0.250 and 0.417 hr, respectively. Following SC dosing, exposure of SOPH-120 based on the C_{max} increased in a greater than dose-proportional manner. As the dose increased in a ratio of 1:3.33:10, the C_{max} of SOPH-120 increased in a ratio of 1:4.06:16.06.

Table 5. Mean and SD PK parameters of SOPH-120 following a single subcutaneous (SC; 0.3, 1, or 3 mg/kg), intramuscular (IM; 3.0 mg/kg), or intravenous (IV; 3 mg/kg) dose of SOPH-110S to male Beagle dogs.

Route	Dose (mg/kg)	Animal ID	T_{max} (h)	C_{max} (ng/mL)	T_{last} (h)	$AUC_{0\text{-}t}$ (h.ng/mL)	$t_{1/2}$ (hr)
SC	0.3	**Mean**	0.833	52.9	2.00	68.6	NE
		SD	0.289	8.36	0.00	4.51	N/A
SC	1.0	**Mean**	0.667	215	4.00	346	0.799
		SD	0.289	30.8	0.00	70	N/A
SC	3.0	**Mean**	0.667	850	4.00	1180	0.769
		SD	0.289	137	0.00	165	N/A
IM	3.0	**Mean**	0.417	1460	4.00	1620	0.735
		SD	0.144	85	0.00	396	0.0508
IV	3.0	**Mean**	0.25	1510	4.00	1340	0.775
		SD	0.00	219	0.00	174	0.0551

NE: not estimated (insufficient characterization of the terminal phase to estimate this parameter); N/A: not applicable ($N \leq 2$).

The $AUC_{0\text{-}t}$ increased in a greater than dose-proportional manner in a ratio of 1:5.04:17.2. The terminal phase of the PK profiles was insufficiently characterized to estimate the $T_{1/2}$ of SOPH-120 following SC administration of 0.3 mg/kg of SOPH-110S. The $T_{1/2}$ was observed at 0.799 and 0.769 hr following a single SC administration of 1 and 3 mg/kg of SOPH-110S, respectively. Similarly, the $T_{1/2}$ was observed at 0.735 and 0.775 hr following a single IM and IV administration of 3 mg/kg of SOPH-110S, respectively. The mean concentration vs. time plots dose comparison for SQ dosing is provided in Figure 3.

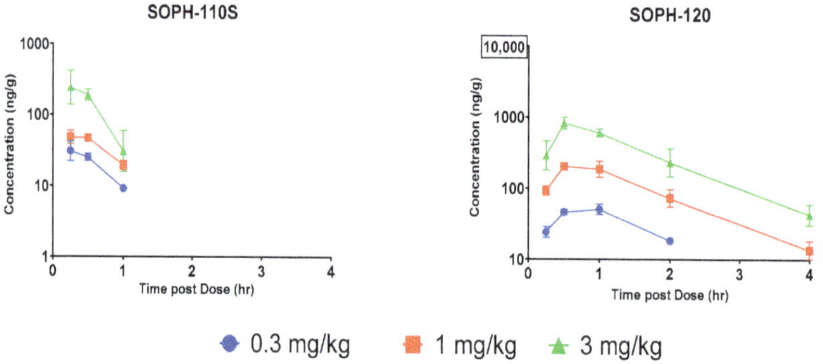

Figure 3. Mean (+SD) concentration vs. time plots for SOPH-110S and SOPH-120 in Beagle dogs following a single SC dose of SOPH-110S at 0.3, 1, or 3 mg/kg (n = 3).

3.8. In Vivo Study of Liver ALDH2 Inhibition in the Presence of Alcohol in Rats

After 10 days' dosing followed by challenge with alcohol, SOPH-110S was found to be a highly potent, dose-dependent inhibitor of ALDH2, comparable to DETC-MeSO, and numerically more potent than the R-enantiomer of SOPH-110S (Figure 4a,b). Several experiments were conducted using variations on the design described, and there was relatively high variability in negative control values: 0.79, 2.35, 3.73, 3.13, and 2.35 mOD/min (mean: 2.47 mOD/min ± 1.10). Thus, the mean and SD of these values are plotted in Figure 3.

https://doi.org/10.3390/cells15020123

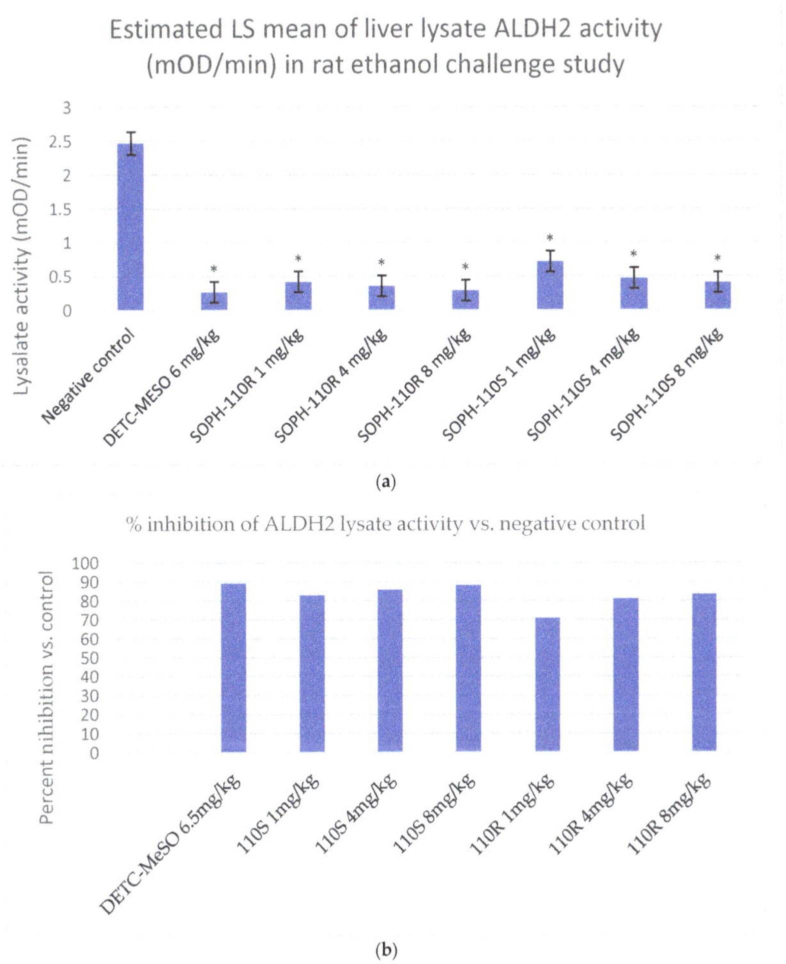

Figure 4. (**a**) Liver lysate ALDH2 activity (mOD/min) measurement in the rat ethanol challenge study. Analysis of variance (ANOVA) with Dunnett's test was used to determine statistical significance between negative control and treated groups. * indicates significant differences with p-value < 0.0001 vs. negative control. Error bars represent the standard error of the LS means. (**b**) Percent of inhibition of ALDH2 lysate activity.

Assay sensitivity was demonstrated in that the positive control, DETC-MeSO, showed significant inhibition of ALDH2 activity compared with the negative control group (Figure 4a; $p < 0.0001$ for all comparisons). SOPH-110S at doses of 1–8 mg/kg produced a dose-dependent 83.0–88.2% reduction in lysate activity (Figure 4b).

4. Discussion

4.1. Preclinical Data Including Proof of Mechanism

The purpose of this series of experiments was to identify and characterize a novel ALDH2 inhibitor to be brought forward into clinical studies. Both in vitro and in vivo experiments have shown that SOPH-110S is a potent inhibitor of ALDH2. SOPH-110S showed greater inhibitory activity than the R enantiomer, and, thus, SOPH-110S was chosen for further development.

https://doi.org/10.3390/cells15020123

SOPH-110S shows very high selectivity for the target enzyme, with no clinically relevant activity in a standard panel of 84 binding targets, which includes receptors, ion channels, transporters, and other enzymes of potential concern with respect to off-target activity. Although unlikely, it could bind to other clinically relevant targets that may be detected in long-term nonclinical or clinical studies. Future studies should evaluate activity across the ALDH superfamily relative to ALDH2 to further characterize the selectivity of SOPH-110S.

Urinary excretion is considered the major excretion pathway of SOPH-110S.

SOPH-110S showed no hERG liability and no cardiovascular or temperature effects and was detectable in plasma with a relatively short half-life in rats and dogs.

In the in vivo proof of mechanism study, SOPH-110S inhibited ALDH2 after 5 days of dosing at 83–88.2% of the negative control in doses ranging from 1 to 8 mg/kg and was comparable to DETC-MeSO (89.1% inhibition) at all doses, demonstrating that SOPH-110S is a highly potent ALDH2 inhibitor in the presence of alcohol. Future studies should evaluate behavioral responses, such as reduction in alcohol self-administration or relapse in animal models, as well as acetaldehyde levels.

4.2. Limitations of Current Treatments for AUD

There has been little innovation in developing drugs to treat AUD in the last several decades. Only three classes of drugs are approved: acamprosate, an N-methyl-D-aspartate (NMDA) receptor modulator; oral naltrexone and long-acting injectable naltrexone (Vivitrol), an opioid receptor antagonist; and disulfiram, an inhibitor of ALDH2.

Acamprosate is approved for the maintenance of abstinence from alcohol in patients with alcohol dependence who are abstinent at treatment initiation. It must be taken three times daily, a schedule that is associated with poor adherence for any oral medication [29]. In the meta-analysis of Maisel et al. (2013) [19], the overall effect size for maintaining abstinence was modest (0.359), and very small for outcomes of heavy drinking (0.072) and reduction in craving (0.034). Naltrexone has a very narrow FDA label, as it is indicated for the treatment of alcohol dependence in patients able to abstain from alcohol in an outpatient setting prior to initiation of treatment. Its effect sizes are even smaller for maintaining abstinence (0.116), reducing heavy drinking (0.189), and reducing craving (0.144) [19,20].

Moreover, both acamprosate and naltrexone can be taken while consuming alcohol and, thus, do not optimally assist the patient in maintaining abstinence.

For large numbers of persons with AUD, long-term abstinence is the necessary and/or desired treatment goal. This group can include individuals with multiple prior relapses, repeated rehabilitation stays, legal mandates (e.g., DWI/DUI), severe psychosocial consequences (such as threatened employment, marital conflict, or parenting failures), and alcohol-related liver disease.

Disulfiram treatment provides a powerful disincentive to continued drinking, via its active metabolite DETC-MeSO [23,24,30]. The clinical effect is to block the reinforcing properties of alcohol and provide an aversive conditioning experience if alcohol is consumed. Patients are educated as to the consequences of drinking while taking disulfiram, and controlled trials have shown that merely the threat of the disulfiram–alcohol reaction (DAR) is sufficient to allow maintenance of abstinence. Petrakis et al. (2005) [21] conducted a large trial in patients with AUD and comorbid psychiatric disorders (n = 254), which compared the efficacy of oral naltrexone alone, placebo, disulfiram open treatment, and naltrexone + disulfiram (open treatment). There was a high rate of adherence (82.7%) and a high overall rate of abstinence (69.7%), regardless of the treatment assignment. Naltrexone + disulfiram was not superior to disulfiram alone but was associated with

https://doi.org/10.3390/cells15020123

more adverse effects than either treatment alone, which included abdominal pain, nausea, vomiting, numb limbs, pins and needles, irregular heartbeat, and restlessness [21].

However, disulfiram has multiple limitations as a treatment for AUD. Its potential adverse effects in the absence of alcohol include drowsiness, headache, cognitive impairment, nausea, vomiting, abdominal pain, and rashes. It has multiple CYP-P450 interactions and generates multiple metabolites and, thus, has multiple potential drug–drug interactions. Moreover, disulfiram is associated with the rare but serious adverse effects of peripheral neuropathy, believed to be due to one of several metabolites [26] and potentially fatal liver toxicity [25]. This is of major concern in this population, with an approximately 50% rate of comorbid liver disease [5,6].

In addition, the amount of DETC-MeSO generated from a given dose is believed to be highly variable. Since the intensity of the DAR is roughly proportional not only to the amount of total amount of alcohol consumed but also to the total inhibition of ALDH2, it is difficult to carefully titrate the amount of inhibition desired with disulfiram to provide the intended disincentive but maintain a safe level of ALDH2 inhibition in the event of a relapse. In decades past, when the dose of disulfiram was as high as 1 gm, severe DARs have been reported, including respiratory depression, cardiovascular adverse effects, seizures, and death.

4.3. Patient Adherence Is Critical to Clinical Outcomes

Adherence to prescribed medication is a major general problem in medicine [31,32]. Poor adherence to treatment is especially problematic in addiction treatment [33]. The nature of the condition is such that craving can drive nonadherence, and brief lapses in oral dosing of disulfiram permit relapse.

Controlled trial data demonstrate that adherence to oral disulfiram can be extremely poor. In the largest controlled trial of disulfiram to date (Fuller et al., 1986 [22]; $n = 605$), there was 80% nonadherence, as demonstrated by urine testing for riboflavin, which was incorporated into the disulfiram tablets. No significant differences over the 1-year period were found on the outcomes of total abstinence, time to first drink, employment, or social stability between the disulfiram group and the two control groups (1 mg of disulfiram and no disulfiram psychosocial treatment). Given the very high rates of nonadherence, this is considered a failed trial. However, there was a highly significant relationship between adherence to dosing and successfully achieved abstinence: among subjects judged adherent, 43% were abstinent vs. only 8% of the nonadherent group [22].

4.4. Effects of ALDH2 Inhibition on Psychiatric Symptoms and Craving

In the large trial of patients with AUD and comorbid psychiatric disorders ($n = 254$), disulfiram was found to be superior to placebo and was associated with dramatically improved symptoms of general psychopathology [21]. In addition, the investigational ALDH2 inhibitor ANS-6637 demonstrated dose-dependent improvements in abstinence-related outcomes as well as reduced craving in a small randomized, controlled trial [34]. Selective ALDH2 inhibition suppresses cue-induced reinstatement of alcohol seeking [35], reduces anxiety-like behaviors [36], and attenuates alcohol-induced dopamine release in the nucleus accumbens [37,38]. These findings suggest that ALDH2 inhibition may act through dual pathways, reinforcing abstinence via aversive conditioning while modulating craving-related circuits. We hypothesize that SOPH-110S may also reduce craving as well as reduce heavy drinking and increase rates of abstinence, similar to disulfiram and ANS-6637, and will evaluate these hypotheses in future clinical trials.

https://doi.org/10.3390/cells15020123

4.5. Long-Acting Injectables as a Proven Strategy to Improve Adherence and Successful Outcomes

It is Sophrosyne's intention to develop a depot long-acting formulation of SOPH-110S if it is demonstrated to be safe and well-tolerated in phase 1. An alcohol challenge study will also be conducted in phase 1, which can potentially demonstrate proof of concept in humans after consuming graded amounts of alcohol.

Long-acting injectable formulations have been used successfully for other psychiatric indications associated with poor adherence, such as schizophrenia and opiate use disorder. The use of a long-acting injectable vs. an oral treatment for schizophrenia is associated with improved adherence, fewer 60-day gaps in medication use, and reduced rates of re-hospitalization [39]. In a very large national cohort study in Sweden (n = 29,823), long-acting injectable antipsychotics were associated with substantially lower risk of rehospitalization (hazard ratio: 0.78; 95% CI: 0.72–0.84) [40]. In a study of opioid use disorder, the proportion of opioid-negative urine samples was superior for subcutaneous buprenorphine vs. sublingual buprenorphine (35.1% vs. 28.4%) [41].

For AUD medications, the level and quality of supervised dosing are believed to be important to treatment success, especially for disulfiram, since it is an oral drug [42]. Observation or administration of an injectable treatment at longer intervals by a healthcare professional or family member is simple and more convenient than daily supervision of an oral medication, potentially improving outcomes. The use of an LAI can potentially prevent the impulsive discontinuation of an ALDH2 inhibitor, allowing the patient more time to thoughtfully consider the advantages of continued abstinence.

4.6. Limitations of the Current Study

This study has several limitations. Nonclinical studies were conducted per FDA guidance and were sufficient for FDA IND approval, but some were conducted with relatively small numbers of animals and in male-only rats. The PK in dogs is short; PK will be evaluated in humans in a planned phase 1 study. Since SOPH-110S is believed to be an irreversible ALDH2 inhibitor, the relationship between pharmacokinetic and pharmacodynamic properties will also be evaluated. Based on the rat PoM study, it is hypothesized that a short PK will still be sufficient to irreversibly inhibit ALDH2. Future studies should also evaluate activity in ALDH2 isoforms, behavioral measures with acetaldehyde kinetics in animal models of AUD, and the feasibility of a long-acting depot injectable with SOPH-110S. Given the known ALD2*2 deficiency in a subpopulation of humans, future evaluation of the safety of SOPH-110S should be explored in subsequent clinical studies.

5. Conclusions

Based on data to date, SOPH-110S is a safe, potent in vitro/in vivo ALDH2 inhibitor intended to be developed for AUD treatment as a long-acting injectable. Future studies are planned to explore the development of a depot formulation with a sufficient duration of action to improve adherence in this difficult-to-treat condition. An LAI treatment for AUD, with the options of dosing once weekly, once monthly, or longer intervals, would support sustained abstinence for the many AUD patients who wish to or must maintain abstinence. Clinical and preclinical evidence suggest that an ALDH2 inhibitor may act through the dual pathway of reinforcing abstinence via aversive conditioning while modulating craving-related circuits. An LAI for AUD may offer the opportunity for reduced rates of relapse, better health and quality of life, and generally improved outcomes.

Author Contributions: Conceptualization, R.D.M.; methodology and performance of experiments, A.S. and A.F.; writing—original draft preparation, R.D.M.; writing—review and editing, A.S. and A.F. All authors have read and agreed to the published version of the manuscript.

https://doi.org/10.3390/cells15020123

Funding: This research was funded by DRW Holdings, Inc.

Institutional Review Board Statement: All animal studies were conducted according to approved protocols in an AAALAC-approved laboratories under the oversight of an Institutional Animal Care and Use Committee.

Informed Consent Statement: Not applicable.

Data Availability Statement: The datasets presented in this article are not readily available because they were extracted from numerous large and proprietary study reports. Requests to access the datasets should be directed to the corresponding author.

Acknowledgments: The authors thank Neil Chapman for administrative and medical writing support.

Conflicts of Interest: All authors are consultants to Sophrosyne Pharmaceuticals. This study was funded by DRW Holdings, Inc. The sponsor had no role in the design or data analysis of the studies; in the collection, analyses, or interpretation of the data; or in the writing of this manuscript. All data were independently reviewed by the FDA in support of the approved IND in September 2025. Publication of these data was approved by DRW Holdings, Inc.

Abbreviations

The following abbreviations were used in this manuscript:

ALDH2	Aldehyde dehydrogenase 2
AUD	Alcohol use disorder
BT	Body temperature
CYP	Cytochrome P450
DAR	Disulfiram alcohol reaction
DBP	Diastolic blood pressure
DETC-MeSO	S-methyl-N, N-diethylthiocarbamate sulfoxide
hERG	Human ether-à-go-go-related gene
HR	Heart rate
IC_{50}	Half-maximal inhibitory concentration
IM	Intramuscular
IV	Intravenous
MAP	Mean arterial pressure
mM	Millimolar
μM	Micromolar
mOD	Rate of change in optical density
mRNA	Messenger ribonucleic acid
NAD, NADH	Nicotinamide adenine dinucleotide (hydride)
nM	Nanomolar
Pk	Pharmacokinetics
pM	Picomolar
SBP	Systolic blood pressure
SD	Standard deviation
SQ	Subcutaneous
WHO	World Health Organization

References

1. World Health Organization. *Global Status Report on Alcohol and Health and Treatment of Substance Use Disorders*; World Health Organization: Geneva, Switzerland, 2024.
2. SAMHSA; Center for Behavioral Health Statistics and Quality. *2022 National Survey on Drug Use and Health*; SAMHSA: Rockville, MD, USA, 2022.

https://doi.org/10.3390/cells15020123

3. SAMHSA Substance Abuse and Mental Health Services Administration. *Treatment Episode Data Set (TEDS) 2022: Ad-Missions to and Discharges from Substance Use Treatment Services Reported by Single State Agencies*; Center for Behavioral Health Statistics and Quality, Substance Abuse and Mental Health Services Administration: Rockville, MD, USA, 2024.
4. Zhao, J.; Stockwell, T.; Naimi, T.; Churchill, S.; Clay, J.; Sherk, A. Association Between Daily Alcohol Intake and Risk of All-Cause Mortality: A Systematic Review and Meta-Analyses. *JAMA Netw. Open* **2023**, *6*, e236185. [CrossRef] [PubMed]
5. Amonker, S.; Houshmand, A.; Hinkson, A.; Rowe, I.; Parker, R. Prevalence of Alcohol-Associated Liver Disease: A Systematic Review and Meta-Analysis. *Hepatol. Commun.* **2023**, *7*, e0133. [CrossRef] [PubMed]
6. Niu, X.; Zhu, L.; Xu, Y.; Zhang, M.; Hao, Y.; Ma, L.; Li, Y.; Xing, H. Global Prevalence, Incidence, and Outcomes of Alco-Hol-Related Liver Diseases: A Systematic Review and Meta-Analysis. *BMC Public Health* **2023**, *23*, 859. [CrossRef]
7. Verhulst, B.; Neale, M.C.; Kendler, K.S. The Heritability of Alcohol Use Disorders: A Meta-Analysis of Twin and Adoption Studies. *Psychol. Med.* **2015**, *45*, 1061–1072. [CrossRef]
8. Deak, J.D.; Miller, A.P.; Gizer, I.R. Genetics of Alcohol Use Disorder: A Review. *Curr. Opin. Psychol.* **2019**, *27*, 56–61. [CrossRef] [PubMed]
9. Liu, M.; Jiang, Y.; Wedow, R.; Li, Y.; Brazel, D.M.; Chen, F.; Datta, G.; Davila-Velderrain, J.; McGuire, D.; Tung, J.Y. Association Studies of up to 1.2 Million Individuals Yield New Insights into the Genetic Etiology of Tobacco and Alcohol Use. *Nat. Genet.* **2019**, *51*, 237–244. [CrossRef]
10. Daviet, R.; Aydogan, G.; Jagannathan, K.; Spilka, N.; Koellinger, P.D.; Kranzler, H.R.; Nave, G.; Wetherill, R.R. Associations between Alcohol Consumption and Gray and White Matter Volumes in the UK Biobank. *Nat. Commun.* **2022**, *13*, 1175. [CrossRef]
11. Ponomarev, I. Epigenetic Control of Gene Expression in the Alcoholic Brain. *Alcohol Res.* **2013**, *35*, 69–76. [CrossRef]
12. Kalivas, P.W.; Volkow, N.D. The Neural Basis of Addiction: A Pathology of Motivation and Choice. *Am. J. Psychiatry* **2005**, *162*, 1403–1413. [CrossRef]
13. Myrick, H.; Anton, R.F.; Li, X.; Henderson, S.; Drobes, D.; Voronin, K.; George, M.S. Differential Brain Activity in Alcoholics and Social Drinkers to Alcohol Cues: Relationship to Craving. *Neuropsychopharmacol. Off. Publ. Am. Coll. Neuropsychopharmacol.* **2004**, *29*, 393–402. [CrossRef]
14. Salinas, A.G.; Mateo, Y.; Carlson, V.C.; Stinnett, G.S.; Luo, G.; Seasholtz, A.F.; Grant, K.A.; Lovinger, D.M. Long-Term Alcohol Consumption Alters Dorsal Striatal Dopamine Release and Regulation by D2 Dopamine Receptors in Rhesus Macaques. *Neuropsychopharmacology* **2021**, *46*, 1432–1441. [CrossRef] [PubMed]
15. Zahr, N.M. Structural and Microstructural Imaging of the Brain in Alcohol Use Disorders. In *Alcohol and the Nervous System*; Sullivan, E.V., Pfefferbaum, A., Eds.; Elsevier: Amsterdam, The Netherlands, 2014; Volume 125, pp. 275–299. [CrossRef]
16. Lai, C.; Lin, H.L.; Chen, Y.C.; Lu, C.Y.; Lee, H.S.; Yin, S.J. Dominance of the Inactive Asian Variant over Activity and Protein Contents of Mitochondrial Aldehyde Dehydrogenase 2 in Human Liv-Er. *Alcohol. Clin. Exp. Res.* **2014**, *38*, 44–50. [CrossRef] [PubMed]
17. Han, B.; Jones, C.M.; Einstein, E.B.; Powell, P.A.; Compton, W.M. Use of Medications for Alcohol Use Disorder in the US: Results from the 2019 National Survey on Drug Use and Health. *JAMA Psychiatry* **2021**, *78*, 922–924. [CrossRef] [PubMed]
18. Bradizza, C.M.; Stasiewicz, P.R.; Paas, N.D. Relapse to Alcohol and Drug Use among Individuals Diagnosed with Co-Occurring Mental Health and Substance Use Disorders: A Review. *Clin. Psychol. Rev.* **2006**, *26*, 162–178. [CrossRef]
19. Maisel, N.C.; Blodgett, J.C.; Wilbourne, P.L. Meta-Analysis of Naltrexone and Acamprosate for Treating Alcohol Use Disorders: When Are These Medications Most Helpful? *Addiction* **2013**, *108*, 275–293. [CrossRef]
20. Garbutt, J.C.; Kranzler, H.R.; O'Malley, S.S. Efficacy and Tolerability of Long-Acting Injectable Naltrexone for Alcohol Dependence: A Randomized Controlled Trial. *JAMA* **2005**, *293*, 1617–1625. [CrossRef]
21. Petrakis, I.L.; Poling, J.; Levinson, C.; Nich, C.; Carroll, K.; Rounsaville, B.; VA New England VISN I MIRECC Study Group. Naltrexone and Disulfiram in Patients with Alcohol Dependence and Comorbid Psychiatric Disorders. *Biol. Psychiatry* **2005**, *57*, 1128–1137. [CrossRef]
22. Fuller, R.K.; Branchey, L.; Brightwell, D.R.; Derman, R.M.; Emrick, C.D.; Iber, F.L.; James, K.E.; Lacoursiere, L.; Lee, K.K.; Lowenstam, I.; et al. Disulfiram Treatment of Alcoholism: A Veterans Administration Cooperative Study. *JAMA* **1986**, *256*, 1449–1455. [CrossRef]
23. Hart, B.W.; Faiman, M.D. In Vitro and In Vivo Inhibition of Rat Liver Aldehyde Dehydrogenase by S-Methyl N,N-Diethylthiolcarbamate Sulfoxide, a New Metabolite of Disulfiram. *Biochem. Pharmacol.* **1992**, *43*, 403–406. [CrossRef]
24. Hart, B.W.; Faiman, M.D. In Vivo Pharmacodynamic Studies of the Disulfiram Metabolite S-Methyl N,N-Diethylthiolcarbamate Sulfoxide: Inhibition of Liver Aldehyde Dehydrogenase. *Alcohol. Clin. Exp. Res.* **1994**, *18*, 340–345. [CrossRef]
25. *LiverTox: Clinical and Research Information on Drug-Induced Liver Injury [Internet]*; National Institute of Diabetes and Digestive and Kidney Diseases: Bethesda, MD, USA, 2012.
26. Filosto, M.; Tentorio, M.; Broglio, L.; Buzio, S.; Lazzarini, C.; Pasolini, M.P.; Cotelli, M.S.; Scarpelli, M.; Mancuso, M.; Choub, A.; et al. Disulfiram Neuropathy: Two Cases of Distal Axonopathy. *Clin. Toxicol.* **2008**, *46*, 314–316. [CrossRef]

https://doi.org/10.3390/cells15020123

27. Water, A.; Verheyen, J.; Xhonneux, R.; Reneman, R.S. An Improved Method to Correct the QT Interval of the Electrocardiogram for Changes in Heart Rate. *J. Pharmacol. Methods* **1989**, *22*, 207–217. [CrossRef]
28. Aylott, M.; Bate, S.; Collins, S.; Jarvis, P.; Saul, J. Review of the Statistical Analysis of the Dog Telemetry Study. *Pharm. Stat.* **2011**, *10*, 236–249. [CrossRef]
29. Claxton, A.J.; Cramer, J.; Pierce, C. A Systematic Review of the Associations between Dose Regimens and Medication Com-Pliance. *Clin. Ther.* **2001**, *23*, 1296–1310. [CrossRef]
30. Madan, A.; Parkinson, A.; Faiman, M.D. Identification of the Human and Rat P450 Enzymes Responsible for the Sulfoxida-Tion of S-Methyl-N,N-Diethylthiocarbamate Sulfoxide (DETC-ME): The Terminal Step in the Bioactivation of Disulfiram. *Drug Metab. Dispos.* **1995**, *23*, 1153–1162. [CrossRef] [PubMed]
31. Osterberg, L.; Blaschke, T. Adherence to Medication. *N. Engl. J. Med.* **2005**, *353*, 487–497. [CrossRef]
32. McLellan, A.T.; Lewis, D.C.; O'Brien, C.P.; Kleber, H.D. Drug Dependence, a Chronic Medical Illness: Implications for Treatment, Insurance, and Outcomes Evaluation. *JAMA* **2000**, *284*, 1689–1695. [CrossRef] [PubMed]
33. Rinn, W.; Desai, N.; Rosenblatt, H.; Gastfriend, D.R. Addiction Denial and Cognitive Dysfunction: A Preliminary Investigation. *J. Neuropsychiatry Clin. Neurosci.* **2002**, *14*, 52–57. [CrossRef]
34. O'Malley, S.S.; Miranda, R.; Book, S.W.; Chun, T.H.; Liss, T.; Malcolm, R.J.; Muvvala, S.B.; Padovano, H.T.; Schacht, J.P.; Blackburn, B.; et al. Preliminary Effects of Orl ANS-6637, an ALDH2 Inhibitor, on Cue-Induced Craving, Safety, and Alcohol Consumption among Adults with Alcohol Use Disorder: A Proof-of-Concept, Randomized, Human Laboratory Trial. *Alcohol. Alcohol.* **2025**, *60*, agaf001. [CrossRef] [PubMed]
35. Arolfo, M.P.; Overstreet, D.H.; Yao, L. Suppression of Heavy Drinking and Alcohol Seeking by a Selective ALDH-2 Inhibitor. *Alcohol. Clin. Exp. Res.* **2009**, *33*, 1935–1944. [CrossRef]
36. Overstreet, D.H.; Knapp, D.J.; Breese, G.R. A Selective ALDH-2 Inhibitor Reduces Anxiety in Rats. *Pharmacol. Bio-Chem. Behav.* **2009**, *94*, 255–261. [CrossRef]
37. Yao, L.; Fan, P.; Arolfo, M.; Jiang, Z.; Olive, M.F.; Zablocki, J.; Sun, H.-L.; Chu, N.; Lee, J.; Kim, H.-Y.; et al. Inhibition of Aldehyde Dehydrogenase-2 Suppresses Cocaine Seeking by Generating THP, a Cocaine Use-Dependent Inhibitor of Dopamine Synthesis. *Nat. Med.* **2010**, *16*, 1024–1028. [CrossRef]
38. Volkow, N.D.; Koob, G.F.; McLellan, A.T. Neurobiologic Advances from the Brain Disease Model of Addiction. *N. Engl. J. Med.* **2016**, *374*, 363–371. [CrossRef] [PubMed]
39. Marcus, S.C.; Zummo, J.; Pettit, A.R.; Stoddard, J.; Doshi, J.A. Antipsychotic Adherence and Rehospitalization in Schizophrenia Patients Receiving Oral Versus Long-Acting Injectable Antipsychotics Following Hospital Discharge. *J. Manag. Care Spec. Pharm.* **2015**, *21*, 754–768. [CrossRef] [PubMed]
40. Tiihonen, J.; Mittendorfer-Rutz, E.; Majak, M.; Mehtala, J.; Hoti, F.; Jedenius, E.; Enkusson, D.; Level, A.; Sermon, J.; Tanskanen, A.; et al. Real-World Effectiveness of Antipsychotic Treatments in a Nationwide Cohort of 29,823 Patients with Schizophrenia. *JAMA Psychiatry* **2017**, *74*, 686–693. [CrossRef] [PubMed]
41. Lofwall, M.R.; Walsh, S.L.; Nunes, E.V.; Bailey, G.L.; Sigmon, S.; Kampman, K.M.; Frost, M.; Tiberg, F.; Linden, M.; Sheldon, B.; et al. Weekly and Monthly Subcutaneous Buprenorphine Depot Formulations vs. Daily Sublingual Buprenorphine with Naloxone for Treatment of Opioid Use Disorder. *JAMA Intern. Med.* **2018**, *178*, 764–773. [CrossRef]
42. Brewer, C.; Meyers, R.J.; Johnson, J. Does Disulfiram Help to Prevent Relapse in Alcohol Abuse? *CNS Drugs* **2000**, *14*, 329–341. [CrossRef]

https://doi.org/10.3390/cells15020123

cells

MDPI

Article

Biomarker-Based Pharmacological Characterization of ENX-102, a Novel α2/3/5 Subtype-Selective GABAA Receptor Positive Allo-Steric Modulator: Translational Insights from Rodent and Human Studies

Pauline Nettesheim [1,2,†], Krishna C. Vadodaria [3,†], Kimberly E. Vanover [3], Laura G. J. M. Borghans [1,2], Estibaliz Arce [3], William Brubaker [3], Stephen Cunningham [3], Stephanie Parks [3], Jordi Serrats [3], Vikram Sudarsan [3], Eve Taylor [3], Erica Klaassen [1], Frederik E. Stuurman [2] and Gabriel E. Jacobs [1,4,*]

[1] Centre for Human Drug Research, 2333 CL Leiden, The Netherlands; p.nettesheim@lumc.nl (P.N.); lborghans@chdr.nl (L.G.J.M.B.)

[2] Department of Clinical Pharmacy and Toxicology, Leiden University Medical Center, 2333 ZG Leiden, The Netherlands

[3] Engrail Therapeutics, Inc., San Diego, CA 92130, USA; krishna@engrail.com (K.C.V.); bill.brubaker@engrail.com (W.B.); lurgan@optonline.net (S.C.); stephanie.parks@engrail.com (S.P.); jordiserrats@gmail.com (J.S.); vikram@engrail.com (V.S.); eve.taylor@engrail.com (E.T.)

[4] Department of Psychiatry, Leiden University Medical Center, 2333 ZG Leiden, The Netherlands

* Correspondence: gjacobs@chdr.nl; Tel.: +31-71-5246-400

† These authors contributed equally to this work.

Abstract

Gamma-aminobutyric acid type A receptors ($GABA_A$Rs) are pentameric ligand-gated ion channels essential for inhibitory neurotransmission in the central nervous system. Subtype-specific expression patterns of $GABA_A$R subunits underlie their diverse roles in regulating anxiety, motor function, and sedation. While non-selective $GABA_A$R positive allosteric modulators (PAMs), such as benzodiazepines, are clinically effective anxiolytic drugs, their non-selective activity across α1/2/3/5 subunit-containing $GABA_A$Rs leads to sedation, cognitive impairment, and risk of dependence. To address this, we evaluated ENX-102, a novel $GABA_A$R PAM, which exhibits selectivity for α2/3/5 subunits. In rodents, ENX-102 demonstrated dose-dependent anxiolytic-like activity following acute and sub-chronic administration, without sedation. ENX-102 exhibited a dose-dependent quantitative electroencephalography (qEEG) spectral signature in rodents that was distinct from that of benzodiazepines. In a double-blind, placebo-controlled, multiple-ascending dose study in healthy human volunteers, ENX-102 was evaluated using the NeuroCart, a CNS test battery including saccadic peak velocity (SPV), adaptive tracking, pupillometry, body sway, the Bond and Lader Visual Analog Scale (VAS), the Visual Verbal Learning Task (VVLT), and qEEG. ENX-102 produced reductions in SPV that were indicative of central target engagement, with minimal effects on alertness and motor coordination, which is consistent with subtype-selective $GABA_A$R targeting. Notably, qEEG revealed increased β-band power and decreased δ- and θ-band activity, which were distinct from the spectral profile of non-selective PAMs, supporting translational alignment with preclinical findings. Across dose levels, ENX-102 was well tolerated and exhibited favorable pharmacokinetics. These results support further clinical development of ENX-102 as a next-generation $GABA_A$R subtype-selective anxiolytic drug.

Keywords: $GABA_A$ receptor; benzodiazepine; subtype-selective; ENX-102; translational

1. Introduction

Gamma-aminobutyric acid (GABA) represents the primary inhibitory neurotransmitter in the human central nervous system (CNS). It is essential for maintaining neurophysiological and behavioral homeostasis in typically dynamic and potentially emotionally challenging environments, and acts as an endogenous ligand at $GABA_A$, $GABA_B$, and $GABA_C$ receptors. $GABA_A$ receptors ($GABA_A$Rs) are expressed across diverse neural circuits, where they crucially regulate phasic GABA-mediated CNS inhibition [1]. As a consequence, dysregulation of GABAergic signaling through $GABA_A$Rs has been implicated in hyperaroused states and/or cortical hyperexcitability, including anxiety, trauma-related disorders, and epilepsy [2,3].

$GABA_A$Rs are ion-gated channels composed of five transmembrane subunits expressed in different configurations in diverse neurocircuits. Although 19 distinct $GABA_A$R subunits (α1–6, β1–3, γ1–3, δ, ε, θ, π, and ρ1–3) have been identified [4], its predominant isoform consists of a single γ, two α, and two β subunits. Such structural and anatomical heterogeneity allows for diverse GABA-mediated physiological effects, making $GABA_A$Rs essential in various CNS functions, depending on the subtype combination and receptor expression site. In both rodents and humans, α1 subunit-containing receptors display a wide expression pattern with relatively greater density in the cerebral cortex, where they contribute to non-specific CNS suppression, which leads to reduced alertness or sedation [5–7]. In contrast, α2 and possibly α3 subunits display a different expression pattern with relatively greater expression in subcortical limbic structures such as the amygdala, potentially contributing to attenuation of hyperarousal and emotional response. Although α5 subunit-containing $GABA_A$Rs expression is generally lower in the CNS, with a less restricted expression in the hippocampus in rodents, they have been implicated in memory function [8–10]. Due to the widespread CNS distribution of $GABA_A$Rs, non-selective positive allosteric modulators (PAMs) such as benzodiazepines have shown broad efficacy in diverse neuropsychiatric disorders, particularly as anxiolytics [11]. However, their non-selective targeting of α1/2/3/5 subunit-containing $GABA_A$Rs is associated with unwanted side-effects including sedation, cognitive impairment, tolerance, and the risk for dependence with chronic use [11]. Consequently, current drug development efforts focus on developing $GABA_A$R PAMs with subtype-selective activity-enhancing efficacy and minimizing activity at subtypes linked with adverse effects [12,13].

Despite the complexity of $GABA_A$R, several subtype-selective PAMs have been developed. For example, α1-selective modulators such as zolpidem and zopiclone have been developed as effective sedative-hypnotics [14]. Other examples include α2-/3-/5-preferring compounds such as PF-06372865 [15], AZD6280 [16], AZD7325 [17], SL65.1498 [18], NS11821 [19], L–838,417 [20,21] and its analogs, TPA-023 [22,23], and TPA023B [24,25]. However, many of these compounds failed to progress clinically, due to issues such as poor bioavailability, insufficient subtype selectivity *in vivo*, and/or toxicological concerns. As a result, there remains a significant unmet need for novel $GABA_A$R subtype-selective PAMs, particularly those targeting α2/3/5 subunits, which may offer safer and more targeted treatment options for hyperarousal-related neuropsychiatric conditions, such as anxiety disorders. ENX-102 (C21H18F2N5O5P) is a novel phosphate-associated version of the previously reported α2-/3-/5-selective $GABA_A$R PAM TPA023B [24]. The pharmacodynamic (PD) characteristics of non-selective and subtype-selective $GABA_A$R agonists have been studied in humans using the NeuroCart, a CNS test battery that combines neuropsychological and neurophysiological assessments with quantitative electroencephalography (qEEG) [26]. Non-selective $GABA_A$R PAMs such as lorazepam consistently reduce the saccadic peak velocity (SPV), reduce alertness (via the Visual Analogue Scale (VAS)), impair adaptive tracking, and increase body sway [27]. In contrast, α2/3 subtype-selective

$GABA_AR$ PAMs (TPA023, NS11821, AZD6280, PF-06372865) produce comparable SPV reductions at equipotent doses with minimal effects on VAS alertness, adaptive tracking, or body sway [16,18,19,23,28]. Regression analyses confirm the distinct PD profiles of α2-/3-selective to non-selective PAMs [29]. The SPV, VAS alertness, body sway, and adaptive tracking can be considered putative biomarkers for arousal, subjective alertness, postural stability, and sustained attention, respectively. Thus, α2/3 subtype-selective modulation may attenuate arousal while avoiding the sedative and motor-impairing effects seen with non-selective compounds [29].

qEEG offers a temporally precise and translational measure of electrophysiological responses to CNS-agents [26,30]. Non-selective $GABA_AR$ PAMs (propofol, lorazepam, midazolam) reliably increase β-band power [17,28,31–34], a biomarker of $GABA_AR$-mediated neurotransmission [35]. They also elevate δ-band activity [16,17,28,36–38], which is associated with non-rapid eye movement (NREM) sleep [39], sedation, and anesthesia-induced unconsciousness [40,41], and reduce α-band activity via benzodiazepine-site modulation [17,32,33,35,42]. Emerging evidence suggests that α2-/3-selective $GABA_AR$ PAMs may produce distinct qEEG spectral signatures. In rodent, compounds like TPA023, TP003, AZD6280, AZD7235, and PF-06372865 dose-dependently increased β- and γ-band power [22,43,44]. In humans, NS11821, AZD6280, and PF-06372865 consistently increased β power and decreased δ and θ power following a single dose [15,16,19,28], unlike lorazepam, which increased both β and δ power in humans and rodents [15–17]. These findings suggest that α2-/3-selective PAMs, following single doses, can be distinguished from non-selective PAMs by a qEEG spectral profile characterized by increased β power (and possibly γ), reduced δ and θ power, and variable α activity [35,45]. Together, this PD signature may reflect anxiolytic efficacy without sedation, offering a promising therapeutic profile that is distinct from that of benzodiazepines.

The current study reports the preclinical and clinical pharmacology of ENX-102, across doses predicted to achieve >80% receptor occupancy at α2-/3 subunit-containing $GABA_ARs$. In rodents, sedative and anxiolytic-like effects were investigated following acute and sub-chronic oral dosing. In healthy human volunteers, CNS biomarkers, observed using the NeuroCart, were employed to characterize PD effects following single and repeated dosing over 12 days. Furthermore, qEEG spectral changes in both rodents and humans were also examined to support translational alignment. Finally, pharmacokinetics and safety were also evaluated in healthy human volunteers.

2. Materials and Methods

2.1. *In Vitro Experiments*

ENX-102 was evaluated for positive allosteric modulation against four $GABA_A$ subtypes ($GABA_A$ α1β3γ2L, α2β3γ2L, α3β3γ2L, and α5β3γ2L) using the SyncroPatch automated patch-clamp platform (SyncroPatch 384i for data recording, Nanion Technologies GmBH, Munic, Germany). Assays were conducted at room temperature in PAM mode across 10 concentrations (0.01 nM, 0.1 nM, 1 nM, 3 nM, 10 nM, 30 nM, 100 nM, 300 nM, 1 μM, 10 μM), with each condition tested in duplicate. GABA was applied at $EC_{5–20}$ levels (5–20% of maximal response). Diazepam served as positive control (0.01, 0.1, 1, 10, 100μM) and 0.2% DMSO as a negative control. Modulatory activity was expressed as the percentage of activation relative to the GABA-alone condition, and was calculated using the following equation: $\%activation = [(Icomp/Icontrol) - 1] \times 100$, where Icomp is the current amplitude in the presence of both ENX-102 and GABA and Icontrol is the current amplitude in the presence of GABA alone.

2.2. *In Vivo Experiments in Rats*

2.2.1. Elevated Plus Maze

Adult male SD rats from Envigo (Indianapolis, IN, USA) were used to assess anxiety-like behavior via the elevated plus maze (EPM). The EPM relies on rats' instinct to explore while avoiding open areas; increased time spent in the open arms is interpreted as reduced anxiety following acute dosing [46,47]. Rats (~260–300 g prior to treatment) were randomly assigned to one of five treatment groups: vehicle, chlordiazepoxide (5 mg/kg, intraperitoneally (i.p.), positive control), or ENX-102 (0.1, 0.3, or 1.0 mg/kg, per os (p.o.)). Each group consisted of 10 rats, a sample size that was selected to ensure adequate statistical power. ENX-102 was formulated in 0.5% Tylose MH 300 solution and administered at a dose volume of 1.0 mL/kg; chlordiazepoxide was dissolved in water and administered at a dose volume. Behavioral testing was conducted 30 min post-dose for chlordiazepoxide and at 1- and 24-h post-dose for ENX-102 under acute (single dose) conditions. For sub-chronic testing, ENX-102 was administered once daily for 14 days, with EPM testing being conducted on the final day. Due to the limited reliability of the EPM in detecting anxiolytic effects under sub-chronic benzodiazepine treatment [48–50], chlordiazepoxide was intentionally excluded as a positive control in this condition. At the designated time post-treatment, each rat was placed in the center of the EPM, facing an open arm, for a 5-min test session. Movement metrics including distance traveled, time spent in each arm, and entries into each arm were automatically recorded by the computer. The experiment remained blinded throughout the active testing period to minimize bias. Unblinding was performed only after all testing was completed. Additional methodological details are provided in Appendix A.1.

2.2.2. Quantitative EEG

Young adult male Sprague Dawley (SD) rats (~275–325 g on arrival) from Envigo (Indianapolis, IN, USA) were used in this study (rats used for qEEG recordings were distinct from those used in the EPM experiments). A sample size of 10–12 animals per group was determined to provide adequate statistical power. Rats were implanted with electrodes, and EEG was recorded wirelessly while they remained in their home cages. Vehicle and 4 doses of ENX-102 were assessed in a cross-over design to ensure N = 12 rats were included in each treatment group, with a four-day washout. ENX-102 and vehicle (0.5% Tylose MH 300® solution) were administered p.o. Afterwards, all rats (N = 12) received lorazepam and 1.0 mg/kg, i.p., dissolved in saline. Each rat underwent six separate 48-h continuous EEG recordings, each of which was preceded by a 2-h baseline EEG (Figure 1). A more detailed technical description of the qEEG recordings is given in Appendix A.2.

2.2.3. Sleep Analysis

Sleep parameters in rats were assessed as part of the qEEG recordings following administration of ENX-102 at four doses (0.03, 0.1, 0.3, 1.0 mg/kg) or lorazepam. Sleep stages analyzed included active wake, quiet wake, NREM, and rapid eye movement (REM) sleep. A more detailed technical description of the sleep analysis is given in Appendix A.3. Unblinding occurred after sleep scoring during the creation of the decoder used for data processing and final analysis.

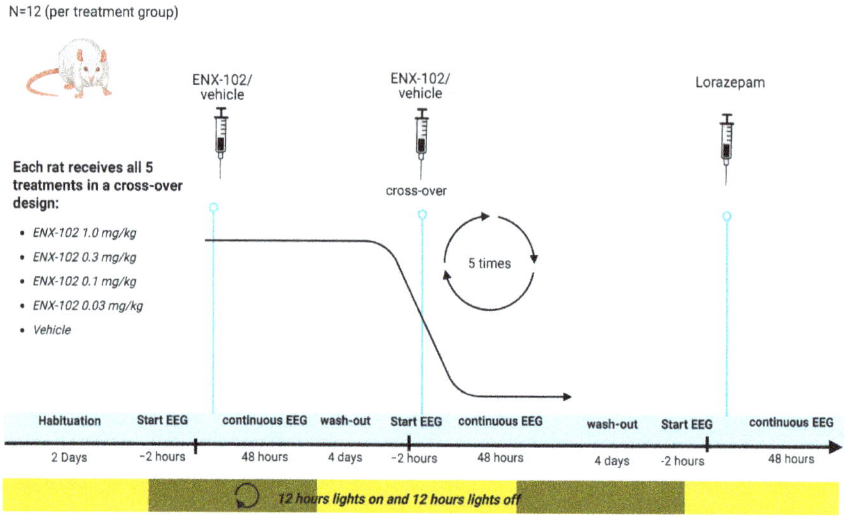

Figure 1. Schematic representation of the qEEG recordings in rats. Twelve rats were used in a cross-over design to assess vehicle and four dose levels of ENX-102. Each dose was followed by a four-day wash-out period. Dosing and EEG recordings were performed on a Monday and Friday schedule, starting during the lights-off period, with dosing occurring 4 h after lights were turned off. EEG recordings began 2 h before dosing and continued for 48 h post-dose. After completion of ENX-102 dosing, all rats (N = 12) received a single dose of lorazepam and EEG was recorded. All animals were habituated to dosing (vehicle dosing) for two days before any data collection took place. Recordings were started 2 h prior to compound administration and recorded continuously for 48 h after administration.

2.3. Multiple Ascending Dose (MAD) Study in Healthy Human Participants

2.3.1. Study Design

A randomized, double-blind, placebo-controlled, multiple ascending dose study was conducted at Centre for Human Drug Research (Leiden, The Netherlands). A total of 40 healthy participants—both male and female—were enrolled in the MAD study. The sample size was based on precedents from comparable clinical studies [13,24], providing adequate power to assess safety, tolerability, pharmacokinetics, and pharmacodynamic effects. Participants were admitted to the Centre for Human Drug Research (CHDR) in Leiden, the Netherlands, from day −1 to day 13. Safety, pharmacokinetic (PK), and pharmacodynamic (PD) assessments were performed during this period. Follow-up visits occurred on day 19 and day 26, during which only safety and PK assessments were conducted.

2.3.2. Study Population

Healthy females and males aged 18 to 55 years, with a body mass index of 18 to 35 kg/m^2, were included. To evaluate eligibility, participants underwent eligibility screening, which included review of their medical and psychiatric history, physical examination, vital sign measurements, an electrocardiogram (ECG), blood chemistry and hematology laboratory tests, and urinalysis. Participants were excluded if they had a clinically significant condition within two years prior to screening that could pose a risk to or interfere with study outcomes (e.g., cardiac, renal, neurologic, gastrointestinal, pulmonary, endocrine, hematologic, immunologic disorders, or malignancy), a current or past psychiatric disorder as defined by Diagnostic and Statistical Manual-5, suicidal ideation within 30 days before screening, or suicidal behavior within the past two years. In addition, participants with a history of convulsions were excluded. Participants were not allowed to have ingested any concomitant medication (excluding hormonal birth control) within 5 half-lives or 30 days (whichever was longer) prior to day 1. In total, ENX-102 was administered across five cohorts, each consisting of eight healthy participants in a 6:2 ratio of active treatment to placebo. This resulted in six participants receiving ENX-102 at each dose level, while the 10 placebo participants across all cohorts were pooled for analysis.

2.3.3. Participant Disposition and Demographics

A total of 104 potential participants underwent screening, of whom 40 (18 female and 22 male) were randomized and completed the study. The consort diagram of randomized participants is shown in Figure S1. The mean (range) age and BMI were 30.5 years (19–52) and 25.2 kg/m^2 (18.5–34.7), respectively. Find more details on the demographics in Table S2.

2.3.4. Dosing Scheme

All doses were administered once daily as p.o. capsules for 12 days, with a fixed amount of 240 mL water intake in a fasted state ($\geq$10 h fasting). In the current MAD study, dose escalation from 0.5 mg to 1.0 mg, 1.5 mg, 2.0 mg, and 5.0 mg was conducted during dedicated dose-escalation meetings, following blinded review of all available PK, PD, and safety data of each cohort. The starting dose of 0.5 mg ENX-102 and anticipated dose escalation scheme were selected based on physiologically based pharmacokinetic (PBPK) modeling of preclinical PK data, unpublished clinical data from the single ascending dose (SAD) study, and published work with the TPA-023B free base, in combination with available non-clinical toxicology data obtained in animals and converted to a human equivalent dose [12]. The SAD study investigated ENX-102 0.5, 1.0, 1.5, 2.0, 3.0, 5.0, and 10.0 mg, in which the observed T_{max} ranged from 1.3 to 4 h and the elimination half-life ranged from 43 to 73 h.

2.3.5. Safety Assessments

Adverse events (AEs), serious adverse events (SAEs), vital signs, electrocardiogram (ECG), and clinical laboratory tests were assessed throughout the study. The definition of an SAE used in this study is provided in Appendix B. Clinical laboratory tests included urinalysis, hematology, chemistry, and coagulation analysis. Supine blood pressure and heart rate were measured after at least 10 min of rest in supine position; standing blood pressure was measured at 1, 3, and 5 min after rising from the supine position to standing to assess orthostasis (i.e., systolic blood pressure decrease > 20 mmHg or diastolic blood pressure decrease > 10 mmHg). Vital signs were assessed pre-dose and at 1- and 5-h post-dosing on days 1 through 12. In addition, the Modified Observer's Assessment Alertness/Sedation score (MOAA/S) and Columbia Suicide Severity Rating Scale (C-SSRS) were administered. The MOAA/S, a rater-administered scale ranging from 0 (unresponsive) to 5 (awake), with values of 0 (unresponsive to pain), 1 (responds only to painful stimulus), 2 (responds to mild prodding), 3 (responds to loud/repeated name calling), 4 (lethargic response to name calling), and 5 (awake), was used to assess and quantify sedation [22]. The MOAA/S was recorded on each study day pre-dose and at 0.5, 1, 1.5, 2, 3, 6, 8, and 10 h post-dosing, as well as within 15 min of any reported somnolence or sedation on days 1 through 12. The C-SSRS evaluated emergent suicidal ideation and behavior and was administered pre-dose and on days 1, 6, 12, 19, and 26.

2.3.6. Pharmacokinetic Assessments

Blood samples (6 mL) for PK analysis were drawn on day 1 pre-dose and at 0.5, 1, 1.5, 3, 6, 8, 10, and 12 h post-dosing and on day 12 pre-dose and at 0.5, 1.5, 3, 6, 8, 10, and 12 h post-dosing. Additional samples were taken on every dosing day, approximately

24 h after the previous day's dose and prior to the subsequent dose, and 24, 72 (day 13), 168 (day 19), and 336 h (day 26) after the last dose on day 12. Assays were performed by Alturas Analytics, Moscow, United States. PK samples were centrifuged within 60 min (2000 G, 10 min, 4 °C), aliquoted in 2.0 mL Sarstedt tubes, and stored at −80 °C within 2 h of collection. ENX-102 analysis was accomplished using a validated high-performance liquid chromatography/liquid chromatography-mass spectrometry/mass spectrometry method. The limit of quantification was 0.500 ng/mL.

2.3.7. Pharmacodynamic (PD) Assessments

NeuroCart CNS Test Battery

The NeuroCart CNS test battery was applied for PD profiling of ENX-102. Saccadic peak velocity (SPV), saccadic reaction time, and smooth pursuit eye movements were assessed to evaluate sedation. Adaptive tracking assessed hand-eye coordination and sustained attention, while body sway measured postural stability. Pupil/iris ratio was measured as a marker of autonomic nervous system modulation. The Bond and Lader visual analog scale (VAS) was used to quantify subjective effects commonly associated with a variety of CNS active drugs. The visual verbal learning task (VVLT) was performed, containing 15 words (in 3 alternate versions) covering basic aspects of verbal learning and memory: acquisition, consolidation, storage, and retrieval.

Baseline measurements were obtained one day before the first administration and were repeated twice according to CHDR Standard Operational Procedures. On day 1 and day 12, at expected steady state, NeuroCart was performed at 0.5, 1, 1.5, 3, 6, 8, 10, and 24 h post-dose. The NeuroCart endpoints included the average values of latency (s), degrees per second (degrees/s) of all correct saccades, percentage of time the participant's eyes smoothly followed the target (%),average performance percentage (%), pupil/iris diameter ratio (mm), anteroposterior movement (mm), VAS alertness (mm) for reaction time, SPV, smooth pursuit, adaptive tracking, pupil size (pupillometry), and body sway, respectively. The endpoints for VVLT were defined as immediate recall: number correct, delayed recall: number correct, and delayed recognition: number correct. See Appendix C for a more detailed technical description of the NeuroCart assessments including the VVLT.

Quantitative EEG

Resting-state qEEG recordings were performed with participants in eyes open and eyes closed conditions 0.5, 1.5, 3, 6, 8, 10, and 24 h post-dose. These recordings were conducted in a sequence with other NeuroCart assessments. qEEG recordings were obtained using a 40-channel recording system (Refa-40, TMSi B.V., Oldenzaal, The Netherlands). All signals were sampled at a rate of 1024 Hz and were filtered prior to storage using a first-order recursive high-pass filter with a cut-off frequency of 0.1 Hz to eliminate low-frequency drift. Digital markers were recorded by the amplifier to indicate the start and end of each eye state (eyes open and eyes closed). See Appendix D for a more detailed technical description of the qEEG.

2.4. Statistical Analyses

2.4.1. *In Vivo* Experiments in Rats: EPM, qEEG, and Sleep Analysis

Rat movement metrics from the EPM were analyzed using one-way analysis of variance (ANOVA) followed by Tukey's post hoc test (95% confidence interval (CI); $p < 0.05$). Data were presented as the mean ± standard error of the mean (SEM). Statistical outliers were defined as values ≥ 2 standard deviations from the group mean. Spectral qEEG data were analyzed in MATLAB version 2019 R, with two pre-dose recordings pooled as baseline. Spectral analysis included quantifying raw, relative, and percent change in spectral power across the traditionally defined EEG frequency bands: δ (0.5–3.9 Hz), θ (4–7.9 Hz), α (8–11.9 Hz), β (12–29.9 Hz), low γ (30–49.9 Hz), and high γ (50–100 Hz). The percentage change from baseline was calculated per channel, rat, dose level, and time segment, then averaged by group. Drug effects were evaluated by comparing treatment groups to the vehicle using one-way ANOVA with Dunnett's test applied to each 60-min interval. For sleep analysis, REM latency was defined as the time between first NREM onset and the first REM episode (TREM–TNREM). Time spent in each sleep state was expressed as a percentage of total time in each sleep state (mean ± SEM). Data were organized by treatment group, sleep state, and time episode and exported to GraphPad PRISM.

2.4.2. MAD in Humans: Safety, PK, and PD Analyses

Safety analyses were performed using descriptive statistics. PK parameters were summarized by dose level and day using descriptive statistics (n, arithmetic mean, SD, coefficient of variation percentage (CV%) for the arithmetic mean, median, minimum, maximum, geometric mean, and CV% for the geometric mean). The maximal concentration (C_{max}), time to reach C_{max} (T_{max}), concentration immediately prior to dosing during multiple administrations (C_{trough}), last quantifiable concentration (C_{last}), time of the last quantifiable concentration (T_{last}), apparent terminal elimination rate constant (λ_z), area under the concentration–time curve from time zero to 24 h ($AUC_{0\text{–}24}$) and extrapolated to infinity time after dosing ($AUC_{0\text{-}inf}$), apparent oral clearance (CL/F), apparent volume of distribution (Vz/F), accumulation ratio (RAUC), and dose-normalized $AUC_{0\text{–}24}$, $AUC_{0\text{-}inf}$, and C_{max} were derived directly from individual concentration–time profiles using non-compartmental analysis. A steady state was assumed if the C_{trough} values plotted against administration day formed a line with an approximately zero slope, as assessed by visual inspection. Dose proportionality for ENX-102 $AUC_{0\text{-}inf}$ (day 1), $AUC_{0\text{–}24}$ (day 12), and C_{max} (day 1 and day 12) were examined using a model with fixed effects for the log-dose and time points (day 1 or day 12) and log-dose by day interaction and random participant effect.

For analysis of the NeuroCart CNS test battery assessments, placebo participants from all cohorts were grouped for comparison. Active treatment groups were compared with placebo using mixed-model analysis of covariance (ANCOVA), with treatment, time, and their interaction as fixed factors. Participant was treated as a random factor, with baseline measurements from day −1 as covariates. Kenward–Roger approximation was used to estimate denominator degrees of freedom, and model parameters were determined using restricted maximum likelihood. Contrasts of placebo vs. single doses (0.5–5.0 mg) of ENX-102 were reported on day 1 and day 12 along with 95% CI, least square mean (LSM) estimates, and *p*-values, with no adjustments made for multiple comparisons since PD measures were all considered exploratory in nature. All PD analyses including qEEG were performed using SAS software, version 9.4 or higher.

3. Results

3.1. In Vitro Experiments

ENX-102 demonstrated selective partial PAM activity at α2β3γ2L, α3β3γ2L, and α5β3γ2L GABAA receptors, with no activity at α1β3γ2L. The E_{max} values were 89% (α2), 113% (α3), and 96% (α5), with corresponding EC_{50} values of 0.82 nM, 2.47 nM, and 0.2 nM (Figure 2B). At α1, ENX-102 showed minimal activation (E_{max} 3%), and no EC_{50} was determined. Diazepam, used as a reference, showed robust PAM activity across all subtypes (Emax 140–429%, EC_{50} 6.2–42.6 nM).

A.

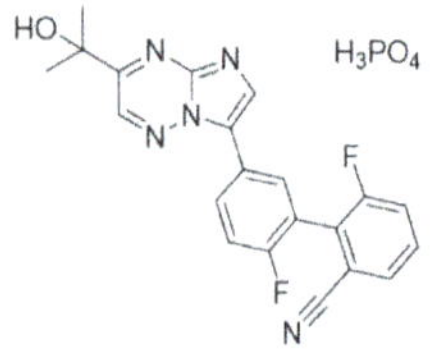

B.

	ENX-102		Diazepam	
GABA$_A$ R	E_{max}	EC_{50} (nM)	E_{max}	EC_{50} (nM)
α1β3γ2L	3% ± 2	N.R.	140% ± 30	22.2
α2β3γ2L	89% ± 5	0.82	385% ± 13	6.19
α3β3γ2L	113% ± 16	2.47	429% ± 27	42.6
α5β3γ2L	96% ± 4	0.20	208% ± 23	15.4

Figure 2. Overview of the chemical structure and *in vitro* pharmacological profile of ENX-102. (**A**) Chemical structure of ENX-102. (**B**) *In vitro* receptor activity of ENX-102 and diazepam at various $GABA_A$ receptor subtypes (α1β3γ2L, α2β3γ2L, α3β3γ2L, α5β3γ2L), showing maximum effect (E_{max}) and half-maximal effective concentration (EC_{50}) values. Data are presented as mean ± SEM. Abbreviations: $GABA_A$ R = γ-aminobutyric acid type A receptor; E_{max} = maximum effect; EC_{50} = concentration for half-maximal response; N.R., not reported.

3.2. *In Vivo Experiments in Rats*

3.2.1. Elevated Plus Maze

In the vehicle-treated group, on average, the rats spent less than 9.4% (±2.5 SE) of the time in the open arms of the EPM, while chlordiazepoxide 5.0 mg/kg significantly increased this to 39.1% (±4.9 SE) (Figure 3). Similarly, ENX-102 dose-dependently increased the percentage of time spent in open arms. While the lowest (0.1 mg/kg, corresponding to a human equivalent dose (HED) ~1.0 mg for a 70 kg individual [51]) did not significantly increase the percentage of time spent in open arms, both 0.3 mg/kg (HED~3.0 mg) and 1.0 mg/kg (~10 mg) showed robust anxiolytic-like effects with significantly greater percentages of time spent in open arms (34.9% (±6.3 SE), 30.9% (±6.7 SE)). Notably, the magnitudes of the anxiolytic-like effects for ENX-102 0.3 mg/kg and 1.0 mg/kg were comparable to that of the positive control, chlordiazepoxide (39.1% (±4.9 SE)).

Anxiolytic-like effects of ENX-102 were maintained with sub-chronic treatment, as rats maintained or increased their time spent in the open arms (32.7% (±5.6 SE) at 0.1 mg/kg, 41.7% (±5.8 SE) at 0.3 mg/kg, and 43.3% (±6.3 SE) at 1.0 mg/kg) for up to 14 days of dosing (Figure 3B). However, these differences were not statistically significant compared to vehicle-treated controls. No animals were excluded from this study or analysis.

3.2.2. Quantitative EEG

Compared to baseline, a single oral dose of ENX-102 in the range of 0.1 mg/kg to 1.0 mg/kg (corresponding to a HED of ~1.0 to 10 mg) dose-dependently increased the β-band power in the quiet wake state between 1.5 and 4.5 h post-dose, aligning with its T_{max} in rats. A single dose of lorazepam i.p. at 1.0 mg/kg, with a T_{max} of approximately 8 min [52], showed a significant increase in β-band power at 1 h post-dose ($p < 0.05$; Figure 4). In addition, ENX-102 at 0.3 and 1.0 mg/kg and lorazepam at 1.0 mg/kg significantly increased the low-γ-band power compared to baseline between 1.5 and 4.5 h post-dose and lorazepam at 1 h post-dose, respectively ($p < 0.05$; Figure 4). Neither ENX-102 nor lorazepam significantly altered the lower frequency bands (δ, θ) or high-γ power, while lorazepam increased the α-band power at a single timepoint, 1 h post-dose, compared to baseline. See Figure S2 for spectral changes observed at 2 h post-dose, which especially highlight the dose-dependency of the γ power increase.

A.

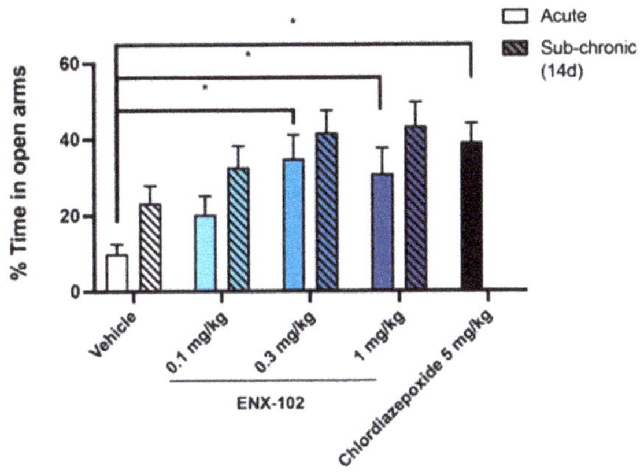

B.

ENX-102 Dose	% Time in Open Arms		
	Acute 1-hour post-dose (% ± SE)	24 hours post-dose (% ± SE)	Sub-chronic 14 days (% ± SE)
Vehicle	9.4 ± 2.5	15.1 ± 3.6	23.3 ± 4.6
Chlordiazepoxide 5 mg/kg	39.1 ± 4.9	N.A.	N.A.
ENX-102 0.1 mg/kg	20.3 ± 4.8	22.0 ± 4.8	32.7 ± 5.6
ENX-102 0.3 mg/kg	34.9 ± 6.3	18.0 ± 2.2	41.7 ± 5.8
ENX-102 1.0 mg/kg	30.9 ± 6.7	28.7 ± 5.6	43.3 ± 6.3

Figure 3. Effects of orally dosed ENX-102 in the rat elevated plus maze (EPM) test. (**A**) ENX-102 was tested in the rat EPM at 0.1, 0.3, and 1 mg/kg dose levels (p.o.) compared with vehicle or the positive control, chlordiazepoxide at 5 mg/kg (i.p.) administered acutely. One hour post-ENX-102 treatment (blue shaded bars) at 0.3 and 1 mg/kg, the percentage of time spent in the open arms of the EPM was significantly increased as compared to vehicle treatment (white bar), and was at a level comparable to that of the positive control chlordiazepoxide (black bar). Data were analyzed by ANOVA followed by post hoc analyses where appropriate. * $p < 0.05$ compared to vehicle. (**B**) Mean percentage of time-spent and standard error (SE) in the open arms at three time points: acute (1 h post-dose), 24 h post-dose, and sub-chronic (after 14 days dosing). N.A. = not applicable.

Compared to baseline during NREM, ENX-102 0.1 mg/kg to 1.0 mg/kg dose-dependently decreased the δ- and θ-band power, and 0.3 mg/kg to 1.0 mg/kg dose-dependently increased the low-γ-band power at 2, 4, 6, 8, 10, and 12 h post-dose and the high-γ-band power at 2, 3, 5 and 6 h post-dose (Figure S3). Lorazepam decreased the θ power at all time points except 7, 10, 11, and 12 h post-dose and the β power at 3, 4, 5, 6, and 8 h post-dose, which seemed comparable to ENX-102 0.3 mg/kg ($p < 0.05$; Figure S3). ENX-102 did not consistently affect the α or β power in rodents, while lorazepam did not consistently affect the δ, α, or γ power (Figure S3).

These data reveal a dose-dependent EEG spectral signature for ENX-102 that is distinct from that of lorazepam.

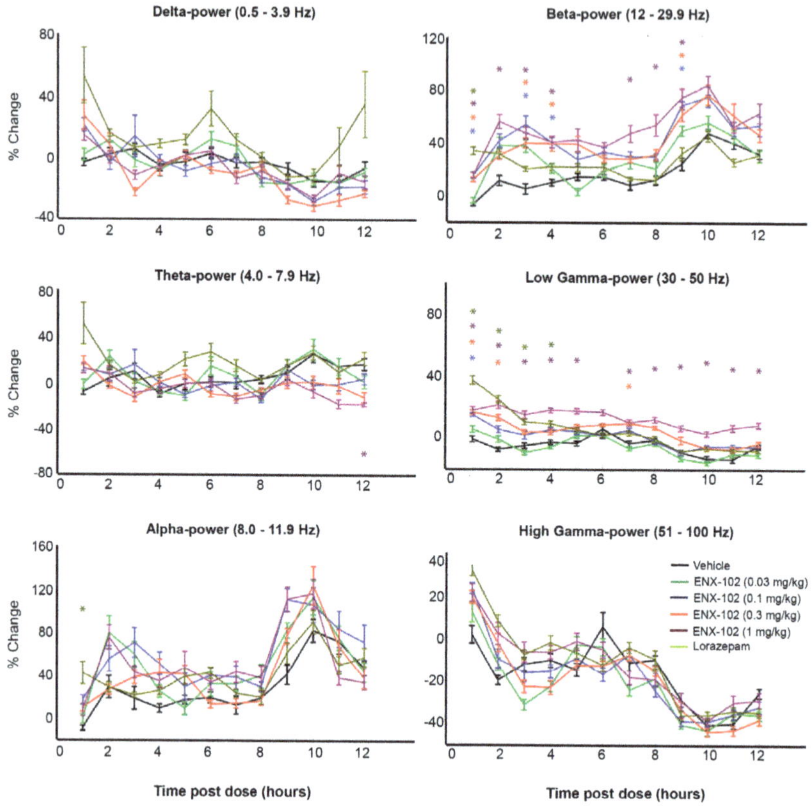

Figure 4. qEEG in rats during quiet wake (0–12 h post-dose)—frontal cortex ENX-12 and lorazepam: Doses tested: vehicle, ENX-102 (0.03, 0.1, 0.3, 1.0 mg/kg), and lorazepam (1.0 mg/kg). Percent change (mean ± SEM) in δ, θ, α, β, low-γ, high-γ bands compared to baseline. Statistical significance indicated with * at different timepoints for each frequency band. Statistically significant increases at T_{max} (1.5–4.5 h) compared to baseline were observed on the β band for ENX-102 0.1, 0.3, and 1.0 mg/kg and in the low-γ band for ENX-102 0.3, 1.0 mg/kg. For lorazepam, statistically significant increases at 1 h post-dose compared to baseline were observed on the β band, α band, and low-γ band. * $p < 0.05$ compared to vehicle.

3.2.3. Sleep Architecture

Lorazepam significantly reduced the percentage of time spent in active wake as compared to vehicle-treated controls ($p < 0.05$; Figure 5A). In contrast, ENX-102 did not significantly impact the percentage of time spent in active wake at any of the doses within the first two hours post-dose, which coincided with the ENX-102 T_{max} in rats approximately 1.5 to 4.5 h post-dose. When examined over a time period of up to 24 and 48 h, hypnograms reveal no significant differences in the proportion of time spent in REM or NREM sleep stages and active or quiet wake stages in ENX-102-treated groups (Figure 5B,C). Based on the onset of REM sleep, the vehicle-treated rats required ~40 min to go into NREM sleep, which was significantly reduced by lorazepam (~20 min), whereas ENX-102 did not impact the latency to onset of NREM sleep at any of the doses tested ($p < 0.05$; Figure S4A).

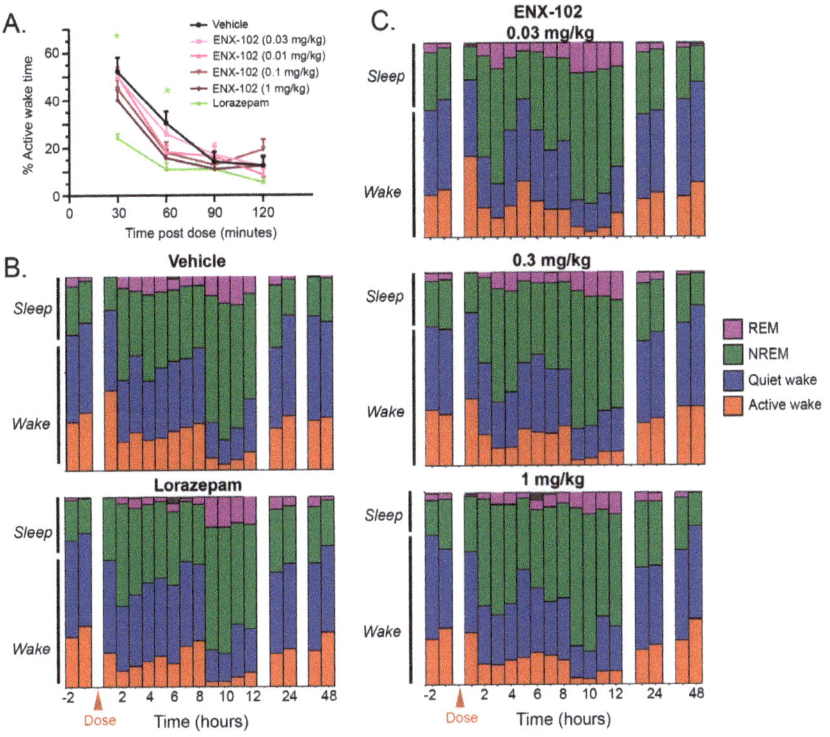

Figure 5. Effect of ENX-102 on sleep architecture in rats. Effect of ENX-102 (at four dose levels) on sleep architecture was examined. (**A**) Data over the first two hours after treatment, with peak plasma levels for ENX-102 and lorazepam, are shown. Lorazepam (green) significantly reduced % time spent in active wake within an hour after administration; ENX-102 (pink) at the different doses tested, 0.01 mg/kg, 0.03 mg/kg, 0.1 mg/kg, or 1 mg/kg, did not significantly impact % time in active wake as compared to vehicle (black)-treated controls. (**B**,**C**) Hypnograms showing proportion of time spent in sleep stages (REM, purple; NREM, green) or awake (quiet wake, blue; active wake, red), for all groups from two hours before dosing up to 48 h after dosing. * $p < 0.05$ compared to vehicle (ANOVA, Dunnett's test). Abbreviations: REM = rapid eye movement; NREM = non-rapid eye movement.

Together, these data show that ENX-102 did not cause sedation or increase the propensity to sleep in rats, even at the highest dose tested (1.0 mg/kg), corresponding to a HED of ~10.0 mg for a 60 kg human.

3.2.4. Pharmacokinetics

Average C_{max} and $AUC_{0\text{–}t}$ increased dose-dependently over the 0.1 to 3.2 mg/kg dose range (C_{max} 35 to 1505 ng/mL; HED = ~3 to 30 mg), with an average T_{max} and $T_{1/2}$ ranging from 1.5 to 4.5 h and 15.5 to 19.1 h, respectively (Table S2).

3.3. Multiple Ascending Dose (MAD) Study in Healthy Human Participants

3.3.1. Safety

AE data are provided in Tables S3 and S4. Over the 12-day dosing period, a total of 408 post-dose AEs were reported, of which 354 (87%) occurred in 30 (100%) participants receiving ENX-102 and 54 (13%) occurred in 9 (90%) participants receiving placebo. The majority of reported AEs (402/408 = 98.5%) in most participants (34/40 = 85%) were mild in intensity and all were self-limiting (data on file). No serious or severe AEs occurred. Five participants experienced moderate AEs, accounting for one event each of gastroenteritis (reported 6 days after final dosing on day 12), headache (reported 9 days after final dosing on day 12), balance disorder (reported on day 1 of dosing), and muscle pain (reported on day 7 of dosing) and two events of somnolence (reported on day 1 and day 6 of dosing in two different participants). These six moderate AEs represent approximately 1.5% of all post-dose AEs (6 of 408). There was no clear dose-dependent increase in incidence of AEs. On day 1, the ENX-102-treated groups reported relatively more AEs than the placebo-treated group. On Day 12, the number of AEs reported was generally lower than on day 1, and they were approximately equally distributed across ENX-102 and placebo groups.

The most commonly reported AEs over the 12-day dosing period were somnolence and fatigue in both the ENX-102- and placebo-treated groups. In the ENX-102-treated group, somnolence and fatigue accounted for 28.5% (101/354) and 22.5% (80/354) of the total AEs, respectively. Similarly, in the placebo-treated group, somnolence and fatigue were the most commonly reported AEs, with somnolence occurring in 13% (7/54) and fatigue occurring in 27.8% (15/54) of participants, respectively. The incidence of somnolence for ENX-102 0.5 mg, 1.5 mg, 2.0 mg, and 5.0 mg was lower on day 12 compared to day 1 (Table S3, Figure 6B,C). Somnolence and fatigue occurred at a median time of 2.5 h (range: 2.25–23 h for somnolence and 2.38 to 7 days) after administration of ENX-102 (Figure 6A). No clinically relevant changes were observed in laboratory evaluations, vital signs, or ECGs. None of the participants reported suicidal ideation or behavior on the C-SSRS. No deaths or serious adverse events (SAEs) occurred.

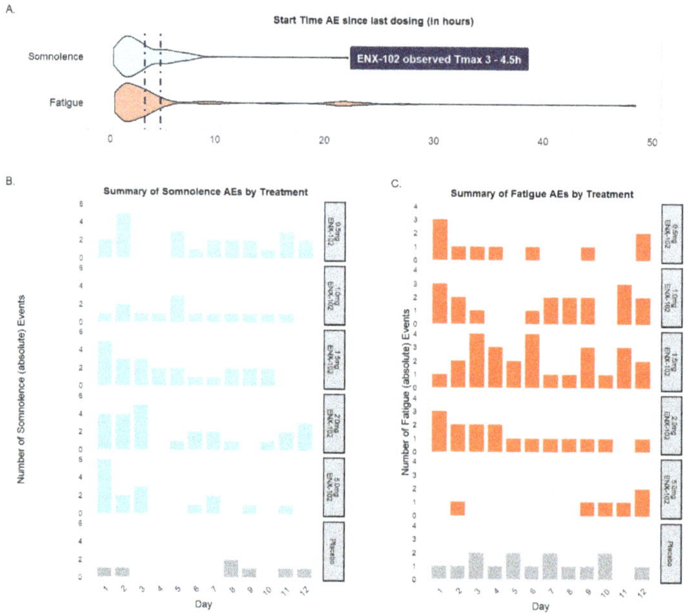

Figure 6. Summary of the most commonly reported adverse events (AEs), somnolence and fatigue, in the human multiple ascending dose (MAD) study. (**A**) Violin plots showing the distribution of onset times of fatigue and somnolence after dosing, relative to the median T_{max} window (3–4.5 h) for ENX-102, across days 1–12. (**B**) Absolute number of somnolence AEs for placebo and ENX-102 0.5, 1.0, 1.5, 2.0, and 5.0 mg per dosing day. (**C**) Absolute number of fatigue AEs for placebo and ENX-102 0.5, 1.0, 1.5, 2.0, and 5.0 mg per dosing day.

These findings demonstrate that repeated dosing of ENX-102 is safe and well tolerated, producing AEs consistent with GABAergic pharmacology, although, importantly, in the absence of sedation, and no evident dose-related increase in somnolence or fatigue was observed. Finally, the occurrence of fatigue and somnolence appeared to coincide with peak plasma exposures and attenuated after repeat administrations.

3.3.2. Pharmacokinetics (PK)

ENX-102 was well-absorbed following oral administration, with peak plasma concentrations being reached within 3–4.5 h, followed by a bi-exponential decrease, with a mean terminal half-life between 39 and 66 h (Figure 7A, Table S5). Table S5 summarizes the PK results on days 1 and 12 after the multiple doses of ENX-102 and shows that mean C_{max} and $AUC_{0–24}$ generally increased dose- and concentration-proportionally over the investigated dose range. On day 1, the C_{max} ranged from 4.6 to 49.7 ng/mL, while, on day 12, it varied from 11.7 to 99.1 ng/mL. The mean half-life ($T_{1/2}$) behaved as anticipated for all dose levels, except for the 5.0 mg group, which exhibited a relatively shorter half-life. Accumulation of ENX-102 at plasma steady-state concentrations (Css) was evident, with a mean accumulation ratio (R_{ac}) ranging from 3.0 to 4.0. For ENX-102 < 2.0 mg, the C_{trough} slope stabilized after days 6 to 12 days of dosing (Figure 7B), indicating that a steady state was achieved. However, the 2.0 mg and 5.0 mg doses did not show the same C_{trough} slope stabilization as lower doses, suggesting the emergence of non-linear PK at higher doses (Figure 7B). The CV% ranged between 30 and 40% for the $T_{1/2}$, C_{max}, and AUC across the investigated dose range.

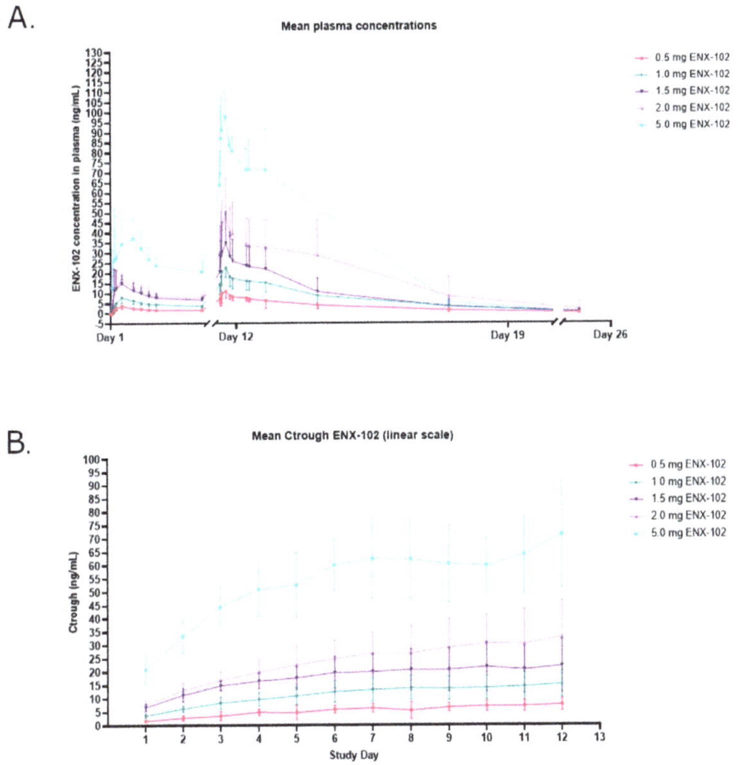

Figure 7. Pharmacokinetic profile of ENX-102. (**A**) Mean plasma concentrations of ENX-102. Mean plasma ENX-102 concentrations in ng/mL for doses of 0.5 mg, 1.0 mg, 1.5 mg, 2.0 mg, and 5.0 mg. (**B**) Mean C_{trough} day 1 to day 12 (linear scale). Mean C_{trough} in ng/mL with standard deviation for doses of 0.5 mg, 1.0 mg, 1.5 mg, 2.0 mg, and 5.0 mg.

Overall, ENX-102 PK demonstrated dose-proportionality, relatively low variability, and a long half-life, with steady-state plasma concentrations being reached by day 12, which make it well-suited for chronic treatment.

3.3.3. Pharmacodynamic (PD) Effects

A detailed summary of the PD results can be found in Table S6. Figures 8 and S5 provide a visual overview of the most informative PD effects of ENX-102 across dose levels and time points. Compared with placebo, the ENX-102 doses ≥ 1.0 mg statistically significantly decreased the SPV on day 1 following a single dose, with estimated differences of −51.46 (95% CI: −80.31 to −22.60), −52.40 (95% CI: −80.83 to −23.97), −65.51 (95% CI: −94.43 to −36.59), and −76.22 (95% CI: −106.35 to −46.09) for the 1.0 mg, 1.5 mg, 2.0 mg, and 5.0 mg doses, respectively (all $p < 0.001$; Table S6). This effect was sustained on day 12, with estimated differences compared with placebo of −51.59 (95% CI: −80.44 to −22.73; $p < 0.001$), −44.11 (95% CI: −73.03 to −15.19; $p = 0.004$), and 5.0 mg −51.39 (95% CI: −81.13 to −21.65; $p = 0.001$) for the 1.0 mg, 2.0 mg, and 5.0 mg doses, respectively (Table S6). Moreover, the SPV reductions for ENX-102 1.0 mg, 2.0 mg and 5.0 mg did not return to baseline by 24 h post-dose on either day 1 or day 12, and the SPV remained reduced prior to the last dose on day 12 (Figure S5A and Figure 8A). Decreases in SPV generally emerged approximately 0.5 h post-dose and reached a peak between 1.5 and 3 h post-dosing on both day 1 and day 12. ENX-102 5.0 mg was the exception, as the peak SPV reductions appeared relatively later, at 6 h post-dose (Figure S5A).

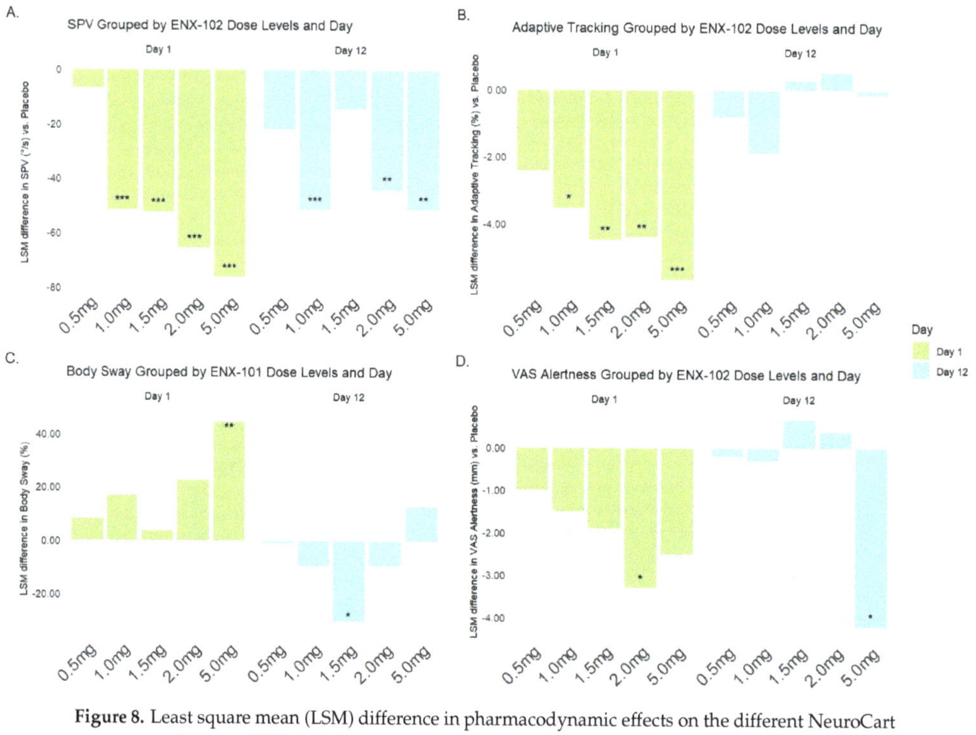

Figure 8. Least square mean (LSM) difference in pharmacodynamic effects on the different NeuroCart assessments between ENX-102 dose groups and placebo after single dose (day 1) and multiple dose (day 12). (**A**) Saccadic peak velocity (SPV) (%/s). (**B**) Adaptive tracking (%) by dose group and day, and panel. (**C**) Body sway (%) by dose group and day, and panel. (**D**) Visual Analogue Scale (VAS) alertness by dose group and day. Statistical significance is indicated by: *** $p < 0.001$; ** $p < 0.01$; * $p < 0.05$ baseline corrected compared with placebo.

In contrast to the SPV, ENX-102 doses ≥ 1.0 mg statistically significantly decreased adaptive tracking compared with placebo on day 1 following a single dose, with estimated differences of −3.532 (95% CI: −6.382 to −0.682; p = 0.0164), −4.470 (95% CI: −7.483 to −1.456; p = 0.0046), −4.399 (95% CI −7.310 to −1.489; p = 0.0040), and −5.660 (95% CI −8.530 to −2.791; p < 0.001) for the 1.0 mg, 1.5 mg, 2.0 mg, and 5.0 mg doses, respectively (Table S6). However, the effect returned to baseline by 12 h post-dose and had disappeared by day 12 (Figure 8; Figure S5B; Table S6). In addition, only ENX-102 5.0 mg statistically significantly increased the SRT and body sway compared to placebo on day 1 following a single dose, with estimated differences of 0.0198 (95% CI: 0.0032 to 0.0365); p = 0.0206) for SRT and 44.8 (95% CI: 10.9 to 89.0; p = 0.0077) for body sway. On day 12, body sway was statistically significantly decreased for ENX-102 1.5 mg, with an estimated difference of −30.2 (95% CI: −46.8 to −8.4; p = 0.0107), but not at any other dose, and no effects were present at any dose for SRT (Figure S5F; Table S6). Furthermore, smooth pursuit eye movements, the pupil/iris ratio, and VAS alertness were comparable to placebo and relatively inconsistent across the investigated dose range on day 1. On day 12, VAS alertness was statistically significantly decreased for ENX-102 5.0 mg, with an estimated difference compared with placebo of −4.2 (95% CI: −7.5 to −0.9); p < 0.0141), but not at any other dose (Figure 8D; Table S6). No effects were present at any dose for smooth pursuit eye movements or the pupil/iris ratio (Table S6).

VVLT performance was comparable to placebo across the investigated dose range on day 1, except for ENX-102 5.0 mg, which showed statistically significant decreases in immediate word recall, delayed word recall, and delayed word recognition, with estimated effects compared with placebo of −3.8 (95% CI: −6.8 to −0.9; p = 0.0119), −4.9 (95% CI: −9.0 to −0.8; p = 0.0189), and −3.0 (95% CI: −6.0 to −0.0; p = 0.0485), respectively. On day 12, neither immediate nor delayed word recall was affected, while delayed word recognition exhibited inconsistent effects across doses with no dose relationship (Table S6).

Together, these data suggest that ENX-102 sustained reduced arousal following multiple dosing, as indicated by a decreased SPV, while undesired CNS effects, such as impairments in sustained attention, memory dysfunction, postural stability, or subjective alertness, diminished over time, were absent by day 12, or presented an inconsistent pattern that would not suggest clinical meaningfulness.

3.3.4. Quantitative EEG

A detailed summary of qEEG results can be found in Table S7. Figure 9 illustrates the direction of EEG power changes induced by ENX-102 in both humans and rats, measured under eyes-closed conditions in the frontal cortex in rats and the fronto-central region in humans. On day 1, compared with placebo and under eyes-closed conditions, ENX-102 1.5 mg statistically significantly reduced both α and γ in the parieto-occipital region (Pz-O1), with estimated differences of −36.0 (95% CI: −57.1 to −4.7; p = 0.0290) and −42.0 (95% CI: −60.2 to −15.4; p = 0.0056), respectively. ENX-102 2.0 mg and 5.0 mg statistically significantly increased the β power in the fronto-central region (Fz-Cz), with estimated differences of 74.9 (95% CI: 29.5 to 136.1; p = 0.0005) and 71.3 (95% CI: 26.9 to 131.4; p = 0.0008), respectively, and reduced the γ power in the parieto-occipital region (Pz-O1), with estimated differences of −34.8 (95% CI: −55.2 to −5.0; p = 0.0266) and −36.3 (95% CI: −56.3 to −7.3; p = 0.0197), respectively (Table S7). On day 12, under eyes-closed conditions, ENX-102 at all dose ranges statistically significantly reduced the δ power in the fronto-central region compared to placebo. The estimated differences range from −29.5 (95% CI: −48.9 to −2.7; p = 0.0340) to −52.4 (95% CI: −65.9 to −33.4; p < 0.0001). In addition, the parieto-occipital regions showed statistically significant reductions in δ power at ENX-102 doses of 1.0 mg and above. Furthermore, ENX-102 at doses ≥ 1.0 mg statistically significantly reduced the θ power in the fronto-temporal region, with estimated differences ranging from −49.2 (95% CI: −64.9 to −26.6; $p = 0.0006$) to −72.9 (CI −81.1; −61.2; $p < 0.0001$). ENX-102 1.5 and 2.0 mg also led to statistically significant reductions in α power across all regions compared to placebo, with estimated differences as follows: Fz-Cz: −56.4 (95% CI: −73.5 to 28.2; $p = 0.0018$) and −58.5 (95% CI: −74.8 to −31.9; $p = 0.0009$), Pz-O1: −49.5 (95% CI: −66.1 to −24.8; $p = 0.0013$) and −46.1 (95% CI: −63.8 to −19.7; $p = 0.0032$), and Pz-O2: −48.0 (95% CI: −64.4 to −24.2; $p = 0.011$) and −53.4 (95% CI: −68.0 to −32.2; $p = 0.0002$). Additionally, ENX-102 at 2.0 mg statistically significantly increased the γ power in the fronto-central region compared to placebo, with an estimated difference of 36.1 (95% CI: 2.0 to 81.7; $p = 0.0366$). Finally, ENX-102 5.0 mg not only statistically significantly reduced the δ, θ, and α power across all regions, but also increased and decreased the γ power at Fz-Cz and Pz-O2, respectively ($p < 0.05$ for all comparisons; see Table S7 for detailed estimates).

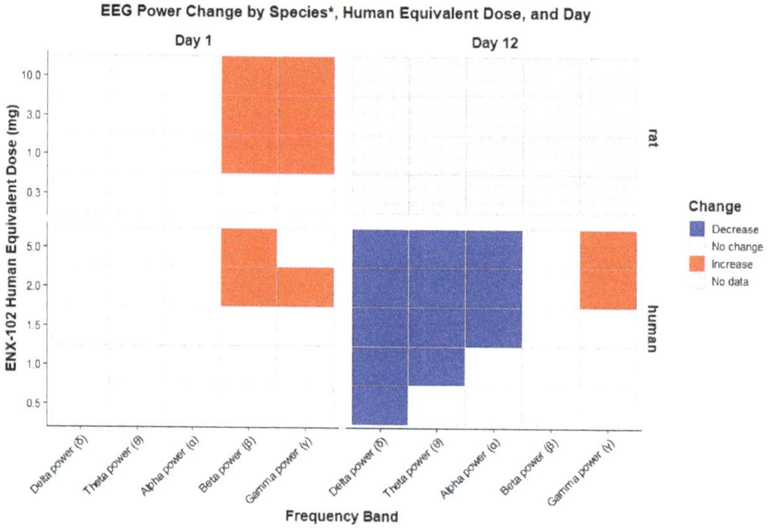

Figure 9. Direction of EEG power change by species, human equivalent dose (HED), and day. (* frontal cortex in rats and Fz-Cz in humans) *Y*-axis shows ENX-102 dose levels: in rats (0.03, 0.1, 0.3, 1.0 mg/kg; approximately equivalent to 0.3, 1.0, 3.0, and 10 mg HED for a 60 kg human, respectively) and in humans (0.5, 1.0, 1.5, 2.0, 5.0 mg). *X*-axis displays EEG power spectral bands (δ, θ, α, β, γ). Red indicates a statistically significant increase in EEG power (vs. placebo in humans, vs. baseline in rats); blue indicates a significant decrease (vs. placebo in humans); white indicates no significant change; grey denotes no data available for rats on day 12. On day 1, significantly increased β and γ power is observed at higher doses (2.0 mg in humans, 1.0 mg/kg in rats). On day 12 in humans, a significant decrease in δ, θ, and α power was observed starting at 0.5 mg, with a significant increase in γ power only at higher doses (2.0 and 5.0 mg).

In summary, the most prominent observed effect following single doses of ENX-102 was increased β power, whereas the most consistent steady-state effects following multiple doses included decreased δ, θ, and α power, across the investigated dose range of 0.5 to 5.0 mg.

4. Discussion

The current report summarizes the preclinical and clinical pharmacological characteristics of ENX-102, a novel $GABA_AR$ α2-/3-/5-selective PAM that is currently under development for anxiety disorders. *In vitro* studies demonstrated that ENX-102 acts as a partial PAM selectively at α2/3/5 subunit-containing $GABA_A$Rs and a functional antagonist at α1 subunit-containing $GABA_A$Rs. This is a distinguishing profile from diazepam, a non-selective $GABA_A$ PAM, as diazepam additionally activates α1 subunit-containing $GABA_A$Rs associated with unwanted side-effects. In humans, ENX-102 dose-dependently decreased the saccadic peak velocity (SPV), a pharmacological biomarker that reflects CNS arousal and is therefore a putative surrogate for anxiolysis, which was sustained at steady-state plasma exposures at doses $\geq$ 1.0 mg, without any indication for CNS suppression, including sedation. These findings are consistent with preclinical data in rodents, demonstrating sustained anxiolytic-like effects in rats at doses $\geq$ 0.3 mg/kg (human equivalent dose (HED) ~3 mg (Figure 3)) that were devoid of sedative effects up to 1.0 mg/kg (HED ~ 10 mg) (Figure 9). Moreover, resting state qEEG revealed a distinct spectral power profile indicative of $GABA_A$R subtype selectivity in both rodents and humans. Finally, ENX-102 exhibited favorable safety and pharmacokinetic (PK) profiles in humans, showing good tolerability and appropriate drug exposure with multiple dosing, which supports its continued clinical development.

The safety profile of ENX-102 in humans appears favorable, particularly compared to non-selective $GABA_A$R PAMs, which are often associated with sedation and adverse events related to CNS suppression following single, subchronic, and/or multiple dosing [11]. Notably, no sedation-related AEs were reported in the human MAD study, and although mild somnolence and fatigue were reported, these effects did not demonstrate dose-dependence. Moreover, the occurrence of somnolence and fatigue was transient, associated with plasma C_{max}, and most pronounced on day 1, yet it was comparable with placebo by day 12 following repeated daily dosing. These findings are consistent with ENX-102's lack of activity or antagonism at the $GABA_A$R α1 subunit, which is typically associated with non-selective $GABA_A$-PAM-mediated sedative effects. Considering that somnolence appeared to be temporally associated with peak plasma concentrations, evening dosing may mitigate the occurrence of somnolence and fatigue when initiating dosing in future studies.

In humans, the PK profile demonstrated dose-proportionality and a relatively long half-life, which makes ENX-102 suited for chronic treatment. With once-daily dosing, a steady state was achieved between 6 and 12 days, with the T_{max} and $T_{1/2}$ ranging from 3.0 to 4.5 h and 39 to 66 h, respectively, and the C_{max} and AUC_{0-24} increasing proportionally with the dose. Accumulation was evident up to 6 days of dosing, due to the long half-life, with no additional accumulation once a steady state was reached. The intersubject PK variability was minimal. Beyond its favorable safety profile, ENX-102 exhibits a pharmacokinetic (PK) profile that supports once-daily oral dosing in future trials.

On day 1 in humans, ENX-102 $\geq$ 1.0 mg not only reduced SPV dose-dependently but also reduced adaptive tracking and/or VAS alertness, while it increased SRT and body sway. In contrast to day 1, however, at steady-state exposures on day 12, reductions in SPV were sustained for ENX-102 1.0 mg, 2.0 mg, and 5.0 mg, while adaptive tracking, VAS alertness, and/or body sway remained unaffected for ENX-102 1.0 mg and 2.0 mg. This suggests no tolerance to the decrease in arousal and potential anxiolytic effects for ENX-102, whereas potentially adverse CNS side effects such as decreased sustained attention, subjective alertness, and postural stability were well tolerated with repeated administration. At a higher, supratherapeutic dose, the VAS alertness remained reduced at 5.0 mg. Furthermore, SPV reductions were sustained beyond the attenuation of adaptive tracking, VAS alertness, and body sway effects for ENX-102 $\geq$ 1.0 mg on day 1, and similarly at steady-state exposures on day 12. SPV reductions were sustained up to 24 h after the final dose for ENX-102 1.0 mg, 2.0 mg, and 5.0 mg. In fact, the mean SPV decreases of −40 to −50 degrees per second for ENX-102 1.0 mg and 2.0 mg on day 12 were comparable to those previously demonstrated by a single dose of lorazepam 2.0 mg administered orally [15,16,18,23]. Single doses of ENX-102 therefore effectively reduced arousal and, in parallel, also undesirably decreased sustained attention, subjective alertness, and postural stability, though to a lesser degree than lorazepam 2.0 mg. The reduction in adaptive tracking, VAS alertness, and body sway following a single dose of ENX-102 were also less pronounced than those inducted by lorazepam 2.0 mg [16,18,23].

Consistent with the related AEs of somnolence and fatigue, the transient undesired PD effects are hypothesized to result from an initial supraphysiological activation of GABA-A-R $\alpha 2/3$ subunits, leading to a rapid increase in neural inhibition (initial GABA burst) at doses of 1.5 mg and higher. These effects are not attributed to loss of subtype selectivity, considering ENX-102's functional antagonism at $\alpha 1$ subunits. At steady-state exposures, however, reduced arousal was robustly sustained for ENX-102 1.0 mg and 2.0 mg, in the absence of deleterious effects on sustained attention, subjective alertness, and/or psychomotor function. Arguably, this is further supported by preclinical data, considering that, in the elevated plus maze (EPM), a well-established rodent model of anxiety, ENX-102 demonstrated anxiolytic-like effects at doses ≥ 0.3 mg/kg (human equivalent dose—HED~3 mg) which were comparable to those of chlordiazepoxide, but no sedative effects were observed up to 1.0 mg/kg (HED~10 mg). These results indicate that ENX-102 1.0 mg to 2.0 mg may effectively sustain reduced arousal over time in the absence of CNS-depressant effects in humans, which was confirmed by the absence of clinical sedation and stable MOAA/S scores from day 1 to day 12 at all dose levels. This is further supported by preclinical data indicating that, at the highest tested dose (HED~10 mg), ENX-102 did not reduce the active wake time in rats, unlike lorazepam. Finally, ENX-102 did not negatively impact memory function in humans, neither following a single dose nor at steady-state exposures. Altogether, these findings align with ENX-102's lack of $\alpha 1$ subunit activity and behavioral effects in preclinical species, differentiating it from non-selective $GABA_AR$ PAMs [17] and supporting ENX-102's potential sustained anxiolytic efficacy without sedation or cognitive impairment, as opposed to non-selective $GABA_AR$ PAMs.

SPV reductions following single doses on day 1 were larger than those demonstrated at steady-state exposures on day 12 for ENX-102 1.5 mg and higher. For instance, ENX-102 at 2.0 mg and 5.0 mg reduced the SPV in the range of −65.5 to −76.2 degrees per second following single doses, while, at steady-state exposures, ENX-102 2.0 mg and 5.0 mg reduced the SPV to −44.1 and −51.4 degrees per second, respectively. However, the reduction from day 1 to day 12 is more consistent with transient effects from an initial burst of GABA on day 1 rather than a true loss of function over time given that the magnitude of SPV decrease on day 12 was still in the range of an acute anxiolytic dose of lorazepam 2.0 mg. It is well established that tolerance develops at differential rates to different behavioral effects of $GABA_AR$ PAMs. For example, in contrast to the development of tolerance for somnolence and sedative effects of non-selective $GABA_AR$ PAMs, tolerance for anxiolytic effects is believed to develop more gradually, to develop partially, or to not occur at all, even following prolonged treatment [53]. Similarly, patients with panic disorder and generalized anxiety disorder demonstrated sustained anxiolytic efficacy that did not diminish with long-term alprazolam or diazepam treatment [53]. Notably, in the MAD study, doses associated with decreased adaptive tracking and VAS alertness and/or increased body sway on both days 1 and 12 were not only associated with relatively greater SPV reductions, but also AEs such as somnolence. Thus, SPV changes following single doses may be conflated by coinciding impairments in attention, alertness, and postural stability on day 1, rather than less sizeable reductions in arousal on day 12 resulting from tolerance.

Resting-state qEEG findings in both humans and rodents seem to indicate an ENX-102 spectral power signature that was not only distinct from non-selective $GABA_AR$ PAMs [44], but also consistent with other α2-/3-selective PAMs [15,16,19,22,43,44]. In both species, single-dose ENX-102 increased the β and, to a certain extent, albeit less consistently, γ power at comparable plasma exposures, as the C_{max} ranged from ~20 ng/mL to ~50 ng/mL (1.5 mg to 5.0 mg) and 35 ng/mL to 362 ng/mL (HED of 1.0 to 10 mg) in humans and rats, respectively, on day 1, while lorazepam, in contrast, has been reported to increase β and δ activity while decreasing the α power in humans [15–17]. In rats, changes in θ band activity have been generally inconsistent [35] and could not readily be distinguished from δ band activity [32,33], while the data on changes in γ band activity are currently too limited to allow for adequate interpretation [17,54]. However, at steady-state plasma exposures for ENX-102 in humans, the β power was unaffected and the γ power remained inconsistent, while, most importantly, decreases in δ, θ, and α power were evident, which was reflective of the single-dose qEEG signatures previously reported for the $GABA_AR$ α2/3-selective PAMs NS11821, AZD6280, and PF-06372865 [15,16,19]. Although the clinical and/or therapeutic implications of the emergence of decreased δ, θ, and α power coinciding with the disappearance of β enhancement following repeated dosing remain to be elucidated, such changes in power spectrum configuration may be indicative of GABAergic adaptation ensuing from sustained $GABA_AR$ α2/3/5 target engagement. This interpretation is supported by the observed changes over time in somnolence in absence of sedation, adaptive tracking, VAS alertness, SRT, and body sway in the MAD. Additionally, regional differences in γ activity were evident, with indications for increased activity in frontal cortical regions that contrasted with predominantly decreased activity in parietal-occipital regions. Considering that the frontal cerebral cortex is primarily involved in processes related to cognitive control, whereas the parieto-occipital regions are responsible for sensory integration [55], the observed reductions in parietal-occipital γ activity may indicate modulation of sensory integration by ENX-102, which has been implicated in functional processes relevant to anxiety-associated hyperarousal such as disrupted valence and/or autonomic nervous system hyperactivation [56]. This notion is supported by recent resting-state functional magnetic resonance imaging (fMRI) findings for alprazolam that show network connectivity alterations in individuals with social anxiety disorder, particularly in regions involved in sensory and motor processing- including visual, somatosensory, and motor cortices [57]. Notably, regions particularly involved in somatosensory and visual processing demonstrated considerable functional-anatomical overlap with parietal-occipital areas [58], which provides further plausibility to the notion that modulation of sensory-related activity may contribute to attenuated arousal. While there are ongoing efforts to correlate qEEG changes with fMRI-based functional connectivity, current qEEG hardware lacks the spatial resolution to reliably distinguish changes in γ power originating from different cortical regions [59]. At any rate, these findings arguably implicate ENX-102-mediated parietal-occipital modulation of γ activity in functional changes from which, at least theoretically, anxiety disorders might benefit. Finally, ENX-102 demonstrated rather inconsistent effects on α power across species since, in humans, it was decreased at higher doses (≥1.5 mg), while, in rats, no significant effects were observed. In previous studies, the same seemed to hold true for non-selective $GABA_AR$ PAMs, since, although these often decreased the α power in humans, such effect was less consistently observed in rodents, as both increases [44] and no changes have previously been found [60]. Similarly, lorazepam increased the α power at a single timepoint in rats in the current study, which possibly resulted from a different α frequency range definition compared to other studies, rather than representing a pharmacological effect. Additionally, qEEG work in rodents has historically focused relatively more on other frequency bands, particularly β bands, which potentially contributes to inconsistent α power findings. Altogether, qEEG demonstrated a distinct spectral power signature in both rodents and humans that was not only similar in both species but also evident at comparable plasma exposures, and, as such, these findings illustrate qEEG to be a potentially valuable and translatable pharmacological biomarker in the early development of subtype-selective $GABA_AR$ PAMs.

Several strengths and limitations of the current report merit mentioning. While the EPM is primarily suited for assessing acute anxiolytic effects due to habituation, ENX-102's effect appeared sustained over time. The primary goal was to evaluate ENX-102's anxiolytic potential alongside chlordiazepoxide; ideally, a follow-up study designed to assess long-term efficacy would strengthen these findings.

In the MAD study, ENX-102's PD profile was characterized using the NeuroCart test battery, which has previously distinguished the PD profiles of several $GABA_AR$ $\alpha2/3$ subtype-selective agonists from non-selective $GABA_AR$ PAMs in humans [18,24]. However, as all previous studies examined single-dose effects, it did not allow for comparison of ENX-102's PD profile following multiple-dose administration. Similarly, while single-dose qEEG comparisons between rats and humans were possible, the lack of multiple-dose qEEG data in rats limits the ability to draw definitive translational conclusions about ENX-102's potential effects at a steady state. ENX-102 1.5 mg revealed rather unexpected findings at steady-state exposures. Unlike the 1.0 mg and 2.0 mg doses, it did not affect the SPV, adaptive tracking, or VAS alertness, although it significantly decreased body sway compared to 1.0 mg and 2.0 mg. This discrepancy appears to stem from individual variability in SPV responses: on day 12, three out of sic participants showed an increase in SPV post-dose at 1.5 mg, while the other three showed a decrease (data on file). Finally, PD effects were demonstrated in rats and healthy, non-anxious, and medication-naive human participants. As dose–response relationships and/or $GABA_AR$ function may differ in patient populations, such as those with anxiety or epilepsy, dose adjustments may be required in subsequent development phases. Finally, although SPV reductions and qEEG changes were sustained up to day 12 without tolerance apparently developing, the impact of dosing for longer periods of time in patient populations needs further investigation in subsequent clinical trials. The fact that 12 days of dosing resulted in maintained SPV reduction is encouraging and supports further evaluation of ENX-102 for use in anxiety disorders, which are often persistent and require ongoing treatment [61]. Lastly, while SPV may serve as a pharmacological biomarker for $GABA_AR$ $\alpha2/3$-mediated changes in arousal, its potential role as a surrogate biomarker for anxiolysis in humans remains to be established.

5. Conclusions

The novel $GABA_AR$ $\alpha2$-/3-/5-selective PAM ENX-102 was safe and well tolerated in healthy human participants following multiple dosing, demonstrated a PK profile supporting once-daily dosing, and, most importantly, reduced arousal that was sustained and not associated with sedative effects typically associated with non-selective $GABA_AR$ PAMs targeting $\alpha1$ subunits. These findings were consistent with both *in vitro* data and *in vivo* behavioral findings at similar plasma exposures in rodents, and, moreover, qEEG revealed a high degree of cross-species consistency and alignment from preclinical models to healthy human participants, illustrating its translational potential as a pharmacological biomarker in the early development of subtype-selective $GABA_AR$ PAMs. Collectively, these results provide compelling evidence that ENX-102 may have the potential for anxiolytic efficacy without causing sedation, and, more broadly, highlight the value of biomarker-based early clinical characterization of novel CNS-active compounds that are in development for the treatment of neuropsychiatric disease.

illustrating its translational potential as a pharmacological biomarker in the early development of subtype-selective $GABA_AR$ PAMs.

Supplementary Materials: The following supporting information can be downloaded at: https://www.mdpi.com/article/10.3390/cells14201575/s1, Figure S1: Consort diagram of randomized participants.; Figure S2: qEEG signature in rats of ENX-102 at 2 h post-dose.; Figure S3: qEEG in rats at non-rapid eye movement sleep (NREM) (0–12 h post dose).; Figure S4: onset of rapid eye movement (REM), non-rapid eye movement (NREM), and latency for REM for vehicle, ENX-102, and lorazepam in rats; Figure S5: mean (standard pharmacodynamic parameters of the NeuroCart saccadic peak velocity (SPV), adaptive tracking, smooth pursuit eye movements, body sway, VAS alertness) measured at 0.5, 1, 1.5, 3, 6, 8, 10, and 24 h on day 1 and on day 12. Table S1: Demographic and baseline characteristics of study participants by treatment; Table S2: pharmacokinetics for ENX-102 in rats; Table S3: summary of the total number of adverse events (AEs), and the five most commonly reported AEs per treatment group on day 1, day 12, and days 1 through 12; Table S4: summary of number of participants with treatment emergent adverse events (AEs) and number of treatment emergent AEs by system organ class and preferred term; Table S5: pharmacokinetic parameters of ENX-102 0.5 mg, 1.0 mg, 1.5 mg, 2.0 mg, and 5.0 mg at day 1 and day 12; Table S6: central nervous system effects of ENX-102 (0.5 mg, 1.0 mg, 1.5 mg, 2.0 mg, and 5.0 mg) compared to placebo on day 1 and day 12; Table S7: qEEG results of the human multiple ascending dose study eyes open and eyes closed states at day 1 and at day 12.

Author Contributions: Conceptualization, P.N., K.C.V., K.E.V., S.C., J.S., V.S. and G.E.J.; formal analysis, E.K.; investigation, P.N., K.E.V., L.G.J.M.B., E.A., W.B., S.C., S.P., J.S., V.S., E.T., E.K., F.E.S. and G.E.J.; manuscript writing and preparation, P.N., K.C.V., K.E.V. and G.E.J.; project administration W.B., S.P., J.S. and E.T. All authors have read and agreed to the published version of the manuscript.

Funding: This research was funded by Engrail Therapeutics, Inc.

Institutional Review Board Statement: All procedures involving animals were conducted in accordance with International Guiding Principles for Biomedical Research Involving Animals (2012) and were reviewed and approved by an Institutional Animal Care and Use Committee associated with the contracted laboratory. For EPM experiments, this study was approved under Institutional Animal Care and Use Committee (IACUC) protocol number 204_0613, dated 8 June 2019. For qEEG experiments, this study was approved under IACUC protocol number 295_0119, titled "EEG in Rats", dated 18 January 2019. The study in human volunteers was approved by the Medical Ethics Committee of Stichting Beoordeling Ethiek Biomedisch Onderzoek (Assen, The Netherlands) under protocol number NL78813.056.21, dated 28 September 2021. This study was conducted according to the Dutch Act on Medical Research Involving Human Subjects (WMO) and in compliance with all International Conference on Harmonisation—Good Clinical Practice (ICH-GCP) guidelines and the Declaration of Helsinki. The trial was registered on the International Clinical Trials Registry Platform under the ID: ISRCTN61386931.

Informed Consent Statement: Informed consent was obtained from all participants involved in the study.

Data Availability Statement: The original contributions presented in this study are included in the article/Supplementary Materials. Further inquiries can be directed to the corresponding author.

Acknowledgments: We would like to express our gratitude to Karen Broekhuizen, for her advice in writing the manuscript.

Conflicts of Interest: This study was funded by Engrail Therapeutics, Inc. KCV, KEV, EA, WB, SC, SP, JS, VS, ET are/were employees of Engrail Therapeutics at the time of this research and may hold stock/equity in the company. Engrail Therapeutics contributed to the study design, data collection, analysis, interpretation, manuscript preparation, and the decision to publish. P.N., L.G.J.M.B., E.K., F.E.S., G.E.J. are/were employees of the Centre for Human Drug Research (CHDR), which received funding from Engrail Therapeutics to conduct this study.

Abbreviations

The following abbreviations are used in this manuscript:

α	Alpha
β	Beta
δ	Delta
γ	Gamma
θ	Theta
AE	Adverse Event
ANCOVA	Analysis of Covariance
AUC	Area Under the Curve
BMI	Body Mass Index
CHDR	Centre for Human Drug Research
CNS	Central Nervous System
C_{max}	Maximum Plasma Concentration
C_{trough}	Trough Plasma Concentration
CV	Coefficent of Variation
EEG	Electroencephalography
EC_{50}	Half Maximal Effective Concentration
E_{max}	Maximum Effect
EPM	Elevated Plus Maze
Fz-Cz	Fronto-Central EEG electrode region
GABA	Gamma-Aminobutyric Acid
$GABA_AR$	$GABA_A$ Receptor
HED	Human Equivalent Dose
ICH GCP	International Conference on Harmonisation—Good Clinical Practise
i.p.	Intraperitoneal
MAD	Multiple Ascending Dose
MoA	Mechanism of Aciton
MOAA/S	Modified Observer's Assessment of Alertness/Sedation
NREM	Non-Rapid Eye Movement
PAM	Positive Allosteric Modulator
PBPK	Physiologically Based Pharmacokinetic
PD	Pharmacodynamic
PK	Pharmacokinetic
qEEG	Quantitative EEG
Rac	Accumulation Ratio
REM	Rapid Eye Movement
SAD	Single Ascending Dose
SAEs	Serious Adverse Events
SD	Sprangue Dawley (Rats)
SEM	Standard Error of the Mean
SPV	Saccadic Peak Velocity
SRT	Saccadic Reaction Time
T_{max}	Time to Maximum Concentration
$T_{1/2}$	Elimination Half-life
VAS	Visual Analogue Scale
Vz/F	Apparaent Volume of Distribution
VVLT	Visual Verbal Learning Task
WMO	Dutch Act on Medical Research Involving Human Subjects "Wet medisch-wetenschappelijk onderzoek met mensen"

Appendix A

Appendix A.1

The elevated plus maze (EPM) test was conducted at PsychoGenics, Inc. (Paramus, NJ, USA). The maze apparatus (Hamilton Kinder) consisted of two closed arms (15.75 cm h × 10.8 cm w × 50.17 cm l) and two open arms (10.8 cm w × 50.17 cm l), forming a cross, with a square center platform (6 × 6 cm). All visible surfaces were made of black acrylic. Each arm of the maze was placed on a support column 85.09 cm above the floor. Antistatic black vinyl curtains (7′ tall) surrounded the EPM to make a 7′ × 7′ enclosure. Rats were housed in ventilated cages and acclimated for up to 1 week prior to commencing the study. All rats were examined and weighed prior to initiation of this study to assure adequate health and suitability. During the course of this study, 12/12 light/dark cycle was maintained. The room temperature was maintained between 20 and 23 °C with a relative humidity maintained around 50%. Chow and water were provided ad libitum for the duration of this study. All testing was performed during the rat's light cycle phase. Each rat was randomly assigned across the treatment groups. Rats were acclimated to the experimental room for at least 1 h prior to the test.

Appendix A.2

Quantitative EEG (qEEG) was conducted at PsychoGenics, Inc. (Paramus, NJ, USA). EEG recordings were obtained using DSI DataQuest ART. Rats were assigned unique identification numbers and housed with 3 rats per cage in polycarbonate cages with micro-isolator filter tops. All rats were examined and weighed prior to initiation of this study to assure adequate health and suitability. During this study, 12/12 light/dark cycles were maintained. The room temperature was maintained between 20 and 23 °C with a relative humidity maintained around 50%. Chow and water were provided ad libitum for the duration of this study. Following surgery, rats were single housed. After a period of recovery (7–10 days), animals were transferred to an EEG recording room and placed over a receiver for recordings. ENX-102 was formulated in 0.5% Tylose MH 300 solution and dosed orally. Lorazepam, the benzodiazepine control, administered i.p. at 1 mg/kg, was dissolved in saline. After four days of washout, recordings were performed on a Monday and Friday dosing/recording schedule. EEG recordings started in the lights-off period and dosing occurred 4 h after lights were turned off. Animals were habituated to dosing (vehicle dosing) for 2 days before any data collection took place.

Appendix A.3

For sleep scoring, raw EEG recordings were manually scored using Neuroscore software version 3.3 (Data Sciences International) to identify sleep stages: active wake, quiet wake, non-rapid eye movement (NREM), and rapid eye movement (REM). Using NeuroScore, artifacts were removed offline from the data and sleep stages assigned manually for every 10 s epoch using EEG, EMG, and locomotor activity (LMA) by conventional methods as previously described [62–65] using the fronto-parietal EEG, LMA, and EMG: active wake (less regular, low-amplitude EEG with high EMG and LMAactivity). REM sleep (stable, low-amplitude waves dominated by theta (4–8 Hz) with near absent EMG and no LMA). Sleep stage data were exported from a Neuroscore report template of sleep state time per each 15 min period (2 h pre-dose to 4 h post-dose). The onset of the first sleep and first REM were also reported directly from a Neuroscore report template.

Appendix B

A serious adverse event (SAE) is any untoward medical occurrence that, at any dose, meets one or more of the following criteria:

- Results in death;
- Is life threatening;
  NOTE: The term "life-threatening" in the definition of "serious" refers to an event in which the subject was at risk of death at the time of the event; it does not refer to an event which hypothetically might have caused death if it were more severe.
- Requires inpatient hospitalization or prolongation of existing hospitalization;
- Results in persistent or significant disability/incapacity;
- Is a congenital anomaly/birth defect;
- Important medical events that may not result in death, be life-threatening, or require hospitalization may be considered serious when, based on appropriate medical judgment, they may jeopardize the subject and may require medical or surgical intervention to prevent one of the outcomes listed in this definition. Examples of such medical events include allergic bronchospasm requiring intensive treatment in an emergency room or at home, blood dyscrasias or convulsions that do not result in subject hospitalization, or the development of study drug dependency or drug abuse.

Appendix C

Appendix C.1. Eye Movements

Saccadic peak velocity (SPV) and smooth pursuit eye movements were assessed to evaluate sedation. SPV is known to be highly sensitive to sedation [27] and can indicate anxiolytic properties of benzodiazepines [35]. Smooth pursuit measures neurophysiological function and is sensitive to benzodiazepines [18], the a1-preferring zolpidem [66], and some α2-/3 subunit-selective GABA-A-R modulators [18]. SPV was measured using a computerized system [67]. Volunteers followed a horizontally jumping light source with their eyes. Smooth pursuit eye movements were recorded using the same set up as for the saccadic eye movements [68].

Appendix C.2. Adaptive Tracking

The adaptive tracking test measures eye-hand coordination and sustained attention and is sensitive to CNS effects of alcohol, psychoactive drugs, and sleep deprivation [26]. In this study, the test followed Borland and Nicholson's specifications [69]. Participants used a joystick to keep a dot inside a randomly moving circle on a screen; the circle's speed increased with success and decreased with failure.

Appendix C.3. Pupillometry

Pupil size was measured to monitor any drug effect, which provided evidence of CNS penetration. Pupillometry was performed as previously described [70]; a simultaneous photo of both eyes measured the pupil/iris diameter ratio (mm).

Appendix C.4. Body Sway

Body sway measures postural (in)stability and can be impaired by muscle relaxation, which can be mediated by GABA-A-R containing α5 subunits [71] as well as central sedation mediated by α1 activation. At CHDR, the method has been used to demonstrate effects of sleep deprivation [72], alcohol [73], benzodiazepines [35], and other psychoactive agents (data on file or cite Groeneveld et al. 2016 [26]). Body sway was assessed using methods previously described [23] via an apparatus similar to the Wright ataxia meter. A string was affixed to the waist to capture all body movements over a 2-min period, quantifying sway in millimeters on a digital display.

Appendix C.5. Bond and Lader Visual Analog Scale

The Bond and Lader VAS, originally described by Norris [74], is commonly used to quantify subjective effects of a variety of sedative agents. It measures subjective drug effects in the domains of mood, self-control, and sedation [75]. The VAS alertness is noted as a sensitive parameter for benzodiazepines and is indicative for α1-modulation [35].

Appendix C.6. Visual Verbal Learning Task

The visual verbal learning task (VVLT) contains three different subtests covering basic aspects of verbal learning and memory—acquisition, consolidation, storage, and retrieval of novel information -and has been previously used to characterize GABA-A PAMs [18]. In total, four versions of the VVLT were used. A training version was used for screening purposes, with separate versions administered on day -1 for baseline measurement, on day 1 at the time of the first administration, and at the final administration on day 12. The VVLT assessments on days 1 and 12 were scheduled so that immediate recall occurred around the Tmax of ENX-102. Participants were visually presented 30 words in three consecutive word trials. Immediately after each trial, participants were invited to orally recall as many words as possible to assess VVLT's immediate recall. VVLT delayed recall and delayed recognition were assessed 3 and 10 h post-dose. For delayed recall, participants were asked to recall as many words as possible from the list presented during the immediate recall task. Delayed recognition, also known as the memory recognition test, involved visually presenting 30 previously shown words and 30 distractors to evaluate memory retention. Participants were required to identify the words presented during the immediate recall task.

Appendix D

Resting-state qEEG recordings were performed with participants in two conditions: eyes open and eyes closed. These recordings were conducted in a sequence with other NeuroCart assessments. Each recording session alternated between 64-s periods of eyes open and eyes closed. During this time, participants were instructed to face a featureless wall, refrain from staring, avoid head and eye movements, and suppress eye blinks. This procedure was implemented to minimize extraneous motion artifacts during the recordings.

The qEEG assessments were conducted in eight consecutive blocks of eight periods each, utilizing standard pharmaco-qEEG lead placement, with the same common ground electrode as used in the eye movement registration. Continuous EEG recordings were obtained using a 40-channel recording system (Refa-40, TMSi B.V., The Netherlands). Electrodes were placed according to the international 10–20 system (using a 24-lead cap, TMSi B.V.), with the exception that electrodes at the earlobes (A1 and A2), which were replaced by electrodes at the mastoid sites (M1 and M2). The scalp electrode impedance was maintained below 5 kΩ to ensure good signal quality. The ground electrode was placed at the anterior frontal midline zero (AFz).

To detect ocular artifacts, electrooculography (EOG) was recorded in addition to the EEG. Two Ag/AgCl electrodes were placed at the outer canthi of both eyes, and two additional electrodes were placed approximately 2 cm above and below the right eye to capture vertical and horizontal eye movements.

All signals were sampled at a rate of 1024 Hz and were filtered prior to storage using a first-order recursive high-pass filter with a cut-off frequency of 0.1 Hz to eliminate low-frequency drift. Digital markers were recorded by the amplifier to indicate the start and end of each eye state (eyes open and eyes closed).

EEG recordings were analyzed for power spectral density (PSD) using fast Fourier transform (FFT) analysis. The raw signals were first band-pass filtered using a third-order Butterworth filter, with cutoff frequencies set at 0.1 Hz (low) and 45.0 Hz (high) to remove both high-frequency noise and low-frequency drift. After filtering, the data were segmented into 4-s epochs to allow for time-domain analysis. Epochs containing ocular artifacts (e.g., eye blinks or movements) were identified and excluded to ensure the data used in the subsequent analyses were artifact-free. For the remaining clean epochs, power spectral density (PSD) was calculated for each segment. The PSD values were then averaged across all epochs for each participant under both eyes open and eyes closed conditions. The resulting PSDs were subdivided into the following frequency bands: δ (0.5–6.0 Hz), θ (6.0–8.5 Hz), α (8.5–12.5 Hz), β (12.5–30.0 Hz), and γ (30.0–40.0 Hz). Fast Fourier transform (FFT) analysis was applied to each epoch to calculate the amplitude sum within each frequency band for the electrodes of interest (Fz-Cz, Pz-O1, and Pz-O2).

References

1. Bormann, J. The 'ABC' of GABA Receptors. *Trends Pharmacol. Sci.* **2000**, *21*, 16–19. [CrossRef]
2. Calhoon, G.G.; Tye, K.M. Resolving the Neural Circuits of Anxiety. *Nat. Neurosci.* **2015**, *18*, 1394–1404. [CrossRef] [PubMed]
3. Williamson, J.B.; Jaffee, M.S.; Jorge, R.E. Posttraumatic Stress Disorder and Anxiety-Related Conditions. *Contin. Lifelong Learn. Neurol.* **2021**, *27*, 1738–1763. [CrossRef] [PubMed]
4. Barnard, E.A.; Skolnick, P.; Olsen, R.W.; Mohler, H.; Sieghart, W.; Biggio, G.; Braestrup, C.; Bateson, A.N.; Langer, S.Z. International Union of Pharmacology. XV. Subtypes of Gamma-Aminobutyric AcidA Receptors: Classification on the Basis of Subunit Structure and Receptor Function. *Pharmacol. Rev.* **1998**, *50*, 291–313. [CrossRef]
5. Bonin, R.P.; Orser, B.A. GABAA Receptor Subtypes Underlying General Anesthesia. *Pharmacol. Biochem. Behav.* **2008**, *90*, 105–112. [CrossRef] [PubMed]
6. Sequeira, A.; Shen, K.; Gottlieb, A.; Limon, A. Human Brain Transcriptome Analysis Finds Region- and Subject-Specific Expression Signatures of GABAAR Subunits. *Commun. Biol.* **2019**, *2*, 153. [CrossRef]
7. Song, J.J.; Curtis, M.A.; Faull, R.L.M.; Waldvogel, H.J. The Regional and Cellular Distribution of GABAA Receptor Subunits in the Human Amygdala. *J. Chem. Neuroanat.* **2022**, *126*, 102185. [CrossRef]
8. Collinson, N.; Kuenzi, F.M.; Jarolimek, W.; Maubach, K.A.; Cothliff, R.; Sur, C.; Smith, A.; Otu, F.M.; Howell, O.; Atack, J.R.; et al. Enhanced Learning and Memory and Altered GABAergic Synaptic Transmission in Mice Lacking the A5 Subunit of the GABA $_A$ Receptor. *J. Neurosci.* **2002**, *22*, 5572–5580. [CrossRef]
9. Jacob, T.C. Neurobiology and Therapeutic Potential of A5-GABA Type A Receptors. *Front. Mol. Neurosci.* **2019**, *12*, 179. [CrossRef]
10. Hu, X.; Rocco, B.R.; Fee, C.; Sibille, E. Cell Type-Specific Gene Expression of Alpha 5 Subunit-Containing Gamma-Aminobutyric Acid Subtype A Receptors in Human and Mouse Frontal Cortex. *Complex. Psychiatry* **2018**, *4*, 204–215. [CrossRef]
11. Edwards, Z.; Preuss, C.V. *GABA Receptor Positive Allosteric Modulators*; StatPearls Publishing: Treasure Island, FL, USA, 2024.
12. Serrats, J.; Vadodaria, K.C.; Brubaker, W.; Barker-Haliski, M.; White, H.S.; Evrard, A.; Roucard, C.; Taylor, E.; Vanover, K.E.; Cunningham, S.; et al. ENX−101, a GABA$_A$ Receptor A2,3,5-selective Positive Allosteric Modulator, Displays Antiseizure Effects in Rodent Seizure and Epilepsy Models. *Epilepsia* **2025**, *66*, 2124–2136. [CrossRef]
13. MacLean, A.; Chappell, A.S.; Kranzler, J.; Evrard, A.; Monchal, H.; Roucard, C. BAER-101, a Selective Potentiator of A2- and A3-containing GABA $_A$ Receptors, Fully Suppresses Spontaneous Cortical Spike-wave Discharges in Genetic Absence Epilepsy Rats from Strasbourg (GAERS). *Drug Dev. Res.* **2024**, *85*, e22160. [CrossRef]
14. Monti, J.M.; Spence, D.W.; Buttoo, K.; Pandi-Perumal, S.R. Zolpidem's Use for Insomnia. *Asian J. Psychiatr.* **2017**, *25*, 79–90. [CrossRef]
15. Nickolls, S.A.; Gurrell, R.; van Amerongen, G.; Kammonen, J.; Cao, L.; Brown, A.R.; Stead, C.; Mead, A.; Watson, C.; Hsu, C.; et al. Pharmacology in Translation: The Preclinical and Early Clinical Profile of the Novel A2/3 Functionally Selective GABA(A) Receptor Positive Allosteric Modulator PF-06372865. *Br. J. Pharmacol.* **2018**, *175*, 708–725. [CrossRef]
16. Chen, X.; Jacobs, G.; de Kam, M.L.; Jaeger, J.; Lappalainen, J.; Maruff, P.; Smith, M.A.; Cross, A.J.; Cohen, A.; van Gerven, J. AZD6280, a Novel Partial γ-Aminobutyric Acid A Receptor Modulator, Demonstrates a Pharmacodynamically Selective Effect Profile in Healthy Male Volunteers. *J. Clin. Psychopharmacol.* **2015**, *35*, 22–33. [CrossRef] [PubMed]
17. Chen, X.; Jacobs, G.; de Kam, M.; Jaeger, J.; Lappalainen, J.; Maruff, P.; Smith, M.A.; Cross, A.J.; Cohen, A.; van Gerven, J. The Central Nervous System Effects of the Partial GABA-Aα2,3 -Selective Receptor Modulator AZD7325 in Comparison with Lorazepam in Healthy Males. *Br. J. Clin. Pharmacol.* **2014**, *78*, 1298–1314. [CrossRef] [PubMed]
18. De Haas, S.; Franson, K.; Schmitt, J.; Cohen, A.; Fau, J.; Dubruc, C.; van Gerven, J. The Pharmacokinetic and Pharmacodynamic Effects of SL65.1498, a GABA-A 2,3 Selective Agonist, in Comparison with Lorazepam in Healthy Volunteers. *J. Psychopharmacol.* **2009**, *23*, 625–632. [CrossRef]

19. Zuiker, R.G.; Chen, X.; Østerberg, O.; Mirza, N.R.; Muglia, P.; de Kam, M.; Klaassen, E.S.; van Gerven, J.M. NS11821, a Partial Subtype-Selective $GABA_A$ Agonist, Elicits Selective Effects on the Central Nervous System in Randomized Controlled Trial with Healthy Subjects. *J. Psychopharmacol.* **2016**, *30*, 253–262. [CrossRef]
20. Morales, M.; Varlinskaya, E.I.; Spear, L.P. Anxiolytic Effects of the GABA(A) Receptor Partial Agonist, L-838,417: Impact of Age, Test Context Familiarity, and Stress. *Pharmacol. Biochem. Behav.* **2013**, *109*, 31–37. [CrossRef] [PubMed]
21. Hofmann, M.; Kordás, K.S.; Gravius, A.; Bölcskei, K.; Parsons, C.G.; Dekundy, A.; Danysz, W.; Dézsi, L.; Wittko-Schneider, I.M.; Sághy, K.; et al. Assessment of the Effects of NS11394 and L-838417, A2/3 Subunit-Selective GABAA Receptor-Positive Allosteric Modulators, in Tests for Pain, Anxiety, Memory and Motor Function. *Behav. Pharmacol.* **2012**, *23*, 790–801. [CrossRef]
22. Atack, J.R.; Wafford, K.A.; Tye, S.J.; Cook, S.M.; Sohal, B.; Pike, A.; Sur, C.; Melillo, D.; Bristow, L.; Bromidge, F.; et al. TPA023 [7-(1,1-Dimethylethyl)-6-(2-Ethyl-2 *H*-1,2,4-Triazol-3-Ylmethoxy)-3-(2-Fluorophenyl)-1,2,4-Triazolo[4,3-*b*]Pyridazine], an Agonist Selective for A2- and A3-Containing $GABA_A$ Receptors, Is a Nonsedating Anxiolytic in Rodents and Primates. *J. Pharmacol. Exp. Ther.* **2006**, *316*, 410–422. [CrossRef]
23. De Haas, S.L.; de Visser, S.J.; van der Post, J.P.; de Smet, M.; Schoemaker, R.C.; Rijnbeek, B.; Cohen, A.F.; Vega, J.M.; Agrawal, N.G.B.; Goel, T.V.; et al. Pharmacodynamic and Pharmacokinetic Effects of TPA023, a $GABA_A$ $\alpha_{2,3}$ Subtype-Selective Agonist, Compared to Lorazepam and Placebo in Healthy Volunteers. *J. Psychopharmacol.* **2007**, *21*, 374–383. [CrossRef]
24. Atack, J.; Hallett, D.; Tye, S.; Wafford, K.; Ryan, C.; Sanabria-Bohórquez, S.; Eng, W.; Gibson, R.; Burns, H.; Dawson, G.; et al. Preclinical and Clinical Pharmacology of TPA023B, a $GABA_A$ Receptor A2/A3 Subtype-Selective Partial Agonist. *J. Psychopharmacol.* **2011**, *25*, 329–344. [CrossRef]
25. Van Laere, K.; Bormans, G.; Sanabria-Bohórquez, S.M.; de Groot, T.; Dupont, P.; De Lepeleire, I.; de Hoon, J.; Mortelmans, L.; Hargreaves, R.J.; Atack, J.R.; et al. In Vivo Characterization and Dynamic Receptor Occupancy Imaging of TPA023B, an A2/A3/A5 Subtype Selective γ-Aminobutyric Acid–A Partial Agonist. *Biol. Psychiatry* **2008**, *64*, 153–161. [CrossRef] [PubMed]
26. Groeneveld, G.J.; Hay, J.L.; Van Gerven, J.M. Measuring Blood–Brain Barrier Penetration Using the NeuroCart, a CNS Test Battery. *Drug Discov. Today Technol.* **2016**, *20*, 27–34. [CrossRef]
27. Van Steveninck, A.L.; Schoemaker, H.C.; Pieters, M.S.M.; Kroon, R.; Breimer, D.D.; Cohen, A.F. A Comparison of the Sensitivities of Adaptive Tracking, Eye Movement Analysis, and Visual Analog Lines to the Effects of Incremental Doses of Temazepam in Healthy Volunteers. *Clin. Pharmacol. Ther.* **1991**, *50*, 172–180. [CrossRef]
28. Amerongen, G.V. Integrated Assessment of Neurocognitive, Neurophysiological and Pain Processing in Early Clinical Drug Development. Ph.D. Thesis, Leiden University, Leiden, The Netherlands, 2020.
29. Chen, X.; Broeyer, F.; de Kam, M.; Baas, J.; Cohen, A.; van Gerven, J. Pharmacodynamic Response Profiles of Anxiolytic and Sedative Drugs. *Br. J. Clin. Pharmacol.* **2017**, *83*, 1028–1038. [CrossRef]
30. Jobert, M.; Wilson, F.J.; Ruigt, G.S.F.; Brunovsky, M.; Prichep, L.S.; Drinkenburg, W.H.I.M. Guidelines for the Recording and Evaluation of Pharmaco-EEG Data in Man: The International Pharmaco-EEG Society (IPEG). *Neuropsychobiology* **2012**, *66*, 201–220. [CrossRef] [PubMed]
31. Seifert, H.A.; Blouin, R.T.; Conard, P.F.; Gross, J.B. Sedative Doses of Propofol Increase Beta Activity of the Processed Electroencephalogram. *Anesth. Analg.* **1993**, *76*, 976–978. [CrossRef]
32. Link, C.; Leigh, T.; Fell, G. Effects of Granisetron and Lorazepam, Alone and in Combination, on the EEG of Human Volunteers. *Br. J. Clin. Pharmacol.* **1991**, *31*, 93–97. [CrossRef] [PubMed]
33. Link, C.; Leigh, T.; Dennison, J. The Effects of BRL 46470A, a Novel 5-HT3 Receptor Antagonist, and Lorazepam on Psychometric Performance and the EEG. *Br. J. Clin. Pharmacol.* **1993**, *35*, 395–399. [CrossRef]
34. Breimer, L.T.M.; Hennis, P.J.; Burm, A.G.L.; Danhof, M.; Bovill, J.G.; Spierdijk, J.; Vletter, A.A. Quantification of the EEG Effect of Midazolam by Aperiodic Analysis in Volunteers. *Clin. Pharmacokinet.* **1990**, *18*, 245–253. [CrossRef]
35. De Visser, S.J.; Van Der Post, J.P.; De Waal, P.P.; Cornet, F.; Cohen, A.F.; Van Gerven, J.M.A. Biomarkers for the Effects of Benzodiazepines in Healthy Volunteers. *Br. J. Clin. Pharmacol.* **2003**, *55*, 39–50. [CrossRef]
36. Yamakura, T.; Bertaccini, E.; Trudell, J.R.; Harris, R.A. Anesthetics and Ion Channels: Molecular Models and Sites of Action. *Annu. Rev. Pharmacol. Toxicol.* **2001**, *41*, 23–51. [CrossRef]
37. Purdon, P.L.; Pierce, E.T.; Mukamel, E.A.; Prerau, M.J.; Walsh, J.L.; Wong, K.F.K.; Salazar-Gomez, A.F.; Harrell, P.G.; Sampson, A.L.; Cimenser, A.; et al. Electroencephalogram Signatures of Loss and Recovery of Consciousness from Propofol. *Proc. Natl. Acad. Sci. USA* **2013**, *110*, E1142–E1151. [CrossRef] [PubMed]
38. Murphy, M.; Bruno, M.-A.; Riedner, B.A.; Boveroux, P.; Noirhomme, Q.; Landsness, E.C.; Brichant, J.-F.; Phillips, C.; Massimini, M.; Laureys, S.; et al. Propofol Anesthesia and Sleep: A High-Density EEG Study. *Sleep* **2011**, *34*, 283–291. [CrossRef]
39. Uygun, D.S.; Basheer, R. Circuits and Components of Delta Wave Regulation. *Brain Res. Bull.* **2022**, *188*, 223–232. [CrossRef] [PubMed]
40. Purdon, P.L.; Sampson, A.; Pavone, K.J.; Brown, E.N. Clinical Electroencephalography for Anesthesiologists. *Anesthesiology* **2015**, *123*, 937–960. [CrossRef]

41. Xi, C.; Sun, S.; Pan, C.; Ji, F.; Cui, X.; Li, T. Different Effects of Propofol and Dexmedetomidine Sedation on Electroencephalogram Patterns: Wakefulness, Moderate Sedation, Deep Sedation and Recovery. *PLoS ONE* **2018**, *13*, e0199120. [CrossRef] [PubMed]
42. Veselis, R.A.; Reinsel, R.; Alagesan, R.; Heino, R.; Bedford, R.F. The EEC as a Monitor of Midazolam Amnesia. *Anesthesiology* **1991**, *74*, 866–874. [CrossRef]
43. Dias, R.; Sheppard, W.F.A.; Fradley, R.L.; Garrett, E.M.; Stanley, J.L.; Tye, S.J.; Goodacre, S.; Lincoln, R.J.; Cook, S.M.; Conley, R.; et al. Evidence for a Significant Role of A3-Containing $GABA_A$ Receptors in Mediating the Anxiolytic Effects of Benzodiazepines. *J. Neurosci.* **2005**, *25*, 10682–10688. [CrossRef] [PubMed]
44. Christian, E.P.; Snyder, D.H.; Song, W.; Gurley, D.A.; Smolka, J.; Maier, D.L.; Ding, M.; Gharahdaghi, F.; Liu, X.F.; Chopra, M.; et al. EEG-β/γ Spectral Power Elevation in Rat: A Translatable Biomarker Elicited by $GABA_{A\alpha2/3}$-Positive Allosteric Modulators at Nonsedating Anxiolytic Doses. *J. Neurophysiol.* **2015**, *113*, 116–131. [CrossRef] [PubMed]
45. Visser, S.A.G.; Wolters, F.L.C.; van der Graaf, P.H.; Peletier, L.A.; Danhof, M. Dose-Dependent EEG Effects of Zolpidem Provide Evidence for GABAA Receptor Subtype Selectivity in Vivo. *J. Pharmacol. Exp. Ther.* **2003**, *304*, 1251–1257. [CrossRef]
46. Pellow, S.; Chopin, P.; File, S.E.; Briley, M. Validation of Open: Closed Arm Entries in an Elevated plus-Maze as a Measure of Anxiety in the Rat. *J. Neurosci. Methods* **1985**, *14*, 149–167. [CrossRef]
47. Hogg, S. A Review of the Validity and Variability of the Elevated Plus-Maze as an Animal Model of Anxiety. *Pharmacol. Biochem. Behav.* **1996**, *54*, 21–30. [CrossRef]
48. Brett, R.R.; Pratt, J.A. Chronic Handling Modifies the Anxiolytic Effect of Diazepam in the Elevated Plus-Maze. *Eur. J. Pharmacol.* **1990**, *178*, 135–138. [CrossRef]
49. Sandra, E. Chronic Diazepam Treatment: Effect of Dose on Development of Tolerance and Incidence of Withdrawal in an Animal Test of Anxiety. *Hum. Psychopharmacol. Clin. Exp.* **1989**, *4*, 59–63. [CrossRef]
50. Stock, H.; Foradori, C.; Ford, K.; Wilson, M.A. A Lack of Tolerance to the Anxiolytic Effects of Diazepam on the Plus-Maze: Comparison of Male and Female Rats. *Psychopharmacology* **2000**, *147*, 362–370. [CrossRef] [PubMed]
51. Nair, A.; Jacob, S. A Simple Practice Guide for Dose Conversion between Animals and Human. *J. Basic. Clin. Pharm.* **2016**, *7*, 27. [CrossRef]
52. Walton, N.Y.; Treiman, D.M. Lorazepam Treatment of Experimental Status Epilepticus in the Rat. *Neurology* **1990**, *40*, 990. [CrossRef]
53. Vinkers, C.H.; Olivier, B. Mechanisms Underlying Tolerance after Long-Term Benzodiazepine Use: A Future for Subtype-Selective Receptor Modulators? *Adv. Pharmacol. Sci.* **2012**, *2012*, 416864. [CrossRef] [PubMed]
54. Muthukumaraswamy, S.D. High-Frequency Brain Activity and Muscle Artifacts in MEG/EEG: A Review and Recommendations. *Front. Hum. Neurosci.* **2013**, *7*, 138. [CrossRef]
55. Misselhorn, J.; Schwab, B.C.; Schneider, T.R.; Engel, A.K. Synchronization of Sensory Gamma Oscillations Promotes Multisensory Communication. *eNeuro* **2019**, *6*, ENEURO.0101-19.2019. [CrossRef]
56. Wei, M.; Roodenrys, S. A Scoping Review on the Extent and Nature of Anxiety-Related Research within the Research Domain Criteria (RDoC) Framework: Limited Coverage Using Non-Disorder-Specific Search Terms. *New Ideas Psychol.* **2021**, *63*, 100901. [CrossRef]
57. Blanco-Hinojo, L.; Pujol, J.; Macià, D.; Martínez-Vilavella, G.; Martín-Santos, R.; Pérez-Sola, V.; Deus, J. Mapping the Synchronization Effect of Gamma-Aminobutyric Acid Inhibition on the Cerebral Cortex Using Magnetic Resonance Imaging. *Brain Connect.* **2021**, *11*, 393–403. [CrossRef]
58. Klautke, J.; Foster, C.; Medendorp, W.P.; Heed, T. Dynamic Spatial Coding in Parietal Cortex Mediates Tactile-Motor Transformation. *Nat. Commun.* **2023**, *14*, 4532. [CrossRef]
59. Leicht, G.; Herrmann, C.S.; Mulert, C. BOLD-Response and EEG Gamma Oscillations. In *EEG—fMRI*; Springer International Publishing: Cham, Switzerland, 2022; pp. 641–661.
60. Nickel, B.; Szelenyi, I. Comparison of Changes in the EEG of Freely Moving Rats Induced by Enciprazine, Buspirone and Diazepam. *Neuropharmacology* **1989**, *28*, 799–803. [CrossRef]
61. Perna, G.; Alciati, A.; Riva, A.; Micieli, W.; Caldirola, D. Long-Term Pharmacological Treatments of Anxiety Disorders: An Updated Systematic Review. *Curr. Psychiatry Rep.* **2016**, *18*, 23. [CrossRef]
62. Ivarsson, M.; Paterson, L.M.; Hutson, P.H. Antidepressants and REM Sleep in Wistar–Kyoto and Sprague–Dawley Rats. *Eur. J. Pharmacol.* **2005**, *522*, 63–71. [CrossRef]
63. Parmentier-Batteur, S.; OBrien, J.A.; Doran, S.; Nguyen, S.J.; Flick, R.B.; Uslaner, J.M.; Chen, H.; Finger, E.N.; Wil-liams, T.M.; Jacobson, M.A.; et al. Differential Effects of the MGluR5 Positive Allosteric Modulator CDPPB in the Cortex and Striatum Following Repeated Administration. *Neuropharmacology* **2012**, *62*, 1453–1460. [CrossRef]
64. Leiser, S.C.; Pehrson, A.L.; Robichaud, P.J.; Sanchez, C. Multimodal Antidepressant Vortioxetine Increases Frontal Cortical Oscillations Unlike Escitalopram and Duloxetine—A Quantitative EEG Study in Rats. *Br. J. Pharmacol.* **2014**, *171*, 4255–4272. [CrossRef] [PubMed]

65. Leiser, S.C.; Iglesias-Bregna, D.; Westrich, L.; Pehrson, A.L.; Sanchez, C. Differentiated Effects of the Multimodal Antidepressant Vortioxetine on Sleep Architecture: Part 2, Pharmacological Interactions in Rodents Suggest a Role of Serotonin-3 Receptor Antagonism. *J. Psychopharmacol.* **2015**, *29*, 1092–1105. [CrossRef] [PubMed]
66. de Haas, S.; Schoemaker, R.; van Gerven, J.; Hoever, P.; Cohen, A.; Dingemanse, J. Pharmacokinetics, Pharmacodynamics and the Pharmacokinetic/Pharmacodynamic Relationship of Zolpidem in Healthy Subjects. *J. Psychopharmacol.* **2010**, *24*, 1619–1629. [CrossRef] [PubMed]
67. Wilhelmus, M.M.; Hay, J.L.; Zuiker, R.G.; Okkerse, P.; Perdrieu, C.; Sauser, J.; Beaumont, M.; Schmitt, J.; van Gerven, J.M.; Silber, B.Y. Effects of a Single, Oral 60 Mg Caffeine Dose on Attention in Healthy Adult Subjects. *J. Psychopharmacol.* **2017**, *31*, 222–232. [CrossRef] [PubMed]
68. Bittencourt, P.; Wade, P.; Smith, A.; Richens, A. Benzodiazepines Impair Smooth Pursuit Eye Movements. *Br. J. Clin. Pharmacol.* **1983**, *15*, 259–262. [CrossRef]
69. Borland, R.; Nicholson, A. Visual Motor Co-ordination and Dynamic Visual Acuity. *Br. J. Clin. Pharmacol.* **1984**, *18*, 69S–72S. [CrossRef] [PubMed]
70. Twa, M.D.; Bailey, M.D.; Hayes, J.; Bullimore, M. Estimation of Pupil Size by Digital Photography. *J. Cataract. Refract. Surg.* **2004**, *30*, 381–389. [CrossRef]
71. Crestani, F.; Löw, K.; Keist, R.; Mandelli, M.-J.; Möhler, H.; Rudolph, U. Molecular Targets for the Myorelaxant Action of Diazepam. *Mol. Pharmacol.* **2001**, *59*, 442–445. [CrossRef] [PubMed]
72. van Steveninck, A.L.; van Berckel, B.N.M.; Schoemaker, R.C.; Breimer, D.D.; van Gerven, J.M.A.; Cohen, A.F. The Sensitivity of Pharmacodynamic Tests for the Central Nervous System Effects of Drugs on the Effects of Sleep Deprivation. *J. Psychopharmacol.* **1999**, *13*, 10–17. [CrossRef] [PubMed]
73. Van Steveninck, A.L.; Gieschke, R.; Schoemaker, R.C.; Roncari, G.; Tuk, B.; Pieters, M.S.M.; Breimer, D.D.; Cohen, A.F. Pharmacokinetic and Pharmacodynamic Interactions of Bretazenil and Diazepam with Alcohol. *Br. J. Clin. Pharmacol.* **1996**, *41*, 565–573. [CrossRef]
74. Norris, H. The Action of Sedatives on Brain Stem Oculomotor Systems in Man. *Neuropharmacology* **1971**, *10*, 181–191. [CrossRef]
75. BOND, A.; LADER, M. The Use of Analogue Scales in Rating Subjective Feelings. *Br. J. Med. Psychol.* **1974**, *47*, 211–218. [CrossRef]

cells

MDPI

Review

A Comprehensive Overview of Neurophysiological Correlates of Cognitive Impairment in Amyotrophic Lateral Sclerosis

Seyyed Bahram Borgheai [1,2,†], Brie E. Achorn [3,†], Alyssa H. Zisk [3,4], Sarah M. Hosni [1,5], Karl E. G. Richter [6], Frank S. Menniti [6,7,*] and Yalda Shahriari [1,*]

[1] Department of Electrical, Computer & Biomedical Engineering, University of Rhode Island, Kingston, RI 02881, USA; borgheai@emory.edu (S.B.B.); shosni@odu.edu (S.M.H.)
[2] Neurology Department, Emory University, Atlanta, GA 30322, USA
[3] Interdisciplinary Neuroscience Program, University of Rhode Island, Kingston, RI 02881, USA; briennaachorn@uri.edu (B.E.A.); alyssazisk@uri.edu (A.H.Z.)
[4] AssistiveWare B.V., 1016 PL Amsterdam, The Netherlands
[5] Computer Science Department, Old Dominion University, Norfolk, VA 23529, USA
[6] IMS Scientific Consulting, Malden, MA 02148, USA; krichter7.mail@gmail.com
[7] MindImmune Therapeutics, Inc., Kingston, RI 02881, USA
[*] Correspondence: mennitifs@gmail.com (F.S.M.); yalda_shahriari@uri.edu (Y.S.)
[†] These authors contributed equally to this work.

Abstract

Amyotrophic lateral sclerosis (ALS) is a progressive neurodegenerative disease that leads to the gradual loss of motor control, typically resulting in paralysis and death within 3 to 5 years of diagnosis. ALS shares neuropathological and genetic associations with frontotemporal dementia (FTD), a neurodegenerative condition primarily impacting cognitive functions. These two conditions are increasingly viewed as manifestations of a single molecular disease process that affects distinct brain systems, impacting motor neuronal pathways in ALS and fronto-cortical functions in FTD. However, this simple dichotomy belies the complexity of these conditions. In particular, patients with primary motor ALS can also experience significant cognitive deficits. Investigating the pathobiological and neurophysiological underpinnings of these impairments is essential for a comprehensive understanding of ALS and may open avenues for targeted therapies to alleviate these debilitating symptoms. Moreover, the biophysical correlates of cognitive deficits in ALS may serve as sensitive biomarkers for evaluating potential therapeutics. In this narrative review, we begin with an overview of the clinical features and genetics of ALS, followed by a review of the associated cognitive deficits that are adjunctive to motor decline. We then highlight neuroimaging studies from our laboratory and the broader literature, using EEG and other modalities that are beginning to uncover systems-level brain disruptions potentially underlying cognitive impairment in motor-dominant ALS.

Keywords: amyotrophic lateral sclerosis; cognition; neurophysiology; biomarkers

There is emerging evidence that amyotrophic lateral sclerosis (ALS) extends beyond a purely motor disorder to involve significant cognitive and systems-level brain dysfunction. This narrative review situates ALS within the broader disease spectrum, highlighting shared neuropathological, neurophysiological, and genetic mechanisms while emphasizing the substantial heterogeneity present even in motor-dominant ALS phenotypes. We synthesize clinical and genetic insights alongside electrophysiological and neuroimaging evidence, drawing on complementary modalities from our laboratory and the broader literature, to elucidate neural network disruptions that may underlie cognitive impairments in ALS. By integrating these findings, the review underscores the potential of cognitive biomarkers for disease (sub)characterization and therapeutic applications.

1. Search Method

This is a narrative review that draws on literature identified through targeted searches in PubMed, IEEE Xplore, and Google Scholar, as well as relevant book chapters. The search focused on correlates or biomarkers of cognitive dysfunction in ALS with topics related to clinical, genetic, cognitive screens, structural and functional neuroimaging. We also targeted articles adopting multimodal neuroimaging, e.g., EEG-fNIRS integration, particularly with cognitive tasks performed by ALS patients. Given the narrative nature of this review, the search was not exhaustive; instead, studies were selected based on conceptual relevance and contribution to abovementioned major contexts.

2. Clinical Features of Amyotrophic Lateral Sclerosis (ALS)

In 1869, French neurologist Jean-Martin Charcot devised the term "amyotrophic lateral sclerosis" to represent a clinical syndrome of muscle atrophy and weakness (amyotrophic), along with tissue scarring in the lateral spinal cord (lateral sclerosis), resulting in paralysis and death [1]. ALS is now recognized as a progressive neurodegenerative disorder that primarily affects adults between 40 and 70 years of age, with peak onset in the mid- to late-50s [2]. The defining motor symptoms of ALS are caused by the degeneration of upper motor neurons in the motor cortex and corticospinal tracts, along with lower motor neurons in the brainstem and spinal cord, leading to progressive weakness, loss of voluntary motor control, and eventual paralysis [2–4]. The disease typically begins with focal weakness that then spreads to include most muscles, with the failure of eye, sphincter, and diaphragm muscles occurring in the late stages of the disease [2]. A diagnosis requires a comprehensive evaluation of clinical symptoms, neurological examinations, electromyography, and lab tests, such as blood and genetic testing [4]. Classic ALS constitutes 70% of all ALS cases and can be categorized into limb-onset, bulbar-onset, or respiratory-onset, depending on the site of initial symptom manifestation [5]. These categorizations notwithstanding, ALS is a heterogeneous disease, with significant individual variability in initial presentation, disease progression, spreading patterns, and survival [2]. This heterogeneity relates to the diverse pattern of upper and lower motor neuron involvement among patients [1]. Like other complex diseases, there is a diagnosis continuum with ALS; this means that patients may be given a diagnosis of suspected, possible, probable, or definite ALS [4]. Ultimately, a diagnosis of definite ALS is given when the clinical examination and electromyography indicate both upper and lower motor neuron signs in three body regions, and laboratory tests rule out all other diseases that may be mistaken for ALS. Today, it is estimated that approximately 3 to 5 per 100,000 individuals will develop ALS [4]. Patients can be classified based on whether there is a family history of the disease, which accounts for ~10% of cases and is known as familial ALS (fALS). However, for most patients, the cause of ALS is unknown, and no family history is present; these cases are referred to as sporadic ALS (sALS) [2].

3. Genetics

Insight into possible molecular mechanisms of pathogenesis in ALS has come from the discovery of genetic variants that lead to ALS in familial cases. Furthermore, genome-wide, population-based studies suggest that as much as 50% of the risk for "sporadic" ALS may have a genetic basis [6]. To date, variations in over 30 genes have been linked to ALS [3]. Of these, Superoxide dismutase [Cu-Zn] (SOD1), the guanine nucleotide exchange factor C9ORF72, Transactive response DNA-binding protein 43 (TDP-43), and Fused in

https://doi.org/10.3390/cells15010037

sarcoma (FUS) are the most well-documented, accounting for two-thirds of fALS cases and ~10–15% of sALS cases (Table 1) [7,8]. It is now established that a patient's genetic background can greatly influence clinical phenotypes. For instance, individuals with SOD1 mutations are more likely to exhibit limb-onset ALS, while those with C9ORF72 mutations are more likely to present with bulbar-onset ALS and to develop psychiatric disorders [9,10]. What is known about mechanisms relating genetic variations to disease phenotypes is outlined below.

Table 1. Gene-specific ALS phenotypes.

Gene	Frequency	Typical Age of Onset	Phenotypic Tendencies	Comments
SOD1	~20% of fALS	40–60	Limb-onset; variable progression depending on mutation	Some variants are aggressive, others slow
C9ORF72	~34% of fALS	40–70	Bulbar-onset more common; psychiatric symptoms; ALS-FTD overlap	Immune and autophagy dysfunction
TARDBP (TDP-43)	3–5% of fALS	50 s	Slow progression except certain variants (e.g., G376D)	RNA metabolism defects
FUS	4–5% of fALS	Juvenile–adult	Aggressive, early onset; often cognitive symptoms	Nuclear transport defects

SOD1 was the first ALS gene to be identified in 1993 and remains one of the most studied genetic contributors [11]. Individuals with SOD1 mutations typically experience symptom onset in the lower limbs and lower motor neurons, but the clinical phenotype varies across different SOD1 variants [11]. For example, the Gly73Ser and Ser106Leu mutations are associated with a more severe phenotype, while patients homozygous for the Asp91Ala allele have a slowly progressing disease course with urinary incontinence and sensory abnormalities appearing in the late stages [12]. SOD1 is ubiquitously expressed and acts as a potent antioxidant that protects cells from superoxide radicals [13]. SOD1 mutations are thought to result in a gain-of-toxic-function due to protein misfolding and aggregation. Mutations may also alter protein function, leading to aberrant protein–protein interactions. The accumulation of toxic hydroxyl radicals can damage both nuclear and mitochondrial DNA [13,14].

The most common genetic cause of ALS is a G_4C_2 hexanucleotide repeat expansion in the C9ORF72 gene, accounting for ~34% of fALS cases [15]. In the CNS, neurons and microglia express the highest levels of C9ORF72, while myeloid cells, particularly innate immune dendritic cells, show the highest expression in the periphery [16]. The C9ORF72 gene encodes a protein that is believed to function as a Differentially Expressed in Normal and Neoplasia (DENN) domain protein that serves as a GDP/GTP exchange factor (GEF) to activate Rab-GTPases involved in membrane trafficking [17]. C9ORF72 mutations often lead to adult-onset ALS with deficits in RNA metabolism, autophagy, immune modulation, nucleocytoplasmic transport, and protein homeostasis [11].

TDP-43 is a DNA/RNA-binding protein that regulates the expression of thousands of genes and seems to be particularly important for RNA metabolism [15,18]. TDP-43 is ubiquitously expressed and tends to localize to discrete subnuclear structures [19]. In 2006, it was observed that TDP-43, encoded by the TARDBP gene, exhibited a pattern of nuclear depletion and subsequent cytoplasmic accumulation in the motor neurons of ALS patients [20]. Individuals with mutations in TARDBP typically present with adult-onset ALS and demonstrate slow disease progression; however, those with certain variants, such

https://doi.org/10.3390/cells15010037

as G376D, can exhibit rapid progression [11]. These mutations tend to lead to defects in RNA metabolism, stress granule formation, and axon pathology.

FUS is a DNA/RNA-binding protein that plays an important role in DNA repair and RNA metabolism [21]. Mutations in FUS have been linked to ALS, with over 50 different variants observed in ALS patients [22]. Corroborating this, FUS mutations lead to impairments in RNA metabolism, stress granule formation, DNA repair deficits, and axon pathology [11]. Individuals with ALS-linked FUS mutations exhibit juvenile-to-adult onset, with an aggressive disease course and cognitive impairment and postural tremors in some patients [11].

4. Cognitive Deficits in ALS

Although ALS is defined as a disease of motor pathways in the central nervous system (CNS), other brain regions can be affected and other symptoms may be present, notably including deficits in cognitive function [23]. Beyond the motor neurons and muscles, "non-motor" regions of the CNS—including the frontal, parietal, and somatosensory areas—can be affected, contributing to the neuropsychiatric and cognitive symptoms commonly observed in patients with the disease [24]. Determining the nature and frequency of non-motor, specifically cognitive, symptoms, though, can be technically challenging. Modified evaluation tasks have been proposed to make evaluations accessible to patients with motor and speech impairments [25], but this may affect interpretation, as norms for cognitive evaluations typically do not include people with significant motor or speech impairments. Rapid declines make longitudinal studies difficult due to the increasing inaccessibility of evaluations and lack of applicable norms as disability severity increases. Mood disorders—whether stemming from underlying disease biology, coexisting conditions, or psychological reactions to a fatal and currently incurable illness—are known to impact the assessment of cognitive functions. However, studies have varied in their use of cognitive evaluation methods and diagnostic tools. Cognitive function in ALS patients is commonly assessed using general tools such as the Montreal Cognitive Assessment (MoCA) and the Mini Mental State Examination (MMSE). ALS-specific instruments have also been developed. The Edinburgh Cognitive and Behavioural ALS Screen (ECAS) or ALS Cognitive Behavioral Screen (CBS), designed specifically for people with ALS [26], has been validated in several populations, including in the UK, where it was first developed, as well as in Italy [25,27] and China [28]. Additionally, many studies apply the revised Strong criteria to determine the presence or absence of ALS-cognitive impairment (ALSci) [28].

The prevalence of cognitive deficits in ALS has been investigated in a number of studies. A population-based study [29] in an Irish cohort of ALS patients (n = 160 patients, 110 matched controls, excluding those patients with frank dementia) found 34% had cognitive impairment (as defined by ECAS) while 47% had no detectable cognitive deficits. In another comprehensive study [30], in non–speed-dependent cognitive tasks, cognitive impairment was observed in 51% of patients with ALS (n = 279), compared with only 5% of controls (n = 129); notably, 15% of patients met diagnostic criteria for FTD. A similar study [31] in a geographically distinct cohort in Italy found very similar results: of the 188 ALS patients that were evaluated, approximately 30% had at least one category of cognitive impairment, 50% were cognitively normal, and 13% had ALS-FTD. Results were very similar in a study using the ECAS in China: about 35% of ALS patients had cognitive impairment [32]. In a recent, longitudinal study of ALS patients (n = 189 at start of study), Costello et al. [33] found that 23% of all patients met criteria for cognitive impairment at the first assessment (mean time from symptom onset to evaluation was 19 months). On average, overall cognition was relatively spared over the course of three follow-up sessions over one year. However, by eight months, only 42% of starting patients remained

https://doi.org/10.3390/cells15010037

in the study, dropping to 26% by one year. Costello et al. [33] attempted to control for non-random dropout using Cox survival models accounting for age, diagnostic delay, site of onset, and C9orf72 genotype, but with only 49 of 189 patients assessed at the last follow-up, it is unclear whether the statistical analysis was sufficient to correct for the attrition rate. Meta-analysis has also been used to evaluate the presence of cognitive deficits over a greater number of patients and to pool results across geographic regions. A recent meta-analysis [34] of 45 cross-sectional studies and 13 longitudinal studies found results similar to those of Costello with respect to the percentage of patients with cognitive impairment at diagnosis. The number of patients with cognitive impairment at the start of the study ranged from 14% to 54%, with a weighted average of 37%. Thus, despite the different populations and methods, approximately one-third of ALS patients already suffer from some form of cognitive impairment by the time of diagnosis.

While the presence of cognitive deficits in ALS patients is reasonably well established, the natural history of cognitive decline is less clear. An influential study by Elamin and colleagues indicated a decline in cognition over time [35]. This study further noted that the population with cognitive impairment at diagnosis also suffered from a more rapid progression of other symptoms of ALS, including motor deficits. These investigators developed an algorithm using a simple score that included (1) site of onset (bulbar vs. spinal), (2) rate of decline measured using the ALS Functional Rating Scale (Revised), and (3) impairment of executive function (but not ECAS) that had predictive value for the prognosis of disease in two test cohorts [23]. Information on patient genotype for C9orf72, FUS, TARDBP (TDP-43) and optineurin, available for nearly all cases, was incorporated in a subsequent analysis and indicated a higher percentage of C9orf72 expansion carriers in the group of patients suffering the most rapid declines. However, in contrast to Elamin and colleagues' results, two other groups using a similar or greater number of patients did not see a progressive loss of cognitive abilities over time [25,36]. Meta-analyses have also been performed to help clarify progression of cognitive decline in ALS. Benbrika et al. [37] reviewed 190 articles from the English and French language literature and found ample evidence for cognitive deficits at diagnosis, and while there was some evidence of worsening toward the end of the disease, there was not consistent evidence of decline throughout the disease. More recently, a somewhat overlapping meta-analysis by Finsel et al. [34] found similar results. This meta-analysis found some studies in which a progressive decline was present but also many where a decline was not detected despite sufficient numbers of patients to observe a decline had it been present. Both the review and the meta-analysis pointed to several factors that could confound either longitudinal studies or cross-sectional studies with age-matched controls. Depression is worse shortly after diagnosis, and that depression typically reduces measured cognitive performance, thus possibly lowering apparent cognitive power early in the disease, masking any later true decline in cognitive ability. There are also practice effects when using the ECAS repeatedly, possibly artifactually increasing cognition scores in retests [25].

Taking into consideration the limitations of prior work, a recent, large study to assess cognitive function by McHutchison et al. included whole-genome sequencing for each patient and repeated cognitive and behavioral evaluation every 3–6 months for up to six total assessments using ECAS and structured interviews [38]. Three different versions of ECAS were used to reduce practice effects. Latent class growth analysis, a statistical procedure similar to principal component analysis, was used to identify subgroups of cognitive phenotypes within the study population over time [39]. Their outcomes showed that lower ECAS scores at baseline were correlated with a decrease in cognition over time. Furthermore, they found that C9orf72 expansion carriers were over-represented in this group, but C9orf72 status alone did not fully explain progressive deficits in cognition.

https://doi.org/10.3390/cells15010037

In summary, and as concluded by Finsel et al. [34], there is strong evidence for cognitive dysfunction early in the course of ALS, possibly even before diagnosis, and this could be considered a characteristic of ALS or of subtypes of ALS. However, the evidence for a gradual, progressive decline in cognition is weaker [34]. While it does occur in some cases, there is insufficient evidence for a uniform deterioration of cognition in all ALS patients as the disease progresses. Several authors emphasized that testing and following presymptomatic carriers of ALS genes may be a way to determine whether early decreases in cognition may be part of the disease process in those groups [34,40].

5. Brain Structural Correlates of Cognitive Decline in ALS

Postmortem studies of ALS neuropathology have shown that even when only lower motor neuron symptoms were present, mis-localized, aggregated TDP-43 can be found in many parts of the CNS, not only motor pathways [41]. Post-mortem evidence of synapse loss in the prefrontal cortex has been correlated with cognitive decline during life in ALS patients [42]. Thus, both positive (TDP-43 mis-localization and/or accumulation) and negative (synapse loss) elements of neuropathology have been demonstrated in CNS regions beyond motor pathways in ALS.

There is also evidence of degeneration beyond the motor cortex during life, although correlating ALS symptom-defined subtypes with neuroimaging markers remains challenging [43]. Cortical thinning has been detected by structural magnetic resonance imaging (MRI) in ALS patients with or without cognitive impairment or FTD-like symptoms [44]. There is some evidence that thinning of the right fronto-temporal insular cortex may be specifically associated with ALS with cognitive impairment as compared with ALS without cognitive deficits [45]. The Canadian ALS Neuroimaging Consortium (CALSNIC) studied the relationship between ECAS scores and MRI-based brain morphometry [46]. In addition to the expected white matter degeneration in the corticospinal tracts, they found degeneration in the corpus callosum, cingulum and superior longitudinal fasciculus of ALS patients with impaired verbal fluency or executive dysfunction. In one of the largest and most comprehensive studies to date of neuroimaging correlates of cognition in ALS patients (293 patients, 237 controls), Tan and colleagues found significant changes in both white and gray matter that correlated with impaired function in specific cognitive domains [47]. Patients were evaluated a maximum of five times at three-to-six-month intervals. The changes were measured by MRI, using cortical thickness and subcortical structural volume of gray matter and connectome analysis of diffusion tensor imaging of white matter. Language deficits were significantly associated with frontal, temporal, parietal and subcortical gray matter neurodegeneration; memory dysfunction with hippocampal and medial-temporal atrophy; and executive impairment with widespread white matter changes. These results establish that structural changes outside the corticospinal tracts can be analyzed by MRI and are correlated, on a population basis, with the types of cognitive losses that can occur in ALS.

Besides MRI, other techniques have been used to investigate possible correlates of extra-motor degeneration in ALS. Regional cerebral blood flow measurements by positron emission tomography (PET) have been used as a proxy for neural activity. In a small study, Abrahams and colleagues compared cognitively impaired ALS patients with unimpaired patients and controls [48]. They found cognitively affected patients displayed significantly impaired activation in cortical and subcortical regions, including the dorsolateral prefrontal cortex, lateral premotor cortex, medial prefrontal and premotor cortices, insular cortex, and the anterior thalamic nuclei. In a much larger study [49], fluorodeoxyglucose positron emission tomography (FDG-PET) identified changes in the brain associated with cognitive decline in ALS patients [50]. They found hypometabolism in the frontal lobe present in

https://doi.org/10.3390/cells15010037

ALS patients with cognitive impairment and even greater hypometabolism in ALS-FTD patients. ALS patients without cognitive impairment were not similarly affected. They also found hypermetabolism in the cerebellum of patients with ALS-FTD. Foucher et al. [51] also used FDG-PET and found a correlation between ECAS score and metabolic activity. They identified the right orbitofrontal gyrus as a region of reduced metabolic activity that correlated with poorer ECAS scores. They also found regions of hypermetabolism in the occipital cortex and cerebellum, similar to Canosa and colleagues' findings.

Although the FDG-PET method provides less spatial resolution than MRI, both methods have detected differences between ALS patients with cognitive dysfunction and both cognitively healthy ALS patients and similarly aged, cognitively healthy controls. Both techniques have identified changes specifically in white and gray matter in selected regions in the frontal lobes, providing converging evidence that cognitive decline in ALS may be correlated with imaging changes in regions generally associated with language and cognition that lie outside the canonical motor pathways associated with ALS neurodegeneration.

6. Functional Neuroimaging Correlates of Cognitive Impairment in ALS

As ALS clinical symptoms evolve, the corresponding neural dysfunction may be probed through various electrophysiological and neuroimaging techniques. These techniques include electroencephalography (EEG), which measures the electrical signals at the scalp generated by synaptic activity in the underlying cortex [52]; functional near-infrared spectroscopy (fNIRS), which measures hemodynamic responses associated with cortical activity by detecting changes in oxygenated and deoxygenated hemoglobin[5]; and functional magnetic resonance imaging (fMRI), which measures changes in blood oxygen utilization also resulting from cortical synaptic activity [53–55]. While single-modality studies continue to produce significant insights, combining neuroimaging techniques from different modalities is being increasingly employed to provide a more comprehensive view of ALS-related neural alterations [56–60]. Together, such techniques offer promising windows into both the motor and non-motor manifestations of the disease. Here, our focus is on the cognitive neural signatures of ALS as revealed through studies that either used resting-state or task-based paradigms (Table 2) using single or multimodal neuroimaging techniques.

Resting-state functional connectivity (RSFC) has been widely explored in ALS. This approach is attractive because it obviates patients' difficulties with task-based paradigms [61], particularly at advanced stages of the disease. Increased RSFC has been reported in ALS patients during eyes-closed fMRI, particularly in the default mode (DMN) and frontoparietal (FPN) networks involving prefrontal, frontal, and parietal regions [49,62–64]. This hyperconnectivity has been linked to clinical and cognitive impairments [65], disease progression, and areas of structural disconnection [62,63]. Conversely, other eyes-closed fMRI studies have found reduced RSFC in networks associated with cognitive and behavioral functions [66–68]. Using EEG, Fraschini and colleagues [69] reported reduced electrical RSFC, while Kopitzki and colleagues [70] found preserved hemodynamic RSFC using fNIRS. A high-density longitudinal EEG study during eyes-open rest identified elevated gamma coherence between fronto-parietal regions and theta coherence across bilateral motor areas [71]. Another study found reduced EEG power in ALS patients across multiple regions, including delta/theta bands in prefrontal areas, beta in sensorimotor, and delta/alpha/beta in occipital and temporal lobes [72]. In eyes-closed studies [59,73,74], alpha power reduction was associated with disease-related structural degeneration, particularly loss of pyramidal neurons. Thus, there is little consensus on the biomarkers identified using single-mode neuroimaging methods. The apparent discrepancies among these various studies may stem from methodological or imaging technique differences that

https://doi.org/10.3390/cells15010037

affect RSFC estimates [69,75]. This emphasizes the importance of multimodal approaches to fully characterize ALS-related neural alterations.

Connectivity-focused multimodal resting-state paradigms may help to clarify the apparent conflicts among the single-modality studies. An fNIRS-EEG study by Deligani et al. [59] reported significantly increased fronto-parietal EEG connectivity in the alpha and beta bands, along with enhanced interhemispheric and intra-right hemisphere fNIRS connectivity in frontal regions. Their study also observed reduced EEG spectral power in the alpha and theta bands across frontal, central, temporal, parietal, and occipital regions. Moreover, increased hemodynamic spectral power was observed in the frontal (very low frequency oscillations (VLFO)~0.01–0.05 Hz) and parietal regions (VLFO and low frequency oscillations (LFO) ~0.05–0.15 Hz) in ALS patients. Similar findings were reported by Kopitzki et al. [70], who demonstrated altered functional connectivity using DTI and fNIRS during closed-eyes recordings. These changes suggest that, beyond the motor circuits, ALS significantly affects cognitive and extra-motor networks. The default mode network (DMN), which underlies memory, emotional processing, self-referential thought, and aspects of consciousness [76], has shown consistently increased connectivity in ALS, often correlating with greater disability and faster progression [66,68]. Additionally, executive dysfunction in ALS has been linked to disruptions in the frontoparietal (FPN) and dorsal attention networks (DAN), which regulate attention and cognitive control [77–79]. Interpretation of the underlying structural changes that result in the changes in connectivity varies, as these have been linked to both impaired inhibition and compensatory mechanisms for structural decline [63]. While there is yet to be a consensus on resting state correlates of cognitive function in ALS, these analyses clearly point to widespread alteration in functional networks associated with the disease, consistent with clinical findings that ALS patients may experience neurocognitive symptoms beyond motor impairment.

Analysis of EEG temporal changes in response to cognitive-sensory events, specifically in the context of event-related potentials (ERPs), has also been widely explored in patients with ALS. In these activation paradigms, EEG signals are monitored under conditions where subjects either passively or actively process sensory inputs such as a repeated tone. The averaged EEG signal has well characterized temporal components that have been mapped to different epochs of early signal processing (see Figure 1). Abnormalities in ERPs, including the P300 component, as well as earlier components such as P100 and N200, elicited during oddball paradigms, are among the most frequently reported temporal features impaired in ALS patients [58,80–82]. P300 responses correspond to cognitive substrates related to attentional demand and working memory, collectively referred to as mental workload [83–85]. Decreased amplitudes of the P300 response have been linked to markers of ALS progression [80–82]. Given that the P300 component is associated with selective attention, visuospatial filtering, and working memory, the reduced P300 amplitude in ALS may reflect deficits in selective attention. Consistent with this interpretation, Shahriari et al. [86] demonstrated that stronger positive deflections around 220 ms at midline frontal electrodes (Fz, Cz) were associated with better task performance in ALS, suggesting that preserved early attentional engagement contributes to more effective cognitive control in this population.

In addition to attenuated P300 amplitudes, ALS patients have also been shown to exhibit significant trial-to-trial variability in the timing of the P300 component relative to stimulus presentation, referred to as latency jitter [87–90]. Single-trial P300 latencies, which reflect stimulus evaluation time, have long been studied in cognitive neuroscience and are known to correlate with task-related metrics, including reaction times, particularly in tasks emphasizing accuracy over speed [91–93]. However, this association between P300 latencies and task-related metrics can break down in neurotypical individuals exhibiting

https://doi.org/10.3390/cells15010037

higher latency jitter or when task demands prioritize speed [91,93]. Latency jitter can be considered a form of neural variability, which is an essential feature for adaptive learning but is often maladaptively elevated in a range of neurological conditions [94,95]. Elevated latency jitter has been documented not only in ALS but also in autism [94], attention deficit hyperactivity disorder [92], schizophrenia [96], depression [97], traumatic brain injury [98], disorders of consciousness [99], and dementia [97], showing the potential cognitive deficits ALS patients share with non-motor neurocognitive disorders.

Later ERP components have also been found to be attenuated in ALS patients in a visuospatial mental task paradigm [58]. Notably, the N400 component, a marker commonly associated with semantic memory processing, was less pronounced in subjects with ALS compared to healthy controls, suggesting potential impairments in semantic processing. Additionally, a late positive component (~650 ms post-stimulus), typically elicited during mental arithmetic tasks in healthy individuals, was significantly diminished in ALS patients [100,101]. This reduced activation may reflect deficits in computational or executive processing in the ALS population. These later ERP components, which are often reported as absent or significantly weakened in ALS patients [101], could represent promising new neurophysiological markers of disease-related cognitive impairment.

While ERP analyses shed light on both early sensory processing and some higher cognitive deficits in patients with ALS, spectral measures have shown potential to reflect a broader spectrum of cognitive and extra-motor deficits in higher-order cognition using more complex tasks, including motor imagery, the mental execution of a motor action without actual physical execution. Brain systems activated during motor imagery include those associated with physical execution through integration in a system referred to as the mirror neuron system. Some studies [102,103] suggest that brain regions classically associated with physical speech production and language use, namely Broca's and Wernicke's areas, are integrated into the human mirror neuron system [104] and are activated during motor imagery tasks [105,106]. Broca's area has also been found to contain a motor representation of hand actions [107]. Motor imagery is associated with changes in the synchrony of neural activity across brain regions, particularly in the mu (8–12 Hz) and beta (13–25 Hz) frequency bands. Both amplitude suppression (event-related desynchronization, ERD) and enhancement (event-related synchronization, ERS) in these bands have been observed at the onset of a motor imagery task [108–111]. Studies have found correlations between ERD/ERS amplitudes and clinical measures of ALS progression, such as upper limb and respiratory function [112,113]. Dysfunctions in imagery-specific cortical synchronization have been linked to impaired sensory-motor coupling in the somatosensory cortex [114]. This interpretation aligns with earlier fMRI findings [115], which demonstrated reduced cortical activation in ALS patients during motor imagery tasks. One proposed explanation for reduced ERD during motor imagery is the disruption of the prefrontal-parietal network involved in imagery tasks, an effect possibly stemming from ALS-related extra-motor pathology involving the prefrontal cortex, as suggested by previous neuroimaging and neuropathological studies [116–118]. In a separate study, Kasahara and colleagues suggested that reduced ERD in ALS may be a consequence of progression-related neuronal loss based on a quantitative EEG (QEEG) resting-state analysis [112]. The authors speculated that this might reflect an association between bulbar deficits and cognitive dysfunction, potentially due to fatigue or attentional deficits. Further, the reduced ERD patterns observed in ALS patients may reflect mirror neuron system dysfunction, as it has been proposed that suppression in both the mu and beta frequency bands may serve as an index of mirror neuron system activity [119,120].

Other spectral changes have been observed in subjects with ALS in non-motor tasks. In a brain–computer interface (BCI) study employing a visuo-mental paradigm that integrated

https://doi.org/10.3390/cells15010037

a P3Speller with mental arithmetic, Borgheai et al. [58] reported a significantly greater increase in frontal delta power from baseline in healthy controls than in ALS patients. Delta power augmentation during mental calculation tasks is generally interpreted as reflecting heightened internal concentration, facilitating the suppression of external interference during cognitive performance. The attenuated delta power increase observed in ALS patients is therefore speculated to reflect diminished internal concentration during task execution. Notably, this electrophysiological signature was also correlated with lower cognitive scores in the ALS group, further supporting its potential as a marker of cognitive dysfunction in this population. A similar pattern was reported in a longitudinal BCI study of individuals with ALS conducted by Shahriari and colleagues [86], where higher delta activity at occipital locations was negatively correlated with daily BCI performance, reinforcing the link between excessive delta activity and impaired attentional or cognitive processing. In the theta frequency band, healthy participants showed a profound power increase, while this pattern was attenuated significantly in the ALS patient group [58]. A smaller increase in theta power was also associated with lower cognitive scores in ALS patients. Theta oscillations are generally linked to multiple cognitive domains, including attentional engagement and working memory processes [58,121]. Thus, elevated theta activity during the mental task may reflect task-imposed increased cognitive workload and attention. Additionally, a significant reduction in beta power was observed in the healthy control group compared to ALS patients. Frontal beta activity is typically associated with top-down control mechanisms [122,123], particularly related to general task-related processing. Beta power is known to increase during the intended maintenance of a "status quo" mental state [124], while its later suppression is thought to reflect content-specific frontal modulation related to working memory and decision-making, as supported by recent studies [125,126]. In agreement, Shahriari et al. [86] showed that higher beta power at occipital electrodes was positively correlated with successful BCI use in people with ALS, suggesting that preserved beta modulation may contribute to sustaining attentional control and visual processing in demanding tasks. Importantly, significant correlations were found between scores on the CBS- and EEG-derived delta and theta power [118], which show the clinical relevance of these biomarkers, though their causal mechanisms are yet to be known.

Beyond basic spectral-temporal features, neuronal network dysfunction in ALS also appears to be accompanied by dysregulation in neurovascular coupling [57,127,128]. Using a multimodal EEG-fNIRS recording with a visuo-mental protocol, Borgheai et al. [58] showed predominantly positive associations in EEG–fNIRS correlation maps in healthy controls, while participants with ALS exhibited largely negative correlations, indicating a desynchronization between electrical (EEG) and hemodynamic (fNIRS) responses during mental tasks. Interestingly, significant hemodynamic differences were localized to fNIRS locations corresponding to the dorsolateral prefrontal cortex (DLPFC), a region known to be critically involved in working memory processes [129–131]. This finding further supports the hypothesis that individuals with ALS may exhibit executive dysfunctions affecting cognitive workload processing, particularly within the DLPFC. Moreover, a recent multi-modal BCI study [57] demonstrated that integrated EEG–fNIRS features recorded during a brief pre-screening session could predict which workload levels in a subsequent visuo-mental BCI task would yield higher performance. Using multivariate statistical models, the authors showed that combined CBS battery measures (e.g., attention and concentration), hemodynamic features (such as the range of HbO_2 responses across frontal fNIRS channels and fronto-temporal RSFC), and resting-state EEG spectral markers (including temporal–parietal delta changes) collectively predicted variability in BCI performance.

https://doi.org/10.3390/cells15010037

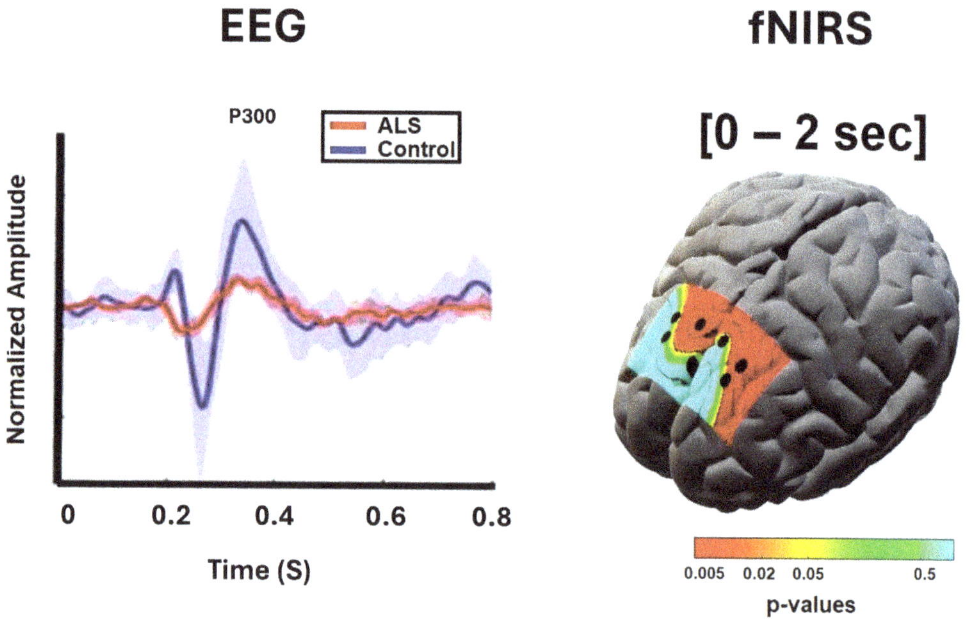

Figure 1. (**Left**): Normalized target ERPs averaged across all ALS participants (red) and all healthy controls (blue) in the 800 ms following stimulus onset in Parietal-occipital region. The shaded bands (light blue and red) indicate the standard deviation within the control and patient groups. (**Right**): *p*-value maps derived from statistical comparison of peak values of hemodynamic concentrations in 0–2 s post-stimulus (adapted from Borgheai et al. [127]).

The data reviewed above indicate that neural signaling abnormalities are observed at multiple levels in ALS. Recent studies have used graph theory to analyze frontal network dysfunctions in ALS patients compared to healthy controls. Graph theory is emerging as a promising framework to analyze the connections between measured network disruptions and observed functional neurocognitive deficits. Graph theoretical analyses model the brain as a network where recording channels or regions of interest (ROIs) act as nodes and their functional or structural connections serve as links [132,133]. These models are then mapped back to the roles of the nodes and their connections to cognitive processes. Sorrentino et al. [133] used magnetoencephalography (MEG)-based minimum spanning tree (MST) analysis to report a shift toward a more centralized and randomized network configuration across all frequency bands as the disease progressed. This pattern has similarly been observed in MST-based analyses of other neuropsychiatric conditions, such as schizophrenia [133] and major depressive disorder [134]. Building on these applications, Hosni and colleagues [60] introduced a nonlinear dynamic graph-based framework using recurrence quantification analysis (RQA) features extracted from EEG and fNIRS. By applying nonlinear graph-derived measures, they demonstrated that these approaches could capture subtle and complex patterns of brain dynamics that are often missed by conventional spectral analyses. This work highlights the potential of integrating nonlinear graph metrics with traditional graph-theoretical models to provide a more comprehensive characterization of network alterations in ALS, particularly in relation to disrupted connectivity and abnormal information flow.

Complementing these findings, an activity-based fNIRS study using a visuo-mental protocol also revealed a more centralized frontal network in ALS patients compared to controls [127]. This star-like organization—marked by increased maximum degree

https://doi.org/10.3390/cells15010037

and betweenness centrality—suggests a shift toward a more random, dysregulated structure [134,135], where information flow becomes overly reliant on a few central hubs. Furthermore, Borgheai et al. [136] performed nodal analyses to characterize key network hubs, identifying the right prefrontal cortex (PFC) as housing the most dominant relay stations in ALS patients—nodes critical for intermodular communication. These hubs are known to be particularly vulnerable to neurodegeneration [137]. Compared to controls, ALS patients exhibited reduced node efficiency and strength in the left PFC, alongside an increase in the right PFC, suggesting a lateralized shift in prefrontal engagement. This rightward asymmetry aligns with earlier neuroimaging findings: PET studies during verbal fluency tasks [48] and ERP studies using dual spatial and working memory tasks [138] also showed altered PFC activity in ALS patients.

These findings converge on the hypothesis that ALS-related executive dysfunction, especially dysfunction involving working memory manipulation, is reflected in altered PFC topology and lateralization. The PFC is a well-established node in executive functioning and working memory networks [139,140], and hemodynamic activity in this region is frequently observed during working memory engagement [58,141,142]. The observed dominance of right PFC hubs in ALS patients may indicate compensatory executive control mechanisms, particularly for functions like inhibitory regulation [143] and memory retrieval monitoring [144]. The right PFC is also critically involved in spatial working memory [145], supporting the notion that hyperconnectivity in this region may reflect difficulty in processing the spatial aspects of dual-task paradigms. Frontal lateralization has been observed in other neuropsychiatric disorders, such as schizophrenia [146] and MDD [134], and the similarity in topological markers may be especially relevant considering the prevalence of depressive symptoms in ALS [147–149].

The functional network alterations observed in ALS putatively reflect structural atrophy localized to frontal regions, especially evident in task-negative resting-state fMRI studies [63,150]. Dimond et al. [150], using resting-state MRI-DTI, reported associations between verbal fluency errors and disrupted clustering coefficients in right frontal and temporal lobes, implicating reduced white matter integrity in executive dysfunction. Similarly, Tedeschi et al. [66] documented suppression in the right fronto-parietal network, potentially driven by right-lateralized frontal atrophy. Shahriari et al. [136] reported a significantly broader distribution of high-beta band connectivity and absence of distinct connections in directivity analysis in ALS patients compared to controls during a P300-speller task. Their results possibly reflect the involvement of inhibitory interneuron networks associated with beta-frequency oscillations [63,151–153]. One interpretation is that such hyperconnectivity may serve a compensatory role in the context of progressive structural degeneration [65], potentially due to the loss of inhibitory interneurons, a hallmark of ALS pathology [154]. These localized atrophies may undermine several key cognitive executive networks, including the executive control network (ECN), fronto-parietal network (FPN), and dorsal attention network (DAN), all of which encompass frontal regions and could contribute to the broad cognitive impairments observed in ALS [66,72,77,155].

https://doi.org/10.3390/cells15010037

Table 2. List of selected neuroimaging studies addressing cognitive dysfunctions in ALS.

	AUTHORS	NEUROIMAGING MODALITY	#ALS/CONTROL	METHOD/TASK	MAJOR FINDINGS IN NON-MOTOR REGIONS IN ALS
REST	Ma et al., 2015 [49]	fMRI	20/20	Functional network analysis	Within the hub regions, higher functional connectivity in the prefrontal cortex with the connectivity strength in the abnormal hub associated with clinical variables, was obtained
REST	Luo et al., 2012 [62]	fMRI	20/20	Functional connectivity	Alteration in low-frequency fluctuations in frontal areas was concluded
REST	Douaud et al., 2011 [63]	fMRI + DWI	25/15	Functional/structural connectivity	Increased functional connectivity spanning sensorimotor, premotor, prefrontal and thalamic regions associated with decreased structural connectivity, was obtained
REST	Agosta et al., 2013 [65]	fMRI	20/15	Functional network analysis	A divergent connectivity pattern in the default mode network (DMN), including decreased connectivity of the right orbitofrontal cortex and increased parietal connectivity, was associated with clinical and cognitive deficits
REST	Li et al., 2017 [67]	fMRI	21/21	Functional connectivity	Decreased functional connectivity in the left and right ventrolateral PFC, along with widespread and frequency-dependent FCS changes, provided evidence that ALS patients exhibit consistent impairments in extra-motor regions even at relatively early stages of the disease
REST	Fraschini et al., 2016 [69]	EEG	21/16	Functional network analysis	A significant group difference in MST dissimilarity and MST leaf fraction in the beta-band was found
REST	Kopitzki et al., 2016 [70]	fNIRS + DTI	31/30	Functional (hemodynamic) /structural connectivity	Anterior-temporal homotopic rs-FC was found to correlate with fractional anisotropy in the central corpus callosum (CC)
REST	Deligani et al., 2020 [59]	fNRIS + EEG	10/9	Functional connectivity	Increased fronto-parietal EEG connectivity in the alpha and beta bands, along with increased interhemispheric and right intra-hemispheric fNIRS connectivity in the frontal and prefrontal regions, was observed
REST	Nasseroleslam et al., 2019 [71]	EEG + MRI	100/34	Functional/structural connectivity	Increased EEG coherence between parietal–frontal scalp regions in the gamma band was observed. EEG signals associated with less extensively involved non-motor regions also showed enhanced structural connectivity on MRI
REST	Dukic et al., 2019 [72]	EEG+ MRI	74/47	Functional connectivity	Increased co-modulation in frontal regions (theta and gamma bands), along with decreased synchrony in the temporal and frontal regions (theta to beta bands), was observed

https://doi.org/10.3390/cells15010037

Table 2. *Cont.*

	AUTHORS	NEUROIMAGING MODALITY	#ALS/CONTROL	METHOD/TASK	MAJOR FINDINGS IN NON-MOTOR REGIONS IN ALS
TASK-BASED	Riccio et al., 2013 [84]	EEG	8/NA	P300-BCI spelling	The temporal filtering capacity in the RSVP task was a predictor of both the P300-based BCI accuracy
	Pinkhardt et al., 2008 [85]	EEG	20/20	ERP/dichotic listening task	A distinct decrease in the fronto-precentral negative difference signal (Nd) was observed in ALS, along with increased processing of non-relevant stimuli, as reflected by the P3 component, suggesting a reduced focus of attention
	Shahriari et al., 2019 [86]	EEG	9/NA	P300-BCI spelling	BCI performance was positively correlated with a positive deflection in EEG amplitude around 220 ms at frontal mid-line locations
	Zisk et al., 2021 [90]	EEG	6/9	P300-BCI spelling	Latency jitter was significantly increased in participants with ALS
	Kasahara et al., 2012 [112]	EEG	8/11	ERD/motor imagery task	The ERD of ALS patients was significantly smaller
	Hosni et al., 2019 [113]	EEG	6/11	ERD/motor imagery task	Decreased ERD features were correlated with ALS clinical scores, specifically disease duration, bulbar, and cognitive functions
	Stanton et al., 2007 [115]	fMRI	16/17	Motor imagery task	Reduced activation during motor imagery was observed in ALS in the left inferior parietal region, as well as in the anterior cingulate gyrus and medial prefrontal cortex
	Borgheai et al., 2020 [127]	fNIRS	9/10	Functional network analysis /visuo-mental task	A more centralized frontal network organization was observed in the ALS group, with the most frequent network hubs showing an asymmetrical pattern predominantly localized in the right prefrontal cortex
	Shahriari et al., 2015 [136]	EEG	9/13	P300/BCI spelling	A significantly broader distribution of high-beta band connectivity, along with an absence of distinct connections in the directivity analysis, was observed in ALS patients
	Hammer et al., 2010 [138]	EEG	17/20	ERP/Go-NoGo task	An anteriorization of the NoGo-P3 was observed, a pattern that has been established as an index of impaired inhibitory function
	Proudfoot et al., 2017 [151]	MEG	11/10	Spectral power/motor task	A relative increase in beta-power was revealed in the frontal lobe and premotor regions

7. Correspondence Between Structural and Functional Neuroimaging and Other Cognitive Biomarkers of ALS

The relationship between structural and functional neuroimaging and other ALS disease biomarkers is an area that is still under investigation, but emerging studies have explored some associations. In a study of 64 ALS patients and 27 controls [156], analyses of structural imaging, diffusion-weighted imaging (DWI), and resting-state fMRI networks revealed a shared pathological pattern; the brain connections exhibiting the greatest structural damage corresponded directly to those showing the greatest functional impairment. Comparing 74 ALS patients with 47 age-matched healthy controls, Dukic et al. [72] reported cortico-cortical EEG connectivity (imaginary coherence [iCoh]) changes (beta-band in motor and delta-band in frontoparietal and frontotemporal channels) associated with structural degeneration as captured by MRI (within the grey matter volume of

https://doi.org/10.3390/cells15010037

motor and frontal cortical channels), as well as the functional (ALSFRS-R) and cognitive scores. This study applied two-way ANOVA on all their EEG measures (spectral power and iCoh) that could categorize ALS patients versus healthy cohorts but did not result in a significant difference between patients' subgroups based on genomic (C9ORF72) status. However, in their follow-up study [155], using clustering methods, the authors categorized four different ALS sub-phenotypes based on distinct EEG network features, where one of the clusters had the highest proportion of C9ORF72-positive patients. Similarly, in another ALS study [157], when graph theory analysis was applied to RSFC between MRI-based subcortical volumes, C9ORF72-positive patients with behavioral variant FTD (with and without comorbid ALS) showed reorganization of network hubs compared to those with ALS only (without behavioral variant FTD). Similarly, Geronimo et al. [158] reported that C9ORF72-positive ALS patients (though low in number: $n = 4$) exhibited greater suppression in visual evoked potential (VEP) and online P300 BCI performance compared to both control and ALS-only groups. A recent resting-state EEG study has also reported alpha and beta band power suppression in ALS patients associated with increased C9orf72 expression and reduced cortical thickness in parietal, occipital and temporal regions [159]. In another task-based EEG study [160] in 87 asymptomatic family members (AFMs) of patients with familial C9orf72 ALS, stimulus-locked evoked potentials were compared between 37 AFM individuals carrying the pathological repeat expansion (C9+) and 50 AFMs without expansion (C9−). Go and NoGo stimulus-locked N2 (180–350 ms) in C9+ AFMs showed an increased negative potential compared to the C9- cohort in the bilateral precuneus and superior parietal regions. Another recent resting-state MEG study [161] could classify symptomatic ALS, asymptomatic C9orf27 and SOD1 carriers compared to healthy controls. In their report, the symptomatic ALS group ($n = 61$) was characterized by widespread reduced beta power across temporal, frontal, and motor regions. This contrasted sharply with asymptomatic C9orf72 carriers ($n = 16$), who showed a mixed pattern of decreased beta power (occipital, temporal) and increased power in other bands (frontal theta, motor/occipital alpha), while asymptomatic SOD1 carriers ($n = 12$) exhibited no significant change in beta power. These findings indicate that functional neuroimaging biomarkers have the potential not only to distinguish patients from healthy controls but also to differentiate between genomic and non-genomic ALS subgroups.

It is noteworthy that there is potential to link neuroimaging biomarkers to cortical hyperexcitability in ALS [162–164]. This includes studies using transcranial magnetic stimulation (TMS) and neuroimaging to probe upper/lower motor neuron dysfunction or hyperexcitability. However, these studies, so far, have focused mainly on the primary motor cortex [162]. Thus, there is a critical need to broaden these neurophysiological investigations to capture the full spectrum of ALS as a multisystem disease. For instance, expanding TMS-EEG studies [163,164] to non-motor cortical regions, such as the prefrontal or parietal cortex, is crucial for identifying patterns of excitability dysfunction linked to common cognitive and behavioral deficits (e.g., FTD comorbidity). Furthermore, assessing excitability changes while subjects perform cognitive tasks (moving beyond the passive resting state) allows researchers to dynamically challenge the functional integrity of both motor and non-motor networks, providing more sensitive, active biomarkers for early cognitive decline and network breakdown that may be missed during simple rest.

8. Summary and Perspectives

ALS is diagnosed as a disease of progressive motor paralysis caused by degeneration of upper and lower motor neurons. However, from a genetic perspective, ALS is closely related to FTD, which has a hallmark of progressive degeneration of frontal cortical regions resulting in neurocognitive dysfunction and dementia. In reality, these related neurode-

https://doi.org/10.3390/cells15010037

generative conditions are heterogeneous in disease presentation, course and underlying neural system dysfunction. There is substantial evidence that patients with ALS also suffer neurocognitive dysfunction, as measured using both generalized clinical cognitive test batteries such as the MoCA and MMSE and batteries specifically designed for testing cognitive function in ALS patients such as the ECAS and ALS-CBS. Approximately one-third of people with ALS show these cognitive deficits, although it is as yet uncertain whether these deficits are progressive. This indeterminacy reflects, in part, the general limitations of such tests and, in particular, the difficulty of measuring cognitive function in the ALS population, i.e., against the backdrop of the progressive motor deterioration and the emotional and affective lability that accompanies this devastating fatal illness. Notwithstanding, the presence of cognitive dysfunction in ALS patients points to broader neural dysfunction beyond the motor system. The ability to sensitively measure this broader neural dysfunction would afford a more fundamental understanding of the pathophysiology underlying ALS while providing biomarkers to follow disease progression and use in testing potential therapeutic agents. EEG and fNIRS are proving valuable tools in investigating the broader neural dysfunction in ALS. These modalities reflect cortical synaptic activity at different and complementary spatial and temporal resolutions. Importantly, these neuroimaging modalities are 'portable', a significant consideration given the limited mobility of many with ALS. While there are currently no definitive neurophysiological signatures of ALS, emerging trends are summarized below.

Neural dysfunction measurable by EEG, fNIRS, and other imaging modalities is evident at the early stages of sensory processing, detected as abnormal ERP responses. This is particularly notable in the P300 and later ERP epochs, which reflect the initial integration of sensory signals into higher internal cognitive processing such as attentional systems. ERP responses in ALS patients are both diminished and temporally more variable, i.e., have higher jitter. Timing is critical to the integration of individual neuronal signals into functional networks. Thus, it is plausible that an increase in jitter, as directly measured in early sensory processing tasks, also contributes to the decay in broader network integrity, evident in analyses when subjects are performing more complex motor imaging tasks. Examples of the latter include the synchrony of neural activity across brain regions in the mu and beta frequency bands that occurs with motor imagery, ERD and ERS, which are consistently found to be abnormal in ALS patients. Deficits in neural synchrony across brain regions in ALS patients are beginning to be investigated using analytical techniques such as graph theory to both reflect and inform on higher-order functional neurocognitive deficits. These analyses point to the degradation of frontal cortical networks related to executive functioning, attention, and working memory manipulation. Frontal network dysfunction is a cardinal feature of FTD, and the finding of such dysfunction in ALS patients is consistent with putatively common underlying neurodegenerative processes, as suggested by the overlapping genetic underpinnings of these diseases. It is also noted that the neurophysiological studies indicating frontal network dysfunction in ALS suggest the neurocognitive domains instantiated by these networks bear further clinical investigation. Thus, while ALS is fundamentally a 'motor' disease, neurophysiology points to broader neural systems dysfunctions affecting multiple cognitive domains. This understanding has consequences in caring for the needs of ALS patients as well as developing therapeutics to treat their underlying neural dysfunctions.

A corollary to an understanding of the neurophysiological dysfunctions underlying ALS is the development of potential biomarkers to follow disease progression and use in clinical trials of ALS therapeutics. In this regard, the hierarchical nature of the neurophysiological abnormalities in people with ALS may be used to clinical advantage. Combined measures may be particularly useful in quantifying potential efficacy in clinical trials of

https://doi.org/10.3390/cells15010037

therapeutic interventions. To the extent to which distortions in electrophysiological and/or hemodynamic features, in temporal, spectral, connectivity, or network domains, reflect the underlying neurodegenerative process, therapies that can be shown to impact these features in early clinical testing may be promoted for clinical evaluation of efficacy against the broader array of ALS symptoms. Finding therapeutics that impact the cognitive and non-motor functional decay in ALS may be a sensitive and accessible avenue towards comprehensive therapies that also address the more aggressive motor degeneration in the disease.

9. Limitations, Gaps, and Future Directions

Neuroimaging studies dissecting cognitive dysfunctions in ALS are characterized by considerable methodological and analytical variability, limiting generalizability, comparability, interpretability, and the development of reliable biomarkers that can be used in practical clinical settings. Key challenges include small sample sizes, limited use of cognitive task–based paradigms, insufficient integration of multimodal neuroimaging data (e.g., combining EEG, fNIRS, MRI, or DTI to capture complementary temporal–spatial signatures), variability in preprocessing pipelines, lack of standardized processing metrics, and inadequate age- and disease-severity matching across cohorts. Furthermore, biological and genetic markers are often under-incorporated in study designs, resulting in reduced ability to account for their influence. Compounding this, many studies insufficiently control for additional confounding factors, such as medications and comorbidities including depression, anxiety, sleep disturbances, and chronic pain, all of which can independently affect cortical oscillatory activity. Moreover, measures of functional connectivity vary widely across studies due to substantial differences in the choice of connectivity metrics and frequency-specific analyses. The aforementioned issues are often compounded by the limited spatial and temporal validity of neuroimaging modalities. For example, EEG source localization for subcortical structures can render inferences about cortico-subcortical connectivity particularly uncertain. Consequently, EEG-derived connectivity findings, especially those involving deep brain networks, are often inconsistent, reflecting methodological constraints rather than true pathophysiological differences. The inherently low spatial resolution of EEG and contamination of frontal channels by facial muscle artifacts (e.g., eye movements, blinks, or jaw tension) further complicate analysis and data interpretation. Therefore, adopting integrative approaches, such as utilizing network analysis based on multimodal neuroimaging (e.g., combining EEG with fNIRS, which offers high frontal capturing potential), especially while patients are performing cognitive tasks, is increasingly essential for overcoming methodological limitations in neuroimaging studies of ALS. The concurrent recording of electrophysiological (EEG) and hemodynamic (e.g., fNIRS) signals allows capture of both temporal dynamics and spatially targeted cortical activation, increasing the likelihood of detecting disease-relevant network dysfunction while reducing reliance on any single modality's limitations. Indeed, methodological reviews highlight the growing use of such dual-modal systems, classify common integration strategies (e.g., EEG-informed fNIRS, fNIRS-informed EEG, parallel analyses), and outline their advantages and challenges [165–168]. Recent studies show that multimodal, graph-based frameworks are increasingly being applied in neurorehabilitation and BCI studies, demonstrating both their feasibility and potential to uncover clinically meaningful patterns of network reorganization [60,169–171]. Adopting graph-theoretical network analysis on multimodal data may thus offer more robust biomarkers. By integrating spectral, hemodynamic, and connectivity features into indices of global and local network organization (e.g., efficiency, modularity, hub disruption), such approaches can capture disease-related

https://doi.org/10.3390/cells15010037

alterations in large-scale brain networks that are less sensitive to modality-specific noise or methodological variability than pairwise connectivity metrics.

The aforementioned methodological limitations, together with intrinsic heterogeneity of ALS, further contribute to divergent findings reported across studies. ALS exhibits pronounced variability in site of onset, balance of upper- and lower-motor-neuron involvement, rate of progression, and the presence of cognitive or behavioral impairment, each of which is likely associated with distinct patterns of cortical and network-level dysfunction. Small, cross-sectional cohorts often fail to capture this phenotypic diversity, leading to overrepresentation of specific disease subtypes and increased inter-study variability. In this context, analyzing patients based on distinct sub-categories can offer a promising and necessary framework to effectively cluster the heterogeneous clinical symptom domains and dissect the unique underlying neural mechanisms that drive disease variability. By combining detailed clinical phenotyping, longitudinal designs, multimodal imaging, and network-level analytical approaches, future studies can better disentangle true ALS-related electrophysiological signatures from variability driven by analytic choices, source-localization limitations, and the underlying biological heterogeneity of the disease.

Author Contributions: All authors contributed to the writing of this manuscript. All authors have read and agreed to the published version of the manuscript.

Funding: This work was supported by the National Science Foundation (NSF-1913492, NSF-2006012) and the Institutional Development Award (IDeA) Network for Biomedical Research Excellence (P20GM103430).

Data Availability Statement: No new data were created or analyzed in this study.

Conflicts of Interest: Alyssa H. Zisk was employed by the company AssistiveWare B.V. Karl E. G. Richter was employed by the company IMS Scientific Consulting. Frank S. Menniti was employed by the company MindImmune Therapeutics, Inc. The remaining authors declare that the research was conducted in the absence of any commercial or financial relationships that could be construed as a potential conflict of interest.

References

1. Grad, L.I.; Rouleau, G.A.; Ravits, J.; Cashman, N.R. Clinical Spectrum of Amyotrophic Lateral Sclerosis (ALS). *Cold Spring Harb. Perspect. Med.* **2017**, *7*, a024117. [CrossRef] [PubMed]
2. Brown, R.H.; Al-Chalabi, A. Amyotrophic Lateral Sclerosis. *N. Engl. J. Med.* **2017**, *377*, 162–172. [CrossRef]
3. Agosta, F.; Chio, A.; Cosottini, M.; De Stefano, N.; Falini, A.; Mascalchi, M.; Rocca, M.A.; Silani, V.; Tedeschi, G.; Filippi, M. The present and the future of neuroimaging in amyotrophic lateral sclerosis. *Am. J. Neuroradiol.* **2010**, *31*, 1769–1777. [CrossRef]
4. Hardiman, O.; Al-Chalabi, A.; Chio, A.; Corr, E.M.; Logroscino, G.; Robberecht, W.; Shaw, P.J.; Simmons, Z.; van den Berg, L.H. Amyotrophic lateral sclerosis. *Nat. Rev. Dis. Primers* **2017**, *3*, 17071. [CrossRef]
5. van Es, M.A.; Hardiman, O.; Chio, A.; Al-Chalabi, A.; Pasterkamp, R.J.; Veldink, J.H.; van den Berg, L.H. Amyotrophic lateral sclerosis. *Lancet* **2017**, *390*, 2084–2098. [CrossRef]
6. Zhang, S.; Cooper-Knock, J.; Weimer, A.K.; Shi, M.; Moll, T.; Marshall, J.N.G.; Harvey, C.; Nezhad, H.G.; Franklin, J.; Souza, C.D.S.; et al. Genome-wide identification of the genetic basis of amyotrophic lateral sclerosis. *Neuron* **2022**, *110*, 992–1008.e11. [CrossRef] [PubMed]
7. Renton, A.E.; Majounie, E.; Waite, A.; Simon-Sanchez, J.; Rollinson, S.; Gibbs, J.R.; Schymick, J.C.; Laaksovirta, H.; van Swieten, J.C.; Myllykangas, L.; et al. A hexanucleotide repeat expansion in C9ORF72 is the cause of chromosome 9p21-linked ALS-FTD. *Neuron* **2011**, *72*, 257–268. [CrossRef] [PubMed]
8. Zou, Z.Y.; Zhou, Z.R.; Che, C.H.; Liu, C.Y.; He, R.L.; Huang, H.P. Genetic epidemiology of amyotrophic lateral sclerosis: A systematic review and meta-analysis. *J. Neurol. Neurosurg. Psychiatry* **2017**, *88*, 540–549. [CrossRef]
9. Li, H.F.; Wu, Z.Y. Genotype-phenotype correlations of amyotrophic lateral sclerosis. *Transl. Neurodegener.* **2016**, *5*, 3. [CrossRef]
10. Breevoort, S.; Gibson, S.; Figueroa, K.; Bromberg, M.; Pulst, S. Expanding Clinical Spectrum of C9ORF72-Related Disorders and Promising Therapeutic Strategies: A Review. *Neurol. Genet.* **2022**, *8*, e670. [CrossRef]
11. Suzuki, N.; Nishiyama, A.; Warita, H.; Aoki, M. Genetics of amyotrophic lateral sclerosis: Seeking therapeutic targets in the era of gene therapy. *J. Hum. Genet.* **2023**, *68*, 131–152. [CrossRef] [PubMed]

https://doi.org/10.3390/cells15010037

12. Akcimen, F.; Lopez, E.R.; Landers, J.E.; Nath, A.; Chio, A.; Chia, R.; Traynor, B.J. Amyotrophic lateral sclerosis: Translating genetic discoveries into therapies. *Nat. Rev. Genet.* **2023**, *24*, 642–658. [CrossRef]
13. Kaur, S.J.; McKeown, S.R.; Rashid, S. Mutant SOD1 mediated pathogenesis of Amyotrophic Lateral Sclerosis. *Gene* **2016**, *577*, 109–118. [CrossRef] [PubMed]
14. Peggion, C.; Scalcon, V.; Massimino, M.L.; Nies, K.; Lopreiato, R.; Rigobello, M.P.; Bertoli, A. SOD1 in ALS: Taking stock in pathogenic mechanisms and the role of glial and muscle cells. *Antioxidants* **2022**, *11*, 614. [CrossRef] [PubMed]
15. Kim, G.; Gautier, O.; Tassoni-Tsuchida, E.; Ma, X.R.; Gitler, A.D. ALS Genetics: Gains, Losses, and Implications for Future Therapies. *Neuron* **2020**, *108*, 822–842. [CrossRef] [PubMed]
16. McCauley, M.E.; Baloh, R.H. Inflammation in ALS/FTD pathogenesis. *Acta Neuropathol.* **2019**, *137*, 715–730. [CrossRef]
17. Smeyers, J.; Banchi, E.G.; Latouche, M. C9ORF72: What It Is, What It Does, and Why It Matters. *Front. Cell. Neurosci.* **2021**, *15*, 661447. [CrossRef]
18. Ghasemi, M.; Brown, R.H., Jr. Genetics of Amyotrophic Lateral Sclerosis. *Cold Spring Harb. Perspect. Med.* **2018**, *8*, a024125. [CrossRef]
19. Sephton, C.F.; Cenik, B.; Cenik, B.K.; Herz, J.; Yu, G. TDP-43 in central nervous system development and function: Clues to TDP-43-associated neurodegeneration. *Biol. Chem.* **2012**, *393*, 589–594. [CrossRef]
20. Neumann, M.; Sampathu, D.M.; Kwong, L.K.; Truax, A.C.; Micsenyi, M.C.; Chou, T.T.; Bruce, J.; Schuck, T.; Grossman, M.; Clark, C.M.; et al. Ubiquitinated TDP-43 in frontotemporal lobar degeneration and amyotrophic lateral sclerosis. *Science* **2006**, *314*, 130–133. [CrossRef] [PubMed]
21. Wang, X.; Arai, S.; Song, X.; Reichart, D.; Du, K.; Pascual, G.; Tempst, P.; Rosenfeld, M.G.; Glass, C.K.; Kurokawa, R. Induced ncRNAs allosterically modify RNA-binding proteins in cis to inhibit transcription. *Nature* **2008**, *454*, 126–130. [CrossRef] [PubMed]
22. Deng, J.; Yang, M.; Chen, Y.; Chen, X.; Liu, J.; Sun, S.; Cheng, H.; Li, Y.; Bigio, E.H.; Mesulam, M.; et al. FUS Interacts with HSP60 to Promote Mitochondrial Damage. *PLoS Genet.* **2015**, *11*, e1005357. [CrossRef]
23. Elamin, M.; Bede, P.; Montuschi, A.; Pender, N.; Chio, A.; Hardiman, O. Predicting prognosis in amyotrophic lateral sclerosis: A simple algorithm. *J. Neurol.* **2015**, *262*, 1447–1454. [CrossRef]
24. Mioshi, E.; Lillo, P.; Yew, B.; Hsieh, S.; Savage, S.; Hodges, J.R.; Kiernan, M.C.; Hornberger, M. Cortical atrophy in ALS is critically associated with neuropsychiatric and cognitive changes. *Neurology* **2013**, *80*, 1117–1123. [CrossRef]
25. Poletti, B.; Solca, F.; Carelli, L.; Faini, A.; Madotto, F.; Lafronza, A.; Monti, A.; Zago, S.; Ciammola, A.; Ratti, A.; et al. Cognitive-behavioral longitudinal assessment in ALS: The Italian Edinburgh Cognitive and Behavioral ALS screen (ECAS). *Amyotroph. Lateral Scler. Front. Degener.* **2018**, *19*, 387–395. [CrossRef]
26. Abrahams, S.; Newton, J.; Niven, E.; Foley, J.; Bak, T.H. Screening for cognition and behaviour changes in ALS. *Amyotroph. Lateral Scler. Front. Degener.* **2014**, *15*, 9–14. [CrossRef] [PubMed]
27. Poletti, B.; Solca, F.; Carelli, L.; Madotto, F.; Lafronza, A.; Faini, A.; Monti, A.; Zago, S.; Calini, D.; Tiloca, C.; et al. The validation of the Italian Edinburgh Cognitive and Behavioural ALS Screen (ECAS). *Amyotroph. Lateral Scler. Front. Degener.* **2016**, *17*, 489–498. [CrossRef]
28. Strong, M.J.; Abrahams, S.; Goldstein, L.H.; Woolley, S.; McLaughlin, P.; Snowden, J.; Mioshi, E.; Roberts-South, A.; Benatar, M.; HortobaGyi, T.; et al. Amyotrophic lateral sclerosis—Frontotemporal spectrum disorder (ALS-FTSD): Revised diagnostic criteria. *Amyotroph. Lateral Scler. Front. Degener.* **2017**, *18*, 153–174. [CrossRef] [PubMed]
29. Phukan, J.; Elamin, M.; Bede, P.; Jordan, N.; Gallagher, L.; Byrne, S.; Lynch, C.; Pender, N.; Hardiman, O. The syndrome of cognitive impairment in amyotrophic lateral sclerosis: A population-based study. *J. Neurol. Neurosurg. Psychiatry* **2012**, *83*, 102–108. [CrossRef]
30. Ringholz, G.M.; Appel, S.H.; Bradshaw, M.; Cooke, N.A.; Mosnik, D.M.; Schulz, P.E. Prevalence and patterns of cognitive impairment in sporadic ALS. *Neurology* **2005**, *65*, 586–590. [CrossRef]
31. Montuschi, A.; Iazzolino, B.; Calvo, A.; Moglia, C.; Lopiano, L.; Restagno, G.; Brunetti, M.; Ossola, I.; Lo Presti, A.; Cammarosano, S.; et al. Cognitive correlates in amyotrophic lateral sclerosis: A population-based study in Italy. *J. Neurol. Neurosurg. Psychiatry* **2015**, *86*, 168–173. [CrossRef]
32. Ye, S.; Ji, Y.; Li, C.; He, J.; Liu, X.; Fan, D. The Edinburgh Cognitive and Behavioural ALS Screen in a Chinese Amyotrophic Lateral Sclerosis Population. *PLoS ONE* **2016**, *11*, e0155496. [CrossRef]
33. Costello, E.; Rooney, J.; Pinto-Grau, M.; Burke, T.; Elamin, M.; Bede, P.; McMackin, R.; Dukic, S.; Vajda, A.; Heverin, M.; et al. Cognitive reserve in amyotrophic lateral sclerosis (ALS): A population-based longitudinal study. *J. Neurol. Neurosurg. Psychiatry* **2021**, *92*, 460–465. [CrossRef]
34. Finsel, J.; Uttner, I.; Vazquez Medrano, C.R.; Ludolph, A.C.; Lule, D. Cognition in the course of ALS—A meta-analysis. *Amyotroph. Lateral Scler. Front. Degener.* **2023**, *24*, 2–13. [CrossRef]
35. Elamin, M.; Bede, P.; Byrne, S.; Jordan, N.; Gallagher, L.; Wynne, B.; O'Brien, C.; Phukan, J.; Lynch, C.; Pender, N.; et al. Cognitive changes predict functional decline in ALS: A population-based longitudinal study. *Neurology* **2013**, *80*, 1590–1597. [CrossRef]

https://doi.org/10.3390/cells15010037

36. Woolley, S.; Goetz, R.; Factor-Litvak, P.; Murphy, J.; Hupf, J.; Lomen-Hoerth, C.; Andrews, H.; Heitzman, D.; Bedlack, R.; Katz, J.; et al. Longitudinal Screening Detects Cognitive Stability and Behavioral Deterioration in ALS Patients. *Behav. Neurol.* **2018**, *2018*, 5969137. [CrossRef] [PubMed]
37. Benbrika, S.; Desgranges, B.; Eustache, F.; Viader, F. Cognitive, Emotional and Psychological Manifestations in Amyotrophic Lateral Sclerosis at Baseline and Overtime: A Review. *Front. Neurosci.* **2019**, *13*, 951. [CrossRef] [PubMed]
38. McHutchison, C.A.; Wuu, J.; McMillan, C.T.; Rademakers, R.; Statland, J.; Wu, G.; Rampersaud, E.; Myers, J.; Hernandez, J.P.; Abrahams, S.; et al. Temporal course of cognitive and behavioural changes in motor neuron diseases. *J. Neurol. Neurosurg. Psychiatry* **2024**, *95*, 316–324. [CrossRef]
39. Muthén, B.; Muthén, L.K. Integrating person-centered and variable-centered analyses: Growth mixture modeling with latent trajectory classes. *Alcohol. Clin. Exp. Res.* **2000**, *24*, 882–891. [CrossRef] [PubMed]
40. Bampton, A.; McHutchison, C.; Talbot, K.; Benatar, M.; Thompson, A.G.; Turner, M.R. The Basis of Cognitive and Behavioral Dysfunction in Amyotrophic Lateral Sclerosis. *Brain Behav.* **2024**, *14*, e70115. [CrossRef]
41. Geser, F.; Stein, B.; Partain, M.; Elman, L.B.; McCluskey, L.F.; Xie, S.X.; Van Deerlin, V.M.; Kwong, L.K.; Lee, V.M.; Trojanowski, J.Q. Motor neuron disease clinically limited to the lower motor neuron is a diffuse TDP-43 proteinopathy. *Acta Neuropathol.* **2011**, *121*, 509–517. [CrossRef]
42. Henstridge, C.M.; Sideris, D.I.; Carroll, E.; Rotariu, S.; Salomonsson, S.; Tzioras, M.; McKenzie, C.A.; Smith, C.; von Arnim, C.A.F.; Ludolph, A.C.; et al. Synapse loss in the prefrontal cortex is associated with cognitive decline in amyotrophic lateral sclerosis. *Acta Neuropathol.* **2018**, *135*, 213–226. [CrossRef]
43. Sun, W.; Liu, S.H.; Wei, X.J.; Sun, H.; Ma, Z.W.; Yu, X.F. Potential of neuroimaging as a biomarker in amyotrophic lateral sclerosis: From structure to metabolism. *J. Neurol.* **2024**, *271*, 2238–2257. [CrossRef]
44. Schuster, C.; Kasper, E.; Dyrba, M.; Machts, J.; Bittner, D.; Kaufmann, J.; Mitchell, A.J.; Benecke, R.; Teipel, S.; Vielhaber, S.; et al. Cortical thinning and its relation to cognition in amyotrophic lateral sclerosis. *Neurobiol. Aging* **2014**, *35*, 240–246. [CrossRef]
45. Consonni, M.; Dalla Bella, E.; Contarino, V.E.; Bersano, E.; Lauria, G. Cortical thinning trajectories across disease stages and cognitive impairment in amyotrophic lateral sclerosis. *Cortex* **2020**, *131*, 284–294. [CrossRef]
46. Chenji, S.; Ishaque, A.; Mah, D.; Fujiwara, E.; Beaulieu, C.; Seres, P.; Graham, S.J.; Frayne, R.; Zinman, L.; Genge, A.; et al. Neuroanatomical associations of the Edinburgh cognitive and Behavioural ALS screen (ECAS). *Brain Imaging Behav.* **2021**, *15*, 1641–1654. [CrossRef]
47. Tan, H.H.G.; Nitert, A.D.; van Veenhuijzen, K.; Dukic, S.; van Zandvoort, M.J.E.; Hendrikse, J.; van Es, M.A.; Veldink, J.H.; Westeneng, H.J.; van den Berg, L.H. Neuroimaging correlates of domain-specific cognitive deficits in amyotrophic lateral sclerosis. *NeuroImage Clin.* **2025**, *45*, 103749. [CrossRef] [PubMed]
48. Abrahams, S.; Goldstein, L.H.; Kew, J.J.; Brooks, D.J.; Lloyd, C.M.; Frith, C.D.; Leigh, P.N. Frontal lobe dysfunction in amyotrophic lateral sclerosis. A PET study. *Brain* **1996**, *119 Pt 6*, 2105–2120. [CrossRef] [PubMed]
49. Ma, X.; Zhang, J.; Zhang, Y.; Chen, H.; Li, R.; Wang, J.; Chen, H. Altered cortical hubs in functional brain networks in amyotrophic lateral sclerosis. *Neurol. Sci.* **2015**, *36*, 2097–2104. [CrossRef]
50. Canosa, A.; Moglia, C.; Manera, U.; Vasta, R.; Torrieri, M.C.; Arena, V.; D'Ovidio, F.; Palumbo, F.; Zucchetti, J.P.; Iazzolino, B.; et al. Metabolic brain changes across different levels of cognitive impairment in ALS: A ^{18}F-FDG-PET study. *J. Neurol. Neurosurg. Psychiatry* **2020**, *92*, 357–363. [CrossRef] [PubMed]
51. Foucher, J.; Oijerstedt, L.; Lovik, A.; Sun, J.; Ismail, M.A.; Sennfalt, S.; Savitcheva, I.; Estenberg, U.; Pagani, M.; Fang, F.; et al. ECAS correlation with metabolic alterations on FDG-PET imaging in ALS. *Amyotroph. Lateral Scler. Front. Degener.* **2024**, *25*, 708–716. [CrossRef]
52. Shahriari, Y.; Besio, W.; Hosni, S.I.; Hillary Zisk, A.; Borgheai, S.B.; Deligani, R.J.; McLinden, J. Electroencephalography. In *Neural Interface Engineering: Linking the Physical World and the Nervous System*; Springer: Berlin/Heidelberg, Germany, 2020; pp. 1–16.
53. Hong, K.S.; Khan, M.J.; Hong, M.J. Feature Extraction and Classification Methods for Hybrid fNIRS-EEG Brain-Computer Interfaces. *Front. Hum. Neurosci.* **2018**, *12*, 246. [CrossRef]
54. Lindquist, M.A.; Meng Loh, J.; Atlas, L.Y.; Wager, T.D. Modeling the hemodynamic response function in fMRI: Efficiency, bias and mis-modeling. *NeuroImage* **2009**, *45*, S187–S198. [CrossRef]
55. Xu, L.; Wang, B.; Xu, G.; Wang, W.; Liu, Z.; Li, Z. Functional connectivity analysis using fNIRS in healthy subjects during prolonged simulated driving. *Neurosci. Lett.* **2017**, *640*, 21–28. [CrossRef] [PubMed]
56. Tulay, E.E.; Metin, B.; Tarhan, N.; Arikan, M.K. Multimodal Neuroimaging: Basic Concepts and Classification of Neuropsychiatric Diseases. *Clin. EEG Neurosci.* **2019**, *50*, 20–33. [CrossRef] [PubMed]
57. Borgheai, S.B.; Zisk, A.H.; McLinden, J.; McIntyre, J.; Sadjadi, R.; Shahriari, Y. Multimodal pre-screening can predict BCI performance variability: A novel subject-specific experimental scheme. *Comput. Biol. Med.* **2024**, *168*, 107658. [CrossRef] [PubMed]

https://doi.org/10.3390/cells15010037

58. Borgheai, S.B.; Deligani, R.J.; McLinden, J.; Zisk, A.; Hosni, S.I.; Abtahi, M.; Mankodiya, K.; Shahriari, Y. Multimodal exploration of non-motor neural functions in ALS patients using simultaneous EEG-fNIRS recording. *J. Neural Eng.* **2019**, *16*, 066036. [CrossRef]
59. Deligani, R.J.; Hosni, S.I.; Borgheai, S.B.; McLinden, J.; Zisk, A.H.; Mankodiya, K.; Shahriari, Y. Electrical and Hemodynamic Neural Functions in People With ALS: An EEG-fNIRS Resting-State Study. *IEEE Trans. Neural Syst. Rehabil. Eng.* **2020**, *28*, 3129–3139. [CrossRef]
60. Hosni, S.M.I.; Borgheai, S.B.; McLinden, J.; Zhu, S.; Huang, X.; Ostadabbas, S.; Shahriari, Y. A Graph-Based Nonlinear Dynamic Characterization of Motor Imagery Toward an Enhanced Hybrid BCI. *Neuroinformatics* **2022**, *20*, 1169–1189. [CrossRef]
61. Menke, R.A.; Proudfoot, M.; Wuu, J.; Andersen, P.M.; Talbot, K.; Benatar, M.; Turner, M.R. Increased functional connectivity common to symptomatic amyotrophic lateral sclerosis and those at genetic risk. *J. Neurol. Neurosurg. Psychiatry* **2016**, *87*, 580–588. [CrossRef]
62. Luo, C.; Chen, Q.; Huang, R.; Chen, X.; Chen, K.; Huang, X.; Tang, H.; Gong, Q.; Shang, H.F. Patterns of spontaneous brain activity in amyotrophic lateral sclerosis: A resting-state FMRI study. *PLoS ONE* **2012**, *7*, e45470. [CrossRef] [PubMed]
63. Douaud, G.; Filippini, N.; Knight, S.; Talbot, K.; Turner, M.R. Integration of structural and functional magnetic resonance imaging in amyotrophic lateral sclerosis. *Brain* **2011**, *134*, 3470–3479. [CrossRef]
64. Verstraete, E.; van den Heuvel, M.P.; Veldink, J.H.; Blanken, N.; Mandl, R.C.; Hulshoff Pol, H.E.; van den Berg, L.H. Motor network degeneration in amyotrophic lateral sclerosis: A structural and functional connectivity study. *PLoS ONE* **2010**, *5*, e13664. [CrossRef]
65. Agosta, F.; Canu, E.; Valsasina, P.; Riva, N.; Prelle, A.; Comi, G.; Filippi, M. Divergent brain network connectivity in amyotrophic lateral sclerosis. *Neurobiol. Aging* **2013**, *34*, 419–427. [CrossRef]
66. Tedeschi, G.; Trojsi, F.; Tessitore, A.; Corbo, D.; Sagnelli, A.; Paccone, A.; D'Ambrosio, A.; Piccirillo, G.; Cirillo, M.; Cirillo, S.; et al. Interaction between aging and neurodegeneration in amyotrophic lateral sclerosis. *Neurobiol. Aging* **2012**, *33*, 886–898. [CrossRef] [PubMed]
67. Li, F.; Zhou, F.; Huang, M.; Gong, H.; Xu, R. Frequency-Specific Abnormalities of Intrinsic Functional Connectivity Strength among Patients with Amyotrophic Lateral Sclerosis: A Resting-State fMRI Study. *Front. Aging Neurosci.* **2017**, *9*, 351. [CrossRef]
68. Mohammadi, B.; Kollewe, K.; Samii, A.; Krampfl, K.; Dengler, R.; Munte, T.F. Changes of resting state brain networks in amyotrophic lateral sclerosis. *Exp. Neurol.* **2009**, *217*, 147–153. [CrossRef]
69. Fraschini, M.; Demuru, M.; Hillebrand, A.; Cuccu, L.; Porcu, S.; Di Stefano, F.; Puligheddu, M.; Floris, G.; Borghero, G.; Marrosu, F. EEG functional network topology is associated with disability in patients with amyotrophic lateral sclerosis. *Sci. Rep.* **2016**, *6*, 38653. [CrossRef]
70. Kopitzki, K.; Oldag, A.; Sweeney-Reed, C.M.; Machts, J.; Veit, M.; Kaufmann, J.; Hinrichs, H.; Heinze, H.J.; Kollewe, K.; Petri, S.; et al. Interhemispheric connectivity in amyotrophic lateral sclerosis: A near-infrared spectroscopy and diffusion tensor imaging study. *NeuroImage Clin.* **2016**, *12*, 666–672. [CrossRef] [PubMed]
71. Nasseroleslami, B.; Dukic, S.; Broderick, M.; Mohr, K.; Schuster, C.; Gavin, B.; McLaughlin, R.; Heverin, M.; Vajda, A.; Iyer, P.M.; et al. Characteristic Increases in EEG Connectivity Correlate With Changes of Structural MRI in Amyotrophic Lateral Sclerosis. *Cereb. Cortex* **2019**, *29*, 27–41. [CrossRef]
72. Dukic, S.; McMackin, R.; Buxo, T.; Fasano, A.; Chipika, R.; Pinto-Grau, M.; Costello, E.; Schuster, C.; Hammond, M.; Heverin, M.; et al. Patterned functional network disruption in amyotrophic lateral sclerosis. *Hum. Brain Mapp.* **2019**, *40*, 4827–4842. [CrossRef]
73. Mai, R.; Facchetti, D.; Micheli, A.; Poloni, M. Quantitative electroencephalography in amyotrophic lateral sclerosis. *Electroencephalogr. Clin. Neurophysiol.* **1998**, *106*, 383–386. [CrossRef]
74. Santhosh, J.; Bhatia, M.; Sahu, S.; Anand, S. Decreased electroencephalogram alpha band [8–13 Hz] power in amyotrophic lateral sclerosis patients: A study of alpha activity in an awake relaxed state. *Neurol. India* **2005**, *53*, 99–101. [CrossRef]
75. Pievani, M.; Filippini, N.; van den Heuvel, M.P.; Cappa, S.F.; Frisoni, G.B. Brain connectivity in neurodegenerative diseases—From phenotype to proteinopathy. *Nat. Rev. Neurol.* **2014**, *10*, 620–633. [CrossRef] [PubMed]
76. Raichle, M.E. The brain's default mode network. *Annu. Rev. Neurosci.* **2015**, *38*, 433–447. [CrossRef]
77. Christidi, F.; Zalonis, I.; Smyrnis, N.; Evdokimidis, I. Selective attention and the three-process memory model for the interpretation of verbal free recall in amyotrophic lateral sclerosis. *J. Int. Neuropsychol. Soc.* **2012**, *18*, 809–818. [CrossRef]
78. Corbetta, M.; Shulman, G.L. Control of goal-directed and stimulus-driven attention in the brain. *Nat. Rev. Neurosci.* **2002**, *3*, 201–215. [CrossRef] [PubMed]
79. Vincent, J.L.; Kahn, I.; Snyder, A.Z.; Raichle, M.E.; Buckner, R.L. Evidence for a frontoparietal control system revealed by intrinsic functional connectivity. *J. Neurophysiol.* **2008**, *100*, 3328–3342. [CrossRef] [PubMed]
80. Kellmeyer, P.; Grosse-Wentrup, M.; Schulze-Bonhage, A.; Ziemann, U.; Ball, T. Electrophysiological correlates of neurodegeneration in motor and non-motor brain regions in amyotrophic lateral sclerosis-implications for brain-computer interfacing. *J. Neural Eng.* **2018**, *15*, 041003. [CrossRef]

https://doi.org/10.3390/cells15010037

81. Raggi, A.; Iannaccone, S.; Cappa, S.F. Event-related brain potentials in amyotrophic lateral sclerosis: A review of the international literature. *Amyotroph. Lateral Scler.* **2010**, *11*, 16–26. [CrossRef]
82. Abiri, R.; Borhani, S.; Sellers, E.W.; Jiang, Y.; Zhao, X. A comprehensive review of EEG-based brain-computer interface paradigms. *J. Neural Eng.* **2019**, *16*, 011001. [CrossRef]
83. Putze, F.; Hesslinger, S.; Tse, C.Y.; Huang, Y.; Herff, C.; Guan, C.; Schultz, T. Hybrid fNIRS-EEG based classification of auditory and visual perception processes. *Front. Neurosci.* **2014**, *8*, 373. [CrossRef] [PubMed]
84. Riccio, A.; Simione, L.; Schettini, F.; Pizzimenti, A.; Inghilleri, M.; Belardinelli, M.O.; Mattia, D.; Cincotti, F. Attention and P300-based BCI performance in people with amyotrophic lateral sclerosis. *Front. Hum. Neurosci.* **2013**, *7*, 732. [CrossRef]
85. Pinkhardt, E.H.; Jurgens, R.; Becker, W.; Molle, M.; Born, J.; Ludolph, A.C.; Schreiber, H. Signs of impaired selective attention in patients with amyotrophic lateral sclerosis. *J. Neurol.* **2008**, *255*, 532–538. [CrossRef]
86. Shahriari, Y.; Vaughan, T.M.; McCane, L.M.; Allison, B.Z.; Wolpaw, J.R.; Krusienski, D.J. An exploration of BCI performance variations in people with amyotrophic lateral sclerosis using longitudinal EEG data. *J. Neural Eng.* **2019**, *16*, 056031. [CrossRef]
87. Arico, P.; Aloise, F.; Schettini, F.; Salinari, S.; Mattia, D.; Cincotti, F. Influence of P300 latency jitter on event related potential-based brain-computer interface performance. *J. Neural Eng.* **2014**, *11*, 035008. [CrossRef]
88. Fjell, A.M.; Rosquist, H.; Walhovd, K.B. Instability in the latency of P3a/P3b brain potentials and cognitive function in aging. *Neurobiol. Aging* **2009**, *30*, 2065–2079. [CrossRef] [PubMed]
89. Jaśkowski, P.; Verleger, R. An evaluation of methods for single-trial estimation of P3 latency. *Psychophysiology* **2000**, *37*, 153–162. [CrossRef] [PubMed]
90. Zisk, A.H.; Borgheai, S.B.; McLinden, J.; Hosni, S.M.; Deligani, R.J.; Shahriari, Y. P300 latency jitter and its correlates in people with amyotrophic lateral sclerosis. *Clin. Neurophysiol.* **2021**, *132*, 632–642. [CrossRef]
91. Kutas, M.; McCarthy, G.; Donchin, E. Augmenting mental chronometry: The P300 as a measure of stimulus evaluation time. *Science* **1977**, *197*, 792–795. [CrossRef]
92. Saville, C.W.N.; Feige, B.; Kluckert, C.; Bender, S.; Biscaldi, M.; Berger, A.; Fleischhaker, C.; Henighausen, K.; Klein, C. Increased reaction time variability in attention-deficit hyperactivity disorder as a response-related phenomenon: Evidence from single-trial event-related potentials. *J. Child. Psychol. Psychiatry* **2015**, *56*, 801–813. [CrossRef] [PubMed]
93. Verleger, R.; Baur, N.; Metzner, M.F.; Smigasiewicz, K. The hard oddball: Effects of difficult response selection on stimulus-related P3 and on response-related negative potentials. *Psychophysiology* **2014**, *51*, 1089–1100. [CrossRef] [PubMed]
94. Magnuson, J.R.; Iarocci, G.; Doesburg, S.M.; Moreno, S. Increased intra-subject variability of reaction times and single-trial event-related potential components in children with autism spectrum disorder. *Autism Res.* **2020**, *13*, 221–229. [CrossRef]
95. Dinstein, I.; Heeger, D.J.; Behrmann, M. Neural variability: Friend or foe? *Trends Cogn. Sci.* **2015**, *19*, 322–328. [CrossRef] [PubMed]
96. Ford, J.M.; White, P.; Lim, K.O.; Pfefferbaum, A. Schizophrenics have fewer and smaller P300s: A single-trial analysis. *Biol. Psychiatry* **1994**, *35*, 96–103. [CrossRef]
97. Patterson, J.V.; Michalewski, H.; Starr, A. Latency variability of the components of auditory event-related potentials to infrequent stimuli in aging, Alzheimer-type dementia, and depression. *Electroencephalogr. Clin. Neurophysiol./Evoked Potentials Sect.* **1988**, *71*, 450–460. [CrossRef]
98. Unsal, A.; Segalowitz, S.J. Sources of P300 attenuation after head injury: Single-trial amplitude, latency jitter, and EEG power. *Psychophysiology* **1995**, *32*, 249–256. [CrossRef]
99. Schettini, F.; Risetti, M.; Aricò, P.; Formisano, R.; Babiloni, F.; Mattia, D.; Cincotti, F. P300 latency Jitter occurrence in patients with disorders of consciousness: Toward a better design for Brain Computer Interface applications. In Proceedings of the 2015 37th Annual International Conference of the IEEE Engineering in Medicine and Biology Society (EMBC), Milan, Italy, 25–29 August 2015; pp. 6178–6181.
100. Van Beek, L.; Ghesquiere, P.; De Smedt, B.; Lagae, L. Arithmetic difficulties in children with mild traumatic brain injury at the subacute stage of recovery. *Dev. Med. Child. Neurol.* **2015**, *57*, 1042–1048. [CrossRef]
101. Xu, H.; Zhang, D.; Ouyang, M.; Hong, B. Employing an active mental task to enhance the performance of auditory attention-based brain-computer interfaces. *Clin. Neurophysiol.* **2013**, *124*, 83–90. [CrossRef]
102. Cooper, D.L. Broca's arrow: Evolution, prediction, and language in the brain. *Anat. Rec. B New Anat.* **2006**, *289*, 9–24. [CrossRef]
103. Shellikeri, S.; Karthikeyan, V.; Martino, R.; Black, S.E.; Zinman, L.; Keith, J.; Yunusova, Y. The neuropathological signature of bulbar-onset ALS: A systematic review. *Neurosci. Biobehav. Rev.* **2017**, *75*, 378–392. [CrossRef]
104. Acharya, S.; Shukla, S. Mirror neurons: Enigma of the metaphysical modular brain. *J. Nat. Sci. Biol. Med.* **2012**, *3*, 118–124. [CrossRef]
105. Garrison, K.A.; Winstein, C.J.; Aziz-Zadeh, L. The mirror neuron system: A neural substrate for methods in stroke rehabilitation. *Neurorehabilit. Neural Repair* **2010**, *24*, 404–412. [CrossRef]
106. Carvalho, D.; Teixeira, S.; Lucas, M.; Yuan, T.F.; Chaves, F.; Peressutti, C.; Machado, S.; Bittencourt, J.; Menendez-Gonzalez, M.; Nardi, A.E.; et al. The mirror neuron system in post-stroke rehabilitation. *Int. Arch. Med.* **2013**, *6*, 41. [CrossRef] [PubMed]

107. Binkofski, F.; Buccino, G. Motor functions of the Broca's region. *Brain Lang.* **2004**, *89*, 362–369. [CrossRef]
108. Jeon, Y.; Nam, C.S.; Kim, Y.-J.; Whang, M.C. Event-related (De) synchronization (ERD/ERS) during motor imagery tasks: Implications for brain–computer interfaces. *Int. J. Ind. Ergon.* **2011**, *41*, 428–436. [CrossRef]
109. Pfurtscheller, G.; Lopes da Silva, F.H. Event-related EEG/MEG synchronization and desynchronization: Basic principles. *Clin. Neurophysiol.* **1999**, *110*, 1842–1857. [CrossRef]
110. Pfurtscheller, G.; Neuper, C. Motor imagery activates primary sensorimotor area in humans. *Neurosci. Lett.* **1997**, *239*, 65–68. [CrossRef]
111. Yuan, H.; He, B. Brain-computer interfaces using sensorimotor rhythms: Current state and future perspectives. *IEEE Trans. Biomed. Eng.* **2014**, *61*, 1425–1435. [CrossRef] [PubMed]
112. Kasahara, T.; Terasaki, K.; Ogawa, Y.; Ushiba, J.; Aramaki, H.; Masakado, Y. The correlation between motor impairments and event-related desynchronization during motor imagery in ALS patients. *BMC Neurosci.* **2012**, *13*, 66. [CrossRef] [PubMed]
113. Hosni, S.M.; Deligani, R.J.; Zisk, A.; McLinden, J.; Borgheai, S.B.; Shahriari, Y. An exploration of neural dynamics of motor imagery for people with amyotrophic lateral sclerosis. *J. Neural Eng.* **2019**, *17*, 016005. [CrossRef]
114. Aguilera, M.; Bedia, M.G.; Santos, B.A.; Barandiaran, X.E. The situated HKB model: How sensorimotor spatial coupling can alter oscillatory brain dynamics. *Front. Comput. Neurosci.* **2013**, *7*, 117. [CrossRef]
115. Stanton, B.R.; Williams, V.C.; Leigh, P.N.; Williams, S.C.; Blain, C.R.; Giampietro, V.P.; Simmons, A. Cortical activation during motor imagery is reduced in Amyotrophic Lateral Sclerosis. *Brain Res.* **2007**, *1172*, 145–151. [CrossRef]
116. Pettit, L.D.; Bastin, M.E.; Smith, C.; Bak, T.H.; Gillingwater, T.H.; Abrahams, S. Executive deficits, not processing speed relates to abnormalities in distinct prefrontal tracts in amyotrophic lateral sclerosis. *Brain* **2013**, *136*, 3290–3304. [CrossRef]
117. Abrahams, S.; Leigh, P.N.; Harvey, A.; Vythelingum, G.N.; Grise, D.; Goldstein, L.H. Verbal fluency and executive dysfunction in amyotrophic lateral sclerosis (ALS). *Neuropsychologia* **2000**, *38*, 734–747. [CrossRef]
118. Woolley, S.C.; York, M.K.; Moore, D.H.; Strutt, A.M.; Murphy, J.; Schulz, P.E.; Katz, J.S. Detecting frontotemporal dysfunction in ALS: Utility of the ALS Cognitive Behavioral Screen (ALS-CBS). *Amyotroph. Lateral Scler.* **2010**, *11*, 303–311. [CrossRef] [PubMed]
119. Hobson, H.M.; Bishop, D.V.M. Mu suppression—A good measure of the human mirror neuron system? *Cortex* **2016**, *82*, 290–310. [CrossRef] [PubMed]
120. Hobson, H.M.; Bishop, D.V. The interpretation of mu suppression as an index of mirror neuron activity: Past, present and future. *R. Soc. Open Sci.* **2017**, *4*, 160662. [CrossRef] [PubMed]
121. Deiber, M.-P.; Missonnier, P.; Bertrand, O.; Gold, G.; Fazio-Costa, L.; Ibanez, V.; Giannakopoulos, P. Distinction between perceptual and attentional processing in working memory tasks: A study of phase-locked and induced oscillatory brain dynamics. *J. Cogn. Neurosci.* **2007**, *19*, 158–172. [CrossRef]
122. Siegel, M.; Donner, T.H.; Engel, A.K. Spectral fingerprints of large-scale neuronal interactions. *Nat. Rev. Neurosci.* **2012**, *13*, 121–134. [CrossRef]
123. Bastos, A.M.; Vezoli, J.; Bosman, C.A.; Schoffelen, J.M.; Oostenveld, R.; Dowdall, J.R.; De Weerd, P.; Kennedy, H.; Fries, P. Visual areas exert feedforward and feedback influences through distinct frequency channels. *Neuron* **2015**, *85*, 390–401. [CrossRef] [PubMed]
124. Engel, A.K.; Fries, P. Beta-band oscillations—Signalling the status quo? *Curr. Opin. Neurobiol.* **2010**, *20*, 156–165. [CrossRef]
125. Wimmer, K.; Ramon, M.; Pasternak, T.; Compte, A. Transitions between Multiband Oscillatory Patterns Characterize Memory-Guided Perceptual Decisions in Prefrontal Circuits. *J. Neurosci.* **2016**, *36*, 489–505. [CrossRef] [PubMed]
126. Spitzer, B.; Haegens, S. Beyond the Status Quo: A Role for Beta Oscillations in Endogenous Content (Re)Activation. *eNeuro* **2017**, *4*. [CrossRef]
127. Borgheai, S.B.; McLinden, J.; Mankodiya, K.; Shahriari, Y. Frontal Functional Network Disruption Associated with Amyotrophic Lateral Sclerosis: An fNIRS-Based Minimum Spanning Tree Analysis. *Front. Neurosci.* **2020**, *14*, 613990. [CrossRef]
128. Yu, X.; Ji, C.; Shao, A. Neurovascular Unit Dysfunction and Neurodegenerative Disorders. *Front. Neurosci.* **2020**, *14*, 334. [CrossRef]
129. Blumenfeld, R.S.; Ranganath, C. Dorsolateral prefrontal cortex promotes long-term memory formation through its role in working memory organization. *J. Neurosci.* **2006**, *26*, 916–925. [CrossRef]
130. Radakovic, R.; Stephenson, L.; Newton, J.; Crockford, C.; Swingler, R.; Chandran, S.; Abrahams, S. Multidimensional apathy and executive dysfunction in amyotrophic lateral sclerosis. *Cortex* **2017**, *94*, 142–151. [CrossRef]
131. Alvarez, J.A.; Emory, E. Executive function and the frontal lobes: A meta-analytic review. *Neuropsychol. Rev.* **2006**, *16*, 17–42. [CrossRef] [PubMed]
132. Stam, C.J.; de Haan, W.; Daffertshofer, A.; Jones, B.F.; Manshanden, I.; van Cappellen van Walsum, A.M.; Montez, T.; Verbunt, J.P.; de Munck, J.C.; van Dijk, B.W.; et al. Graph theoretical analysis of magnetoencephalographic functional connectivity in Alzheimer's disease. *Brain* **2009**, *132*, 213–224. [CrossRef]
133. Stam, C.J. Modern network science of neurological disorders. *Nat. Rev. Neurosci.* **2014**, *15*, 683–695. [CrossRef] [PubMed]

https://doi.org/10.3390/cells15010037

134. Li, X.; Jing, Z.; Hu, B.; Zhu, J.; Zhong, N.; Li, M.; Ding, Z.; Yang, J.; Zhang, L.; Feng, L.; et al. A resting-state brain functional network study in MDD based on minimum spanning tree analysis and the hierarchical clustering. *Complexity* **2017**, *2017*, 9514369. [CrossRef]
135. Fraga Gonzalez, G.; Van der Molen, M.J.W.; Zaric, G.; Bonte, M.; Tijms, J.; Blomert, L.; Stam, C.J.; Van der Molen, M.W. Graph analysis of EEG resting state functional networks in dyslexic readers. *Clin. Neurophysiol.* **2016**, *127*, 3165–3175. [CrossRef]
136. Shahriari, Y.; Sellers, E.W.; McCane, L.M.; Vaughan, T.M.; Krusienski, D.J. Directional brain functional interaction analysis in patients with amyotrophic lateral sclerosis. In Proceedings of the 2015 7th International IEEE/EMBS Conference on Neural Engineering (NER), Montpellier, France, 22–24 April 2015; pp. 972–975.
137. Gleichgerrcht, E.; Kocher, M.; Nesland, T.; Rorden, C.; Fridriksson, J.; Bonilha, L. Preservation of structural brain network hubs is associated with less severe post-stroke aphasia. *Restor. Neurol. Neurosci.* **2016**, *34*, 19–28. [CrossRef]
138. Hammer, A.; Vielhaber, S.; Rodriguez-Fornells, A.; Mohammadi, B.; Munte, T.F. A neurophysiological analysis of working memory in amyotrophic lateral sclerosis. *Brain Res.* **2011**, *1421*, 90–99. [CrossRef] [PubMed]
139. Dehaene, S.; Molko, N.; Cohen, L.; Wilson, A.J. Arithmetic and the brain. *Curr. Opin. Neurobiol.* **2004**, *14*, 218–224. [CrossRef]
140. Klein, E.; Moeller, K.; Glauche, V.; Weiller, C.; Willmes, K. Processing pathways in mental arithmetic—Evidence from probabilistic fiber tracking. *PLoS ONE* **2013**, *8*, e55455. [CrossRef]
141. Sammer, G.; Blecker, C.; Gebhardt, H.; Bischoff, M.; Stark, R.; Morgen, K.; Vaitl, D. Relationship between regional hemodynamic activity and simultaneously recorded EEG-theta associated with mental arithmetic-induced workload. *Hum. Brain Mapp.* **2007**, *28*, 793–803. [CrossRef]
142. Schudlo, L.C.; Chau, T. Dynamic topographical pattern classification of multichannel prefrontal NIRS signals: II. Online differentiation of mental arithmetic and rest. *J. Neural Eng.* **2014**, *11*, 016003. [CrossRef]
143. Monden, Y.; Dan, I.; Nagashima, M.; Dan, H.; Uga, M.; Ikeda, T.; Tsuzuki, D.; Kyutoku, Y.; Gunji, Y.; Hirano, D.; et al. Individual classification of ADHD children by right prefrontal hemodynamic responses during a go/no-go task as assessed by fNIRS. *NeuroImage Clin.* **2015**, *9*, 1–12. [CrossRef] [PubMed]
144. Henson, R.N.; Shallice, T.; Dolan, R.J. Right prefrontal cortex and episodic memory retrieval: A functional MRI test of the monitoring hypothesis. *Brain* **1999**, *122 Pt 7*, 1367–1381. [CrossRef] [PubMed]
145. Causse, M.; Chua, Z.; Peysakhovich, V.; Del Campo, N.; Matton, N. Mental workload and neural efficiency quantified in the prefrontal cortex using fNIRS. *Sci. Rep.* **2017**, *7*, 5222. [CrossRef] [PubMed]
146. Glahn, D.C.; Ragland, J.D.; Abramoff, A.; Barrett, J.; Laird, A.R.; Bearden, C.E.; Velligan, D.I. Beyond hypofrontality: A quantitative meta-analysis of functional neuroimaging studies of working memory in schizophrenia. *Hum. Brain Mapp.* **2005**, *25*, 60–69. [CrossRef]
147. Roos, E.; Mariosa, D.; Ingre, C.; Lundholm, C.; Wirdefeldt, K.; Roos, P.M.; Fang, F. Depression in amyotrophic lateral sclerosis. *Neurology* **2016**, *86*, 2271–2277. [CrossRef]
148. Thakore, N.J.; Pioro, E.P. Depression in ALS in a large self-reporting cohort. *Neurology* **2016**, *86*, 1031–1038. [CrossRef]
149. Wicks, P.; Abrahams, S.; Masi, D.; Hejda-Forde, S.; Leigh, P.N.; Goldstein, L.H. Prevalence of depression in a 12-month consecutive sample of patients with ALS. *Eur. J. Neurol.* **2007**, *14*, 993–1001. [CrossRef]
150. Dimond, D.; Ishaque, A.; Chenji, S.; Mah, D.; Chen, Z.; Seres, P.; Beaulieu, C.; Kalra, S. White matter structural network abnormalities underlie executive dysfunction in amyotrophic lateral sclerosis. *Hum. Brain Mapp.* **2017**, *38*, 1249–1268. [CrossRef] [PubMed]
151. Proudfoot, M.; Rohenkohl, G.; Quinn, A.; Colclough, G.L.; Wuu, J.; Talbot, K.; Woolrich, M.W.; Benatar, M.; Nobre, A.C.; Turner, M.R. Altered cortical beta-band oscillations reflect motor system degeneration in amyotrophic lateral sclerosis. *Hum. Brain Mapp.* **2017**, *38*, 237–254. [CrossRef]
152. Chen, K.S.; Chen, R. Invasive and Noninvasive Brain Stimulation in Parkinson's Disease: Clinical Effects and Future Perspectives. *Clin. Pharmacol. Ther.* **2019**, *106*, 763–775. [CrossRef]
153. Wagner, J.; Wessel, J.R.; Ghahremani, A.; Aron, A.R. Establishing a Right Frontal Beta Signature for Stopping Action in Scalp EEG: Implications for Testing Inhibitory Control in Other Task Contexts. *J. Cogn. Neurosci.* **2018**, *30*, 107–118. [CrossRef]
154. Turner, M.R.; Kiernan, M.C. Does interneuronal dysfunction contribute to neurodegeneration in amyotrophic lateral sclerosis? *Amyotroph. Lateral Scler.* **2012**, *13*, 245–250. [CrossRef] [PubMed]
155. Dukic, S.; McMackin, R.; Costello, E.; Metzger, M.; Buxo, T.; Fasano, A.; Chipika, R.; Pinto-Grau, M.; Schuster, C.; Hammond, M.; et al. Resting-state EEG reveals four subphenotypes of amyotrophic lateral sclerosis. *Brain* **2022**, *145*, 621–631. [CrossRef]
156. Schmidt, R.; Verstraete, E.; de Reus, M.A.; Veldink, J.H.; van den Berg, L.H.; van den Heuvel, M.P. Correlation between structural and functional connectivity impairment in amyotrophic lateral sclerosis. *Hum. Brain Mapp.* **2014**, *35*, 4386–4395. [CrossRef]
157. Smallwood Shoukry, R.F.; Clark, M.G.; Floeter, M.K. Resting State Functional Connectivity Is Decreased Globally Across the C9orf72 Mutation Spectrum. *Front. Neurol.* **2020**, *11*, 598474. [CrossRef]
158. Geronimo, A.; Sheldon, K.E.; Broach, J.R.; Simmons, Z.; Schiff, S.J. Expansion of C9ORF72 in amyotrophic lateral sclerosis correlates with brain-computer interface performance. *Sci. Rep.* **2017**, *7*, 8875. [CrossRef]

https://doi.org/10.3390/cells15010037

159. Dukic, S.; van Veenhuijzen, K.; Michielsen, A.; Westeneng, H.-J.; van den Berg, L.H. Resting-state EEG oscillations are reduced in asymptomatic *C9orf72* repeat expansion carriers. *medRxiv* **2025**. [CrossRef]
160. Dukic, S.; van Veenhuijzen, K.; Westeneng, H.J.; McMackin, R.; van Eijk, R.P.A.; Sleutjes, B.; Nasseroleslami, B.; Hardiman, O.; van den Berg, L.H. Altered EEG Response of the Parietal Network in Asymptomatic C9orf72 Carriers. *Hum. Brain Mapp.* **2025**, *46*, e70275. [CrossRef] [PubMed]
161. Trubshaw, M.; Gohil, C.; Edmond, E.; Proudfoot, M.; Yoganathan, K.; Wuu, J.; Northall, A.; Kohl, O.; Stagg, C.J.; Nobre, A.C.; et al. Divergent Brain Network Activity in Asymptomatic C9orf72 and SOD1 Variant Carriers Compared With Established Amyotrophic Lateral Sclerosis. *Hum. Brain Mapp.* **2025**, *46*, e70345. [CrossRef]
162. Fasano, A.; Vacchiano, V.; Bonan, L.; McMackin, R.; Nasseroleslami, B.; Hardiman, O.; Liguori, R. Neurophysiological biomarkers of networks impairment in amyotrophic lateral sclerosis. *J. Neurol.* **2025**, *272*, 507. [CrossRef] [PubMed]
163. van den Bos, M.A.J.; Menon, P.; Pavey, N.; Higashihara, M.; Kiernan, M.C.; Vucic, S. Direct interrogation of cortical interneuron circuits in amyotrophic lateral sclerosis. *Brain* **2025**, *148*, 1169–1179. [CrossRef]
164. Tremblay, S.; Rogasch, N.C.; Premoli, I.; Blumberger, D.M.; Casarotto, S.; Chen, R.; Di Lazzaro, V.; Farzan, F.; Ferrarelli, F.; Fitzgerald, P.B.; et al. Clinical utility and prospective of TMS-EEG. *Clin. Neurophysiol.* **2019**, *130*, 802–844. [CrossRef]
165. Codina, T.; Blankertz, B.; von Luhmann, A. Multimodal fNIRS-EEG sensor fusion: Review of data-driven methods and perspective for naturalistic brain imaging. *Imaging Neurosci.* **2025**, *3*, IMAG.a.974. [CrossRef]
166. Liu, Z.; Shore, J.; Wang, M.; Yuan, F.; Buss, A.; Zhao, X. A systematic review on hybrid EEG/fNIRS in brain-computer interface. *Biomed. Signal Process. Control* **2021**, *68*, 102595. [CrossRef]
167. Chen, J.; Yu, K.; Bi, Y.; Ji, X.; Zhang, D. Strategic Integration: A Cross-Disciplinary Review of the fNIRS-EEG Dual-Modality Imaging System for Delivering Multimodal Neuroimaging to Applications. *Brain Sci.* **2024**, *14*, 1022. [CrossRef] [PubMed]
168. Li, R.; Yang, D.; Fang, F.; Hong, K.-S.; Reiss, A.L.; Zhang, Y. Concurrent fNIRS and EEG for Brain Function Investigation: A Systematic, Methodology-Focused Review. *Sensors* **2022**, *22*, 5865. [CrossRef]
169. Yu, B.; Zhang, X.; Chen, G.; Li, F. EEG and fNIRS-Based Emotion Recognition Using an Improved Graph Isomorphism Network. In *Neural Computing for Advanced Applications*; Springer: Singapore; pp. 394–408.
170. Zhu, S.; Hosni, S.I.; Huang, X.; Wan, M.; Borgheai, S.B.; McLinden, J.; Shahriari, Y.; Ostadabbas, S. A dynamical graph-based feature extraction approach to enhance mental task classification in brain–computer interfaces. *Comput. Biol. Med.* **2023**, *153*, 106498. [CrossRef] [PubMed]
171. Jiao, Y.; Zhao, K.; Wei, X.; Carlisle, N.B.; Keller, C.J.; Oathes, D.J.; Fonzo, G.A.; Zhang, Y. Deep graph learning of multimodal brain networks defines treatment-predictive signatures in major depression. *Mol. Psychiatry* **2025**, *30*, 3963–3974. [CrossRef] [PubMed]

https://doi.org/10.3390/cells15010037

cells

Systematic Review

Cellular and Molecular Mechanisms of Non-Invasive Brain Stimulation Techniques: A Systematic Review on the Implications for the Treatment of Neurological Disorders

Valerio Sveva [1], Marco Mancuso [1], Alessandro Cruciani [2], Elias Paolo Casula [3], Giorgio Leodori [1], Silvia Antonella Selvaggi [2], Matteo Bologna [1,4], Vincenzo Di Lazzaro [2], Anna Latorre [5,*] and Lorenzo Rocchi [6]

1 Department of Human Neuroscience, University of Rome "Sapienza", Viale dell'Università 30, 00185 Rome, Italy; valerio.sveva@uniroma1.it (V.S.); marco.mancuso@uniroma1.it (M.M.); giorgio.leodori@uniroma1.it (G.L.); matteo.bologna@uniroma1.it (M.B.)
2 Unit of Neurology, Neurophysiology, Neurobiology, and Psychiatry, Department of Medicine and Surgery, Università Campus Bio-Medico di Roma, Via Alvaro del Portillo 21, 00128 Rome, Italy; alessandro.cruciani@unicampus.it (A.C.); sa.selvaggi@unicampus.it (S.A.S.); v.dilazzaro@unicampus.it (V.D.L.)
3 Department of System Medicine, University of Tor Vergata, Via Cracovia 50, 00133 Rome, Italy; elias.casula@gmail.com
4 IRCCS Neuromed, Via Atinense 18, 86077 Pozzilli, Italy
5 Department of Clinical and Movement Neurosciences, UCL Queen Square Institute of Neurology, University College London, London WC1E 6BT, UK
6 Department of Medical Sciences and Public Health, University of Cagliari, Cittadella Universitaria di Monserrato, Blocco I S.S. 554 bivio per Sestu, Monserrato, 09042 Cagliari, Italy; l.rocchi@ucl.ac.uk
* Correspondence: a.latorre@ucl.ac.uk

Highlights

What are the main findings?

- NIBS effects involve multiple, interacting plasticity pathways that extend beyond a simple LTP/LTD framework
- Across studies, NIBS modulates BDNF–TrkB signaling, neurotransmitter receptor activity (NMDA/AMPA, GABA), and calcium-dependent mechanisms supporting synaptic plasticity

What are the implications of the main findings?

- Mechanistic biomarkers—rather than MEP-based LTP/LTD interpretations alone—are needed to guide and optimize NIBS protocols
- The involvement of conserved plasticity pathways supports the translational potential of NIBS for CNS disorders, although further cross-species validation is required

Abstract

Non-invasive brain stimulation (NIBS) techniques—including repetitive transcranial magnetic stimulation (rTMS), theta-burst stimulation (TBS), paired associative stimulation (PAS), transcranial direct current stimulation (tDCS), and transcranial alternating current stimulation (tACS)—have emerged as valuable tools for modulating neural activity and promoting plasticity. Traditionally, their effects have been interpreted within a binary framework of long-term potentiation (LTP)-like and long-term depression (LTD)-like plasticity, largely inferred from changes in motor evoked potentials (MEPs). However, existing models do not fully capture the complexity of the biological processes engaged by these techniques and despite extensive clinical application, the cellular and molecular mechanisms underlying NIBS remain only partially understood. This systematic review,

 https://doi.org/10.3390/cells14241996

conducted in accordance with the PRISMA 2020 guidelines, synthesizes evidence from in vivo, in vitro, and ex vivo studies to delineate how NIBS influences neurotransmission through intracellular signaling, gene expression, and protein synthesis at the cellular level. Emphasis is placed on the roles of classical synaptic models, grounded in Ca^{2+}-dependent glutamatergic signaling and receptor phosphorylation dynamics, as well as broader forms of plasticity involving BDNF–TrkB signaling, epigenetic modifications, neuroimmune and glial interactions, anti-inflammatory pathways, and apoptosis- and survival-related cascades. By integrating findings in humans with those in animal and cellular models, we identify both shared and technique-specific molecular mechanisms underlying NIBS-induced effects, highlighting emerging evidence for multi-pathway, non-binary plasticity mechanisms. Understanding these convergent pathways provides a mechanistic foundation for refining stimulation paradigms and improving their translational relevance for treatment of neurological and psychiatric disorders.

Keywords: rTMS; TBS; PAS; tDCS; tACS; LTP-like; LTD-like plasticity; gene expression; molecular mechanism; cellular mechanism

1. Introduction

Non-invasive brain stimulation (NIBS) techniques have become established tools for modulating brain activity by applying electrical or electromagnetic stimuli to the scalp [1]. Initially developed for research purposes, they are now widely used in both experimental and clinical contexts to probe and influence neural circuits, thereby providing key insights into brain function and plasticity [2]. The term non-invasive brain stimulation (NIBS) encompasses several techniques, such as repetitive transcranial magnetic stimulation (rTMS)—including variants like theta burst stimulation (TBS) and paired associative stimulation (PAS)—as well as transcranial electrical stimulation (tES), most commonly implemented as transcranial direct current stimulation (tDCS) or transcranial alternating current stimulation (tACS) [3]. Over the past decades, interest in these methods has steadily grown, fueled by their relative ease of use, favorable safety profile, and promising translational applications in the treatment of central nervous system (CNS) disorders [4]. Clinical studies have explored their therapeutic potential across a wide spectrum of conditions, ranging from depression [5], Parkinson's disease [6] and Alzheimer's disease (AD) [7,8], to stroke [9], chronic pain [10], epilepsy [11], and migraine [12].

Large, randomized, placebo-controlled studies have shown that NIBS techniques can induce not only immediate but also long-lasting modifications in brain function, with effects persisting well beyond the stimulation period [4]. These sustained outcomes are thought to arise from mechanisms of synaptic plasticity, i.e., the brain's ability to remodel neural connections in response to external stimuli [13]. Synaptic plasticity can be broadly classified into short-term synaptic plasticity (STSP), which supports rapid information processing, and long-term synaptic plasticity (LTSP), which underlies learning, memory, and adaptive behaviors [14]. STSP primarily depends on presynaptic mechanisms such as alterations in neurotransmitter release, whereas LTSP involves enduring structural and molecular modifications, including dendritic spine remodeling [15]. LTSP is commonly conceptualized within a binary framework of long-term potentiation (LTP)—associated with enhanced synaptic efficacy—and long-term depression (LTD)—associated with reduced efficacy [16,17]. Among the key principles governing these processes is Hebbian plasticity, based on the notion "cells that fire together, wire together," which drives activity-dependent strengthening of synapses [18]. To maintain network stability, homeostatic plasticity acts as a counterbalance, globally adjusting synaptic strengths to preserve overall excitatory–inhibitory balance [19]. Metaplasticity, in turn, modulates the synaptic change by altering the thresholds for potentiation or depression, thereby shaping how neurons adapt to future activity [19,20].

In human studies, NIBS-induced changes are typically evaluated through behavioral measures [21–24], such as reaction times [25], or through physiological recordings with electroencephalography (EEG) [26] and electromyography (EMG). Among these, the motor evoked potential (MEP) elicited by single-pulse TMS has emerged as a central biomarker, widely used to quantify NIBS-related plasticity within the LTP/LTD framework [27]. However, MEPs provide an indirect and at times ambiguous readout, as their interpretation is confounded by factors such as spinal excitability and pronounced inter-individual variability [28]. Hence, there is a risk of interpreting NIBS effects within an overly dichotomous framework of LTP-like and LTD-like plasticity, which may obscure the complexity of the underlying neural processes [26].

Despite the widespread use of the LTP-like and LTD-like framework, the underlying NIBS mechanisms remain only partially understood, especially in humans where these cannot be directly verified. Significant evidence suggests that plasticity arises from a dynamic interplay of processes that include calcium (Ca^{2+}) signaling, neurotransmitter release, gene expression, protein synthesis, and neuro-glial interactions [29,30]. Central to these processes are NMDA receptors (NMDARs), which mediate Ca^{2+} influx during membrane depolarization and trigger intracellular cascades required for the induction of LTP [31,32]. AMPA receptors (AMPARs) are equally critical, particularly through their trafficking and clustering at postsynaptic sites, which shape the expression of synaptic plasticity [17]. Beyond glutamatergic transmission, modulatory influences such as neurotrophins—including brain-derived neurotrophic factor (BDNF) [33,34]—and epigenetic mechanisms such as histone acetylation [35] further regulate synaptic strength and neuronal adaptability. Glial cells also play an active role: astrocytes and microglia contribute to synaptogenesis [36,37], synaptic pruning [38,39], and neurotransmitter homeostasis [40,41]. Nevertheless, key questions remain. For example, it is still unclear whether magnetic or electrical stimulation directly modulates excitatory or inhibitory synaptic strength at the single-cell level, and the translational relevance of findings from animal models to human physiology remains limited. These gaps highlight the need for an integrated framework linking cellular and molecular mechanisms to the macroscopic effects observed in humans.

In this review, we present an updated perspective on the cellular and molecular foundation bases of magnetic and electrical NIBS, outlining how these techniques engage shared or distinct pathways of synaptic plasticity, evaluating the current evidence supporting these mechanisms, and considering the implications for future experimental and translational research. While NIBS is increasingly explored as a potential therapeutic tool in neurological disorders, this review focuses specifically on the underlying biomolecular mechanisms of plasticity and does not infer clinical efficacy, which remains to be established. By clarifying these mechanistic underpinnings, our aim is to support the development of more precise research hypotheses and contribute to the responsible advancement of NIBS as a therapeutic strategy for CNS disorders.

2. Materials and Methods

The study is conducted following the PRISMA Statement, and the Review was not registered in any open registry database. A comprehensive search was performed across PubMed/MEDLINE, Scopus, Google Scholar, and Google, independently by three reviewers (V.S., M.M., A.C.), covering all records from inception to January 2025.

Search terms related to rTMS, TBS, PAS, tDCS, and tACS were combined using Boolean operators (OR/AND) as shown in Table 1, and reference lists of eligible papers were screened for additional studies.

Table 1. Search strategy. The left column lists NIBS techniques, which are combined using Boolean operators with the terms in the central (mechanisms) and right (models/populations) columns.

"repetitive transcranial magnetic stimulation"[Title/Abstract] OR "rTMS"[Title/Abstract] "theta burst stimulation"[Title/Abstract] OR "TBS"[Title/Abstract] "paired associative stimulation"[Title/Abstract] OR "PAS"[Title/Abstract] "transcranial direct current stimulation"[Title/Abstract] OR "tDCS"[Title/Abstract] "transcranial alternating current stimulation"[Title/Abstract] OR "tACS"[Title/Abstract]	AND	"cellular effects"[Title/Abstract] OR "molecular effects"[Title/Abstract] OR "mechanisms"[Title/Abstract] OR "cellular mechanisms"[Title/Abstract] OR "molecular mechanisms"[Title/Abstract] OR "neuroplasticity"[Title/Abstract] OR "gene expression"[Title/Abstract] OR "synaptic plasticity"[Title/Abstract] OR "LTP-like"[Title/Abstract] OR "LTD-like"[Title/Abstract])	AND	"human"[Title/Abstract] OR "ex vivo"[Title/Abstract] OR "in vivo"[Title/Abstract] OR "animal model"[Title/Abstract] OR "murine"[Title/Abstract] OR "primate"[Title/Abstract] OR "mammalian"[Title/Abstract] OR "cell line"[Title/Abstract] OR "neuron"[Title/Abstract] OR "in vitro"[Title/Abstract] OR "neurodegenerative disease"[Title/Abstract] OR "Alzheimer's Disease"[Title/Abstract] OR "Parkinson's Disease"[Title/Abstract] OR "stroke"[Title/Abstract] OR "psychiatric diseases"[Title/Abstract]

Eligible studies included in vivo ex vivo, or in vitro investigations of cellular or molecular mechanisms of NIBS in healthy models or models of neurological or psychiatric disease. Reviews, pilot, and retrospective studies were excluded, as were those lacking clear stimulation parameters or relying solely on indirect physiological data.

After removing duplicates, records that were irrelevant to the topic, of low methodological quality, or lacking cellular or molecular outcomes were excluded based on title and abstract screening. Titles and abstracts were screened independently, and full texts assessed for eligibility. Only English-language original studies were included.

Disagreements were resolved by discussion among the reviewers and senior authors. The study-selection process is summarized in Figure 1 (PRISMA flow diagram). For each included study, data on model, stimulation parameters, molecular targets, and main results were extracted and summarized narratively in Tables 2–4. A formal quantitative risk-of-bias analysis was not performed, but methodological strengths and limitations were qualitatively noted.

Table 2. Characteristics of retrieved rTMS studies (listed in alphabetical order). rMS, repeated magnetic stimulation; I/R, ischemic/riperfusion; OGD/R, Oxygen glucose deprivation/reoxygenation; qRT-PCR, quantitative real-time polymerase chain reaction; ICC; immunocytochemistry; CaMKII-CREB, Ca^{2+}-calmodulin-dependent protein-kinase II-cAMP-response element binding protein; ERK, extracellular regulated kinases; BDNF, brain-derived neurotrophic factor; SYN-1, synaptophysin 1; rTMS, repeated transcranial magnetic stimulation; CUMS, chronic unpredicted mild stress; SD, Sprague-Dawley; Ven, venlafaxine; MCAO, middle cerebral artery occlusion; IHC, immunohistochemistry; IF, immunofluorescence; STZ, streptozocin; AD, Alzheimer disease; ELISA, enzyme-linked immunosorbent assay; FC, frontal cortex; DG, dentate gyrus; MSO, maximum stimulator output; RMT, resting motor threshold; HC, healthy controls; CSF, cerebrospinal fluid; CB1R, cannabinoid type 1 receptor; iTBS, intermittent theta-burst stimulation; FLX, fluoxetine; PFC, prefrontal cortex; CB1R, cannabinoid-1 receptor; KAR, kainate receptors.

Authors	Stimulation Technique and Protocol	Parameters/Site of Stimulation	Animal Model/Neural Substrate	N° Groups/Participants	Cellular/Molecular/Genetic Techniques USED	Translation to Human Studies	Results
Baek et al., 2018 [42]	rMS	0.5/10 Hz, 10 min of stimulation	Mouse N2a cell culture of I/R model injury (OGD/R)	3 groups: OGD/R + sham, OGD/R+LF (0.5 Hz) and OGD/R+HF (10 Hz)	qRT-PCR, Western blot, ICC	No	OGD/R+LF: ↓ p-ERK and p-AKT, ↓ BAX and caspase-3, ↑ Bcl-2 and Pro-caspase-3, ↓ NMDAR1, CaMKII–CREB OGD/R+HF: ↑ p-ERK and p-AKT, ↑ BAX and caspase-3, ↓ Bcl-2 and Pro-caspase-3, ↑ NMDAR1, CaMKII–CREB, ↑ BDNF, SYN-1 and PSD-95
Feng et al., 2012 [43]	rTMS	15 Hz, 1000 pulses/d, 3 w, 100% MSO, vertex	Depression model (CUMS) in male SD rats	84 divided in 7 groups: sham, rTMS, Ven, CUMS, CUMS + rTMS, CUMS+Ven, CUMS + rTMS + Ven	ICC, Western blot, ELISA	No	↑ BDNF and pERK1/2 after 3 weeks of rTMS and continued to stay at a stable high level 2 weeks later, after the treatments stopped
Gao et al., 2010 [44]	rTMS	20 Hz, 5 s × 10 times, 7 d, right fronto-parietal cortex (bregma)	SD rat models of MCAO-stroke	30, divided into 3 groups: control, rTMS, sham.	IHC	No	↓ caspase3, ↑ Bcl-2 and ↑ Bcl-2/Bax ratio in the rTMS group
Guo et al., 2017 [45]	rTMS	10 Hz, 10 times (300 pulses/d), 120% RMT, bregma	male SD rat models of MCAO	7- and 14-day-treatment groups, divided into sham, MCAO and rTMS groups.	IF, Western blot, qRT-PCR,	No	↑ BDNF, TrkB, p-AKT and Bcl-2 protein expression, and ↓ Bax expression in hippocampus during rTMS
Hwang et al., 2022 [46]	rTMS	1 Hz single vs. repeated session (20 min/5 d), left hemisphere, 50% MT	male SD rats	16 rats. Single session: 4 real stim, 4 sham Repeated session (5 d): 4 real stim, 4 sham	mRNA-miRNA microarray analysis	No	+ regulation of intracellular transport and synaptic plasticity only with repeated rTMS group. A single session of rTMS primarily induced changes in the early genes.
Kim et al., 2024 [47]	rTMS	1/10 Hz. - Cell lines: 6 cycles, 30 min, 3 times, 2 d - Animal: 20 min, 1 d, 4 w	STZ-induced model of AD in human neuroblastoma cell line. Male SD rats' hippocampi	15 rats divided into 3 groups: control, sham rTMS on STZ-induced AD and real rTMS on STZ-induced AD	IHC, qRT-PCR, Western blotting	No	↑ STAT1, STAT3, STAT5, ERK, JNK, Akt, p70S6K, and CREB in cell lines and in AD's animal model after 10 Hz rTMS ↑ ERK, JNK, Akt, p70S6K in 1 Hz and in 10 Hz rTMS groups at 20 min after stimulation. ↑ CREB only in 10 Hz rTMS after 20 min. Phosphorylation lasts 3 h in 10 Hz and 1 h in 1 Hz rTMS.
McNerney et al., 2022 [48]	rTMS	10 Hz, 10 min/d, 2 times/w, 6 w, bregma	Female 3xTgAD mice and their wild type controls	103 mice divided into real and sham, 2 weeks- and 6 weeks-stimulation groups	IHC, qRT-PCR, ELISA	No	= BDNF in the wild-type group that received 2 weeks of rTMS and ↑ in the 6-week group. ↑ BDNF expression in the 2-week and 6-week rTMS in 3xTgAD groups
Meneses-San Juan et al., 2023 [49]	rTMS	5 Hz, 5 d × 4 w, 1500 pulses/d, FC and DG stimulation	Female BALB/c mice model of depression (CUMS)	40 mice, divided in 2 groups: real rTMS and control group (CUMS+FLX)	IHC, IF, ELISA	No	5 Hz rTMS and FLX reverse the decreased density of the DSs in the FC and DG caused by the CUMS protocol. ↑ SYP in the FC of mice treated with 5 Hz rTMS or FLX. ↑ Histone acetylation and demethylation
Wang et al., 2011 [50]	rTMS	5 Hz, 5 d, 1600 pulses, 50% MSO in rats, 90% RMT in HC in M1, 1200 pulses	Ex vivo male SD rats, HC	12 rats and 8 HC, divided into 2 groups: real vs. sham.	IHC, Western blot	Yes	↑ BDNF, PLC-γ1, shc/N-shc, NMDAR subunits, PSD-95, ERK2, PI3K, Akt in brain slices of rat's PFC and in lymphocytes
Wang et al., 2014 [51]	rTMS	15 Hz, 15 trains of 60 pulses, 100% MSO, 7 d, vertex	Depression model (CUMS) in male SD rats	36 rats, divided in 4 groups: sham, sham+rTMS, CUMS, CUMS+rTMS	Western blotting, ICC	No	↑ CB1R, BDNF and Bcl-2/Bax expression levels in the hippocampus after rTMS ↑ CB1R, abolished after administration of a CB1R antagonist
Weiler et al., 2023 [52]	rTMS (1 Hz), iTBS	15% MSO. rTMs and iTBS: 5 blocks, 600 pulses, repeated at 15 min intervals.	Ex vivo and in vivo male Long–Evans rats. In vitro hippocampal neurons of SD rats	12 Long–Evans rats and 16 SD rats	microarray-based gene expression	No	- In the ex vivo and in vivo Long–Evans model: ↑ Ptk2b, Slc6a13, ↑ Slc5a7, ↑ Ryr2, Chrna5, Grin3a, Glun3a, Arc, Cnp. - In vitro SD rat model: ↑ Gabbr1,2 and Gabra4; ↑ Grik1,4; ↑ Grm3–7.

Table 3. Characteristics of retrieved TBS studies (listed in alphabetical order). TBS, theta-burst stimulation; IHC, immunohistochemistry; IF, immunofluorescence; CREB, cAMP response element-binding protein; iTBS, intermittent theta-burst stimulation; I/R ischemia/reperfusion; MCAO, middle cerebral artery occlusion; SD, Sprague-Dawley; qRT-PCR, quantitative real-time reverse transcription polymerase chain reaction; ELISA, enzyme-linked immunosorbent assay; GFAP, glial fibrillary acidic protein; MSO, maximum stimulator output; cTBS, continuous theta-burst stimulation; AD, Alzheimer Disease; TMT, trimethyltin; ISI, inter-stimulus interval; RMT, resting motor threshold; PT, phototrombotic; PAS, paired-associative stimulation; APP, amyloid precursor protein; PS1, presenilin-1; M1, primary motor cortex; PFC, prefrontal cortex, DG, dentate gyrus; WT, wild-type.

Authors	Stimulation Technique and Protocol	Parameters/Site of Stimulation	Animal Model/Neural Substrate	N° Groups/Participants	Cellular/Molecular/Genetic Techniques Used	Translation to Human Studies	Results
Battaglia et al., 2007 [53]	TBS, HF-rTMS PAS	AD patients: - PAS protocol (200 stimuli, ISI 25 msec); transgenic mice: - TBS (50% MSO) in M1, - HF-rTMS in PFC (300 Hz) and in DG (100 Hz)	AD patients and double transgenic mice (APP/PS1)	10 AD patients; transgenic mice compared with WT	IHC, Western blot	Yes	No increase of MEP amplitude of AD patients after PAS ↓ tyrosine-phosphorylated NR2A/NR2B in APP/PS1 transgenic mice ↑ NR2A subunit only in APP/PS1 prefrontal cortex = PSD-95 in the three cortical areas of APP/PS1 mice
Gandolfi et al., 2017 [54]	TBS	8 bursts of 10 pulses at 100 Hz repeated every 250 ms	Cerebellar slices of Wistar rats	4 groups: controls, 15 min and 120 min from stim, NMDAR antagonist	In situ hybridization, IHC, IF	No	↑ P-CREB at 15 min and 120 min after TBS. ↑ -Fos only at 120 min after TBS. No differences in P-CREB/c-Fos in the presence of an NMDAR antagonist
Hu et al., 2023 [55]	iTBS	10 bursts, 600 pulses, 28 d, 26% MSO	Cerebral I/R injured model (MCAO) in SD rats	4 groups: sham (n = 16), I/R 24 h (n = 19), I/R 28 d (n = 19), I/R + iTBS 28 d (n = 19)	IF, qRT-PCR, ELISA, Western blot, RNA transcriptome sequence analysis	No	↑ GAP-43, MMP9. ↓ GFAP, Iba-1 ↓ CD86, IL-1b, TNF-a; ↑ CD206, IL-10. ↓ CytC, caspase-3, ↑ Bcl-2. ↓ Bax, caspase-3, CytC, caspase-9 after 28 days of iTBS.
Ismail et al., 2024 [56]	iTBS	300 pulses, 25, 50, 75 and 100% MSO	N2A mouse neuroblastoma cells	12-well plate of 10^4 cells divided into 5 groups: after 0.5 h, 3 h, 6 h, 12 h and 24 h	Immunoblotting, ICC	No	↑ NMDAR1, GABBR2, mGluR, TrkB, GAP-43, synapsin-1, BDNF and β-tubulin III at 0.5 h post-iTBS
Labedi et al., 2014 [57]	iTBS	5 blocks, 600 pulses/15 min, 23–25% MSO	male SD rats	16 rats, divided in 4 groups: sham-iTBS, real iTBS, iTBS low dose of ketamine or iTBS high dose of ketamine.	IHC	No	↓ GAD67, CB and PV ketamine largely prevented the loss of PV and GAD67 expression at both low and high dose
Ljubisavljevic et al., 2015 [58]	iTBS, cTBS rTMS (1/5 Hz)	30% MSO, 1 t/d, 4 blocks, 2400 pulse/d	stroke model (MCAO) of male Wistar rats	149 rats divided in: 1 Hz, 5 Hz, iTBS, cTBS, sham	IHC, qRT-PCR	No	↑ BDNF in 5 Hz rTMS, cTBS and iTBS. ↑ Creb1, Gria1–4, Grin2a–2c, Gabbr1, ↑ Gadd45b, Junb, Gls, Bai1. c-Fos and Jun ↑ only after iTBS. ↑ Plat (tPA gene) after cTBS
Stekic et al., 2022 [59]	iTBS	33% MSO, 15 sessions/2 stim per d/600 pulses, frontal	male Wistar rats of AD model induced by TMT intoxication	54 rats divided into 4 groups: - control (n = 12), - TMT (n = 15), - TMT+iTBS (n = 12) - TMT+sham (n = 15)	IHC, Western blot	No	↑ P-ERK 1/2 and PI3K in iTBS group. Restored levels of mTOR and p-Akt/compared to TMT group
Thomson et al., 2020 [60]	iTBS, cTBS	600 pulses, 100% MSO	in vitro SH-SY5Y human neuron model	3 conditions: - cTBS, - iTBS, - sham collected after stim or at 6 h or 24 h	qRT-PCR, IF	No	↑ NTRK2, Bcl2 and MAPK9 after 24 h of iTBS. No effects of cTBS on gene expression
Wu et al., 2024 [61]	iTBS	20 trains, 600 pulses, 14 d, 30% MSO or 80% RMT	male SD rats	97 rats divided into 2 groups: real and sham	IHC, microdyalisis	No	↑ c-Fos. Normalized theta power significantly higher in real iTBS group. ↓ GABA, ↑ Glu, ↓ GABA/Glu ratio in real iTBs group

Table 4. Characteristics of retrieved tDCS studies (listed in alphabetical order). atDCS, anodal transcranial direct current stimulation; SD, Sprague-Dawley; G6P, glucose-6-phosphate; 3-BAIBA, D,L-3-aminoisobutyric acid; TCA cycle, tricarboxylic acid cycle; DCS, direct current stimulation; HA, human astrocytes; EC, endothelial cells; qRT-PCR, quantitative real-time reverse transcription polymerase chain reaction; NOS3, nitric oxide synthase 3; BBB, blood–brain barrier; BDNF, brain derived neurotropic factor; NGS, next-generation sequencing; CREB, cAMP-response element binding protein; CaMKII, Ca^{2+}-calmodulin-dependent protein-kinase II; PT, phototrombotic; M1, primary motor cortex area; WT, wild-type; miRNA, microRNA; lncRNA, long non-coding RNA; circRNA, circular RNA; tACS, transcranial alternating current stimulation; SEP, somatosensory-evoked potential; ctDCS, cathodal transcranial direct current stimulation.

Authors	Stimulation Technique and Protocol	Parameters/Site of Stimulation	Animal Model/Neural Substrate	N° Groups/Participants	Cellular/Molecular/Genetic Techniques Used	Translation to Human Studies	Results
Agrawal et al., 2024 [62]	atDCS	Parietal cortex, 20 min, 5 d, 250 μA	Male SD rats	6 rats	mRNA sequencing and Metabolomic Analysis	No	↑ adenosine, G6P, 3-BAIBA, ↓ sphingosine. ↑ glycolysis and mitochondrial function by the TCA cycle. ↓ Ca^{2+}-related signal.
Cancel et al., 2022 [63]	DCS	0.1–1 mA for 10 min	In vitro HA and mouse brain EC	A monolayer of 3×10^4 HA/cm^2 and 6×10^4 EC/cm^2	qRT-PCR and Western blot immediately and 1 h after stim	No	↑ NOS3 and VEGFR1 (modulate permeability of BBB ↑ c-FOS and BDNF in astrocytes
Holmes et al., 2016 [64]	atDCS	sham, 250- 500- 2000 μA 20 min, sagittal suture, 2.5 mm caudal bregma	Male SD rats	7–8 rats × group stim condition	NGS whole transcriptome RNA-sequencing analysis	No	↑ signaling pathways related to Ca^{2+} ion binding, transmembrane/signal peptide and NLRP3- IL-1β pathway. ↑ Ras signaling pathway
Kim et al., 2017 [65]	atDCS	250 μA, 20 min, 7 d in the right sensorimotor cortex	Male SD rats	19 rats, divided into 3 groups: intact control group(n = 5), sham-operated group (n = 7), real stim group (n = 7)	qRT-PCR after 6 h of stim	No	↑ NMDAR and BDNF, CREB, CaMKII, and synapsin I. ↑ c-Fos and Arc.
Longo et al., 2022 [66]	tDCS delivered with 2 epicranial electrodes	72 h post PT stroke on M1, 3 sessions of 250 μA, 20 min, 3 d	C57BL/6 male mice	Not described. Divided into 2 groups: real or sham tDCS	Western immunoblotting, qRT-PCR, ELISA 24 h after stim	No	↑ ERK1/2-CREB, CaMKII, BDNF. ↑ PSD-95
Magri et al., 2021 [67]	atDCS with unilateral epicranial electrode	HP (3 stim, 20 min, 3 consecutive days), PFC (6 stim, 15 min, 3 d × 2 w)	3xTg-AD mouse vs. age-matched WT mice	14 real vs. 9 sham AD mouse; 7 real vs. 9 sham control WT mice	mRNA sequencing and blood whole transcriptome analysis	No	tDCS is able to modulate the gene expression of peripheral tissues, such as blood, and it suggests that blood gene expression profiles could be used as biomarkers of synaptic plasticity
Podda et al., 2016 [68]	atDCS with unilateral epicranial electrode	left hippocampi, 1 mm left and 1 mm posterior to bregma, 350 μA, 20 min	Male C57 BL/6 mice	18 mice (n = 9 active stim, n = 9 sham control)	qRT-PCR and ELISA, 24 h after stim Western blot 2 h after stim	No	↑ BDNF and pCREB
Sánchez-León et al., 2021 [69]	Anodal or cathodal tDCS, tACS	tDCS: Right-S1, 20 min, 200 μA for cathodal, 150 μA for anodal. tACS: 2,20,200 μA at 1 Hz	adult male C57 mice	10 mice, divided into 2 groups: real and sham stim	IHC	No	↑ GAD65-67 and GABA level imbalance after cathodal stimulation but no changes after anodal stimulation. ↑ SEP amplitude during anodal stimulation and ↓ during cathodal stimulation
Sun et al., 2016 [70]	Cathodal DCS	In vitro: 300 or 400 μA, 10 or 25 min. In vivo: 1 mA, 25 min	In vitro brain slices of male C57BL/6 mice, human cortex in vivo (surgical removal of epileptogenic zone)	Not described	Immunoblot at 0, 15, 30, 60 min in vitro after stim	Yes	mGluR5-mTOR signaling as a novel pathway that neither GABAR nor NMDAR blockade abolished DCS-LTD
Walter et al., 2022 [71]	Anodal or cathodal tDCS	PT, bregma, 2 sessions: 15 min, 250 or 500 μA, 5 d	male C57BL/6JRj mice	62 divided into 2 groups: real or sham stim	IHC	No	ctDCS: ↑ functional recovery, neurogenesis. ↓ microglial activation, and CD16/32. atDCS ↑ neurogenesis.

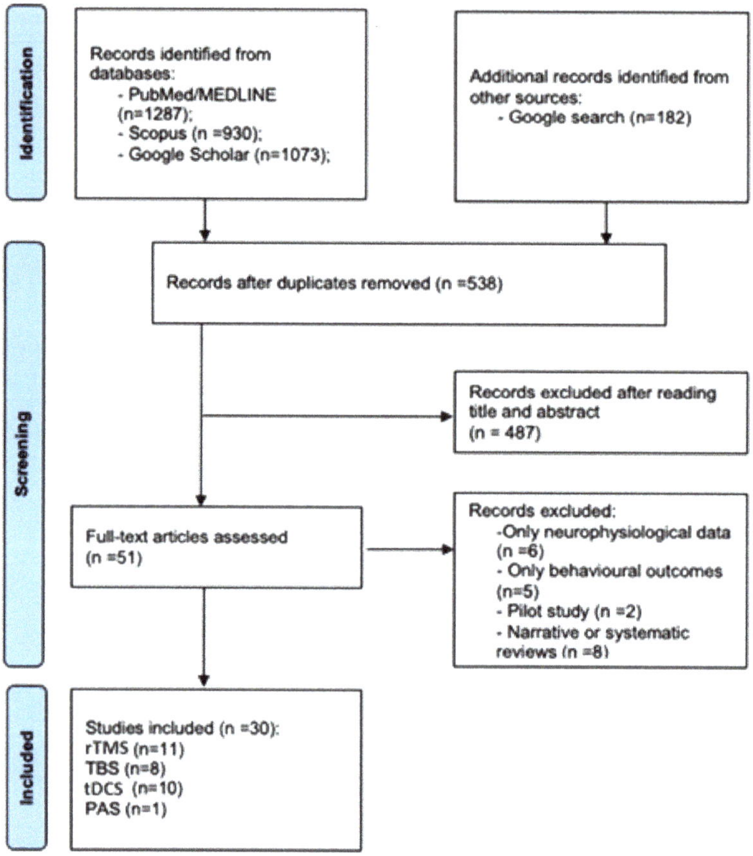

Figure 1. PRISMA flow diagram of retrieved studies.

3. Results

A comprehensive literature search was conducted across PubMed/MEDLINE, Scopus, and Google Scholar, yielding a total of 3290 records (PubMed/MEDLINE: 1287; Scopus: 930; Google Scholar: 1073). An additional 182 records were identified through a supplementary Google search. After removing duplicates, 538 records remained for title and abstract screening. Of these, 487 records were excluded due to lack of relevance to the research question. Fifty-one full-text articles were then assessed for eligibility. After a full-text review, 21 articles were excluded for the following reasons: 6 reported only neurophysiological outcomes, 5 reported only behavioral outcomes, 2 were pilot studies, and 8 were narrative or systematic reviews.

A total of 30 studies met the inclusion criteria and were included in the qualitative synthesis, as summarized in Figure 1 (PRISMA flow diagram).

In this section, the content is organized according to the stimulation techniques employed, each introduced by a brief overview of the methodological principles underlying studies conducted in humans. We then analyze cellular and molecular investigations, highlighting, where possible, translational work in which findings from model systems have been extended to human applications. Table 2 summarizes rTMS studies, Table 3 presents those using TBS and PAS, and Table 4 lists investigations based on tDCS.

Across the studies included, considerable heterogeneity was present in stimulation protocols, model systems, and methodological designs. Stimulation parameters varied widely, ranging from low- (0.5–1 Hz) to high-frequency (10–20 Hz) rTMS/rMS, with large differences in total dose, session duration, pulse number, and intensity (15–120% of motor threshold or MSO). Experimental models also differed substantially, spanning in vitro neuronal and cell-line preparations, ex vivo hippocampal slices, and diverse in vivo rodent models, including ischemia–reperfusion injury, CUMS-induced depression, stroke (MCAO), and Alzheimer's disease. Sample sizes and grouping strategies ranged from small in vitro studies with a few conditions to large multi-group animal experiments exceeding 100 subjects. Methodological variability further contributed to the overall heterogeneity of the evidence base. The most frequently applied cellular and molecular techniques included immunohistochemistry (IHC, n = 13), Western blotting (n = 12), and quantitative real-time polymerase chain reaction (qRT-PCR, n = 11). Additional approaches comprised ELISA (n = 6), immunofluorescence (IF, n = 5), immunocytochemistry (ICC, n = 4), transcriptome analysis (n = 4), microarray analysis (n = 2), in situ hybridization (n = 1), metabolomic profiling (n = 1), and microdialysis (n = 1).

3.1. TMS Protocols

In human studies, TMS delivers a transient magnetic field to the scalp through a coil, inducing depolarization of cortical neurons [72]. When applied repetitively using specific stimulation protocols, it can modulate neural activity, leading to either enhancement or suppression of cortical responses.

3.1.1. rTMS

rTMS protocols are generally categorized into high-frequency (>5 Hz) and low-frequency ($\leq$1 Hz) stimulation, which increase and decrease neuronal excitability, respectively [73]. In human studies, HF-rTMS enhances the size of MEPs during and after stimulation trains, consistent with its design to promote cortical excitation and synaptic potentiation (LTP-like plasticity) [74]. Conversely, LF-rTMS reduces MEP amplitudes and is typically used to induce cortical inhibition and synaptic depression (LTD-like plasticity) in targeted brain regions [75]. Findings from animal models broadly confirm these changes in cortical excitability and have begun to elucidate the underlying cellular and molecular mechanisms.

Several studies converge on the role of BDNF and ERK signaling. Baek and colleagues demonstrated that in N2a cell cultures exposed to ischemia/reperfusion (I/R) injury, 10 Hz—but not 0.5 Hz—stimulation activated ERK, while both frequencies increased BDNF expression [42]. Similarly, in a rat depression model, three weeks of 15 Hz rTMS enhanced BDNF and ERK levels, with effects persisting two weeks beyond stimulation [43]. In AD models, McNerney and coworkers reported that 10 Hz rTMS delivered for two to six weeks upregulated BDNF expression and cholinergic signaling [48], while Kim and colleagues showed that both 1 Hz and 10 Hz protocols promoted proliferation (BDNF), differentiation (CREB), survival (ERK, STAT3, STAT5), and anti-apoptotic signaling (Bcl-2, Akt), albeit with longer-lasting effects at 10 Hz [47].

Other work has described the synaptic and apoptotic pathways. In Sprague Dawley rats, repeated 1 Hz rTMS over five days enhanced intracellular transport and synaptic plasticity compared to a single session, which still induced early gene and miRNA changes related to synaptic function [46]. In models of depression, 15 Hz rTMS for seven days ameliorated depressive-like behaviors and upregulated hippocampal CB1R, BDNF, and Bcl-2/Bax expression [51]. Stroke models have provided converging evidence: Gao and colleagues showed that seven days of 20 Hz rTMS increased the Bcl-2/Bax ratio in a rat middle cerebral artery occlusion (MCAO) model [44], while Guo and colleagues reported enhanced BDNF-TrkB signaling and a higher Bcl-2/Bax ratio following 14 days of 10 Hz stimulation [45].

Epigenetic modulation has also emerged as a possible mechanism. Meneses-San Juan and coworkers found that four weeks of 5 Hz rTMS in a female mouse depression model increased synaptophysin (SYP), histone H3 trimethylation, and reversed stress-induced global DNA hypomethylation [49]. Complementary evidence of neurotransmission modulation comes from Weiler and colleagues, who showed that both 1 Hz rTMS and iTBS altered the expression of genes involved in GABAergic, glutamatergic, cholinergic, and calcium signaling, as well as receptor subunits, ion channels, and mitochondrial proteins in rats and hippocampal brain slices [52].

Finally, translational studies provide further support for the involvement of BDNF-related pathways. Wang and colleagues demonstrated that rTMS enhanced BDNF expression through ERK2 signaling in a rat depression model and found parallel increases in BDNF/NMDAR expression in both rat brain slices and human lymphocytes [50], suggesting conserved mechanisms across species.

3.1.2. Theta Burst Stimulation

In human studies, TBS delivers short high-frequency bursts at a fixed repetition rate to induce lasting changes in cortical or cerebellar excitability [74,76,77]. Theta frequency was initially selected under the assumption that stimulation at the intrinsic rhythm of the hippocampus would promote synaptic plasticity in pyramidal neurons [74]. Depending on the delivery pattern, TBS can produce excitatory or inhibitory effects: intermittent TBS (iTBS) facilitates LTP-like plasticity, whereas continuous TBS (cTBS) induces LTD-like inhibition. iTBS consists of 50 Hz bursts delivered at 5 Hz intervals in 2 s trains every 10 s for 190 s, while cTBS applies the same bursts continuously for 40 s. Both protocols consist of 600 pulses and can elicit effects lasting over an hour, but within a much shorter stimulation time compared to conventional rTMS [74,78,79].

Animal studies have identified calcium signaling as a central mechanism of iTBS-mediated plasticity. Gandolfi and colleagues applied TBS (8 bursts of 10 pulses at 100 Hz every 250 ms) to rat cerebellar slices and observed increased expression of CREB and c-Fos, downstream effectors of NMDAR activation; these effects were abolished by NMDAR antagonists, confirming their involvement [54]. Similarly, Labedi and coworkers demonstrated that iTBS-induced suppression of inhibitory interneurons is mediated by NMDA receptors: ketamine at sub-narcotic doses partially attenuated, and at higher doses completely abolished, the reduction in GAD67 and parvalbumin (PV) expression, while suppression of calbindin (CB) expression persisted, suggesting distinct mechanisms for PV+ versus CB+ interneurons [57].

Additional in vitro work supports broader molecular effects. Ismail et al. applied iTBS to N2a neuroblastoma cells and found upregulation of calcium-, GABA-, and glutamate-related subunits, along with increased BDNF and TrkB expression [56]. Consistently, Stekic and coworkers showed that iTBS enhanced phosphorylation of ERK1/2 and PI3K and restored mTOR and Akt levels in an AD rat model, indicating modulation of the BDNF pathway [59]. In MCAO-induced rats, Ljubisavljevic and colleagues compared iTBS, cTBS, and rTMS (1 Hz and 5 Hz) and found that while BDNF expression increased across several groups, only iTBS upregulated activity-dependent transcriptional regulators of long-term plasticity, glutamatergic and GABA receptor subunits, and glutamate synthesis [58].

iTBS has also been linked to modulation of inflammation and neuroprotection. In a MCAO rat model, Hu et al. showed that 28 days of iTBS reduced glial activation, suppressed proinflammatory molecules, and enhanced anti-inflammatory signaling [55].

Finally, Wu and colleagues reported that 14 days of iTBS over the sagittal suture increased c-Fos-positive neuron density in the hippocampal CA1 region of male SD rats, further supporting the role of iTBS in activating plasticity-related pathways [61].

3.1.3. PAS

PAS combines activation of a sensory pathway, typically via electrical stimulation of a peripheral nerve, with TMS applied over the primary motor cortex (M1) to induce spike timing-dependent plasticity (STDP) [80–82]. When the afferent input from peripheral stimulation reaches the cortex just before the TMS pulse, PAS can enhance cortical excitability in a temporally specific and long-lasting manner, mimicking LTP-like plasticity. In human studies, PAS effects are abolished by NMDA receptor antagonists such as dextromethorphan, memantine, and D-cycloserine, underscoring the critical role of NMDARs in PAS-induced plasticity [83–85]. Conversely, acetylcholinesterase inhibitors such as rivastigmine enhance both LTP- and LTD-like effects of PAS [86].

A notable translational study by Battaglia and colleagues demonstrated altered NMDA-dependent plasticity across M1, the prefrontal cortex, and the hippocampus of APP/PS1 transgenic mice after TBS/HF-rTMS. These findings were paralleled by impaired PAS plasticity in patients with moderate AD, highlighting the conserved mechanisms of NMDAR-dependent modulation across species [53].

3.2. *tES*

3.2.1. tDCS

tDCS delivers constant current across the scalp between an anode and a cathode, with the goal of subthreshold modulation of the resting membrane potential of cortical neurons. This produces after-effects on cortical excitability that can persist from minutes to hours [87]. The direction of current flow is critical: anodal stimulation generally increases excitability, whereas cathodal stimulation reduces it. However, responses also depend on the morphology of the stimulated area [77,88–90].

Glutamatergic signaling is a key mechanism mediating tDCS effects. NMDA receptor activation is required for sustaining tDCS-induced plasticity, and calcium-dependent pathways mediate its non-linear dose–response [62,91–94]. Monoaminergic systems also modulate these effects through dopaminergic, serotonergic, and noradrenergic signaling, although human pharmacological studies mainly confirm these pathways' modulatory influence [95,96].

Evidence from animal and cellular models highlights BDNF as another key mediator. BDNF secretion depends on calcium and NMDA receptor activation [97–99] and supports neuronal regeneration and LTP. In a mouse stroke model, tDCS increased peri-infarct BDNF levels and spike firing, with corresponding increases in circulating BDNF that predicted functional recovery [66]. In vitro, DCS-induced synaptic potentiation was absent in BDNF-deficient brain slices [100], and stimulation of astrocytes alone increased BDNF gene expression [63]. These findings align with computational and experimental evidence implicating glial polarization in tDCS effects [71,101]. Beyond plasticity, tDCS has been shown to modulate autophagy and reduce α-synuclein oligomers in neuroblastoma cell lines [102]. Translationally, BDNF genotype has been linked to tDCS efficacy in stroke-related aphasia recovery [103], though not in depression [104].

tDCS may also support neuroregeneration. In mouse models of cardiac arrest-induced neuronal death, tDCS enhanced expression of proteins related to synaptic function and regeneration, including MAP2, GAP43, PSD95, and synaptophysin [71,105]. Evidence of trans-synaptic neurogenesis in the dorsal hippocampus following frontal cortex stimulation suggests effects extending to interconnected regions [106].

Finally, peripheral contributions cannot be excluded. tDCS modulates tES-evoked potentials, thought to be independent of cortical synaptic activity, as well as compound muscle action potentials from peripheral nerve stimulation, in humans [107]. These effects may reflect arousal-related activation of the ascending reticular activating system. A translational study by Sun and colleagues identified a novel cathodal tDCS-related pathway in mouse brain slices that persisted despite GABAergic and glutamatergic blockade, and found comparable effects in resected human epileptogenic tissue [70].

3.2.2. tACS

Unlike tDCS, tACS does not directly induce synaptic plasticity. Instead, it delivers a weak alternating current (1–2 mA) that modulates brain activity through the "resonance principle," entraining cortical oscillations and influencing neuronal firing patterns without changing the overall firing rate. By using a sinusoidal waveform, tACS can synchronize or desynchronize neuronal networks, with effects determined by stimulation parameters such as frequency, intensity, and phase alignment [87,89,108].

Although not directly plasticity-inducing, alternating electric fields can transiently modify synaptic activity by altering presynaptic calcium dynamics, thereby providing a substrate for downstream neuromodulatory effects. Consistent with this, several studies have shown that tACS interacts with multiple neurotransmitter systems, suggesting potential therapeutic applications in rebalancing excitatory and inhibitory signaling [14,109]. For example, the after-effects of tACS have been linked to NMDA receptor-dependent plasticity [110]. Moreover, it has been shown to restore cholinergic–adrenergic balance after pharmacological perturbation with reserpine (an anti-adrenergic drug) and physostigmine (a reversible cholinesterase inhibitor) [111]. In addition, tACS can influence serotonergic transmission, both by enhancing 5-HT synthesis in the hippocampus, frontal cortex, and hypothalamus [112] and by modulating activity within brainstem nuclei such as the raphe, locus coeruleus, and laterodorsal/pedunculopontine cholinergic nuclei [109].

4. Discussion

This systematic review provides a comprehensive and up-to-date overview of the molecular and cellular underpinnings of the most commonly applied NIBS techniques (Figure 2), in accordance with the PRISMA 2020 guidelines (Supplementary Table S1) [113]. Evidence was gathered from in vivo, in vitro, and ex vivo studies using healthy animals, disease models, and cellular preparations. While many investigations have examined the cellular and molecular effects of TMS and tES, findings are not entirely consistent, and successful translation from animal models to humans remains limited. These inconsistencies likely reflect methodological variability in stimulation parameters, experimental models, and outcome measures, which should be considered when interpreting the overall evidence.

To date, only preliminary and partial cross-species evidence supports the existence of shared molecular mechanisms of NIBS. Wang and colleagues [50] showed that high-frequency rTMS modulates the BDNF–TrkB–NMDAR pathway—including ERK2, PI3K, and Akt activation—in both rat tissue and human lymphocytes; however, these results do not establish downstream engagement of mTOR in humans. Battaglia and coworkers [53] reported altered NMDA-dependent plasticity in APP/PS1 transgenic mice and in patients with AD, with convergent changes in NR2A/NR2B phosphorylation and PSD-95 expression, using PAS and TBS. Sun and coworkers [70] identified mGluR5–mTOR signaling as a novel pathway underlying cathodal DCS-induced LTD, validated in mouse brain slices and resected human cortical tissue, though this work did not show BDNF-dependent initiation of the pathway in human samples. Together, these findings support the possibility that NIBS can modulate partially conserved molecular pathways across species, while also underscoring important gaps: no study has yet validated the full BDNF→mTOR cascade in humans, and systematic translational research is still needed to determine how closely NIBS-induced plasticity mechanisms align between animal models and the human brain.

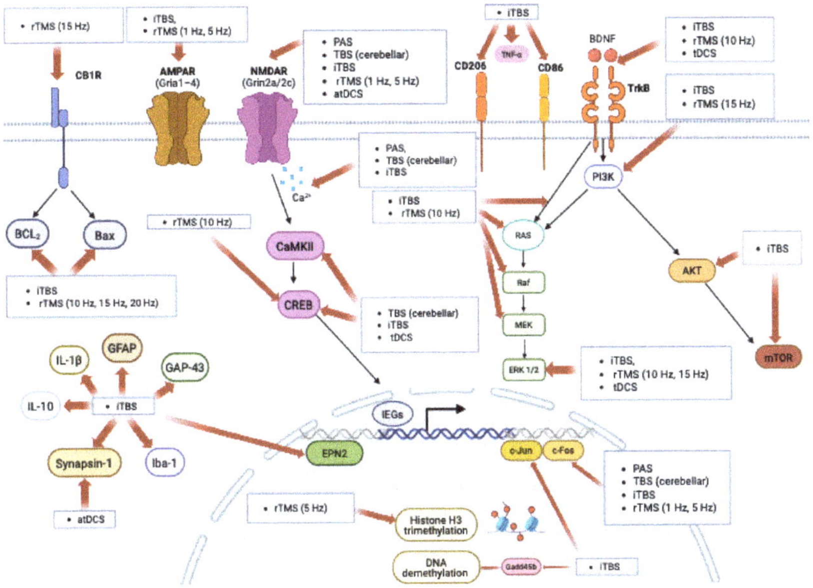

Figure 2. Cellular and molecular brain plasticity pathways and associated brain stimulation techniques.

Across studies, molecular and biochemical modulation induced by NIBS appears to converge on distinct but interconnected pathways. Among these, BDNF emerges as a central hub. Its regulation has been documented with iTBS [56,58], cTBS [58], rTMS at different frequencies (5–15 Hz) [43,45,48,50,51,58], and tDCS [66,71]. In particular, several studies [47,63,66,68] demonstrated that anodal tDCS (atDCS) enhances BDNF expression with downstream activation of CREB and CaMKII. These excitatory protocols rely on calcium influx—mainly through NMDA receptors and voltage-gated calcium channels—initiating signaling cascades that promote BDNF transcription and synaptic remodeling. Supporting this view, Longo and coworkers [66] showed that atDCS in a stroke model increased CaMKII and CREB levels, pointing to a calcium-dependent pathway mediating neuroplasticity.

Further downstream, Stekic and colleagues [59] and Wang and coworkers [50] reported that 15 Hz rTMS and iTBS enhance phosphorylation of PI3K and ERK1/2 and increase Akt and mTOR levels, all downstream effectors of BDNF–TrkB signaling. While these results support BDNF as an endogenous ligand initiating synaptic plasticity, Longo and colleagues [66] suggested that atDCS may instead preferentially engage a BDNF → TrkB → ERK1/2 route. This divergence indicates that although NIBS techniques may converge on BDNF–TrkB signaling, they diverge in their downstream cascades, potentially leading to distinct forms of structural and functional plasticity.

Glutamatergic transmission is another key target of NIBS. Glutamate acts through ionotropic receptors (NMDA, AMPA, kainate) involved in fast excitatory transmission and early synaptic modifications, and through metabotropic receptors (mGluRs) that support long-term plasticity and neuromodulation [56,58,61,114]. Wu and coworkers [61] showed that iTBS increases glutamate levels and reduces the GABA/Glu ratio, suggesting a shift toward excitation. This was accompanied by upregulation of AMPA receptor subunits

(Gria1–4) and NMDA receptor subunits (Grin2a–c) [58], pointing to reinforcement of both fast synaptic transmission and calcium-dependent plasticity. Importantly, AMPA receptor upregulation facilitates rapid excitatory signaling, whereas NMDA receptor modulation supports longer-term synaptic remodeling [42,58,61,115].

Interestingly, BDNF and NMDA signaling appear to converge under certain conditions. Studies on atDCS [47,66] demonstrated increased BDNF expression and ERK1/2 activation, a cascade often linked to NMDA receptor function. Given that NMDA receptors themselves can trigger ERK1/2 activation [116], these findings suggest that BDNF and NMDA pathways may converge on common intracellular mechanisms of plasticity, amplifying each other's effects.

Inhibitory neurotransmission is also shaped by NIBS. GABAergic modulation appears selective and context-dependent: iTBS reduces GABA levels [61], ctDCS increases GAD65/67 with an interhemispheric imbalance [117], and iTBS decreases GAD67 and parvalbumin but not calbindin, with differential NMDA dependence across interneuron subtypes [57]. These findings indicate that NIBS can differentially modulate inhibitory circuits, with implications for cortical reorganization.

At the transcriptional level, NIBS activates multiple genes involved in synaptic plasticity, including c-Fos, Arc [68], CREB [47,68], CaMKII [47,66], PSD-95 [66], Synapsin I [47], and others regulating structural remodeling and neuroprotection [60]. While these results collectively show the capacity of NIBS to engage broad genetic programs, their efficacy may be compromised in pathological contexts. For instance, Battaglia and colleagues [53] reported altered NR2A/NR2B subunit composition and absent PAS-induced potentiation in Alzheimer's patients, highlighting impaired NMDA receptor function as a bottleneck for plasticity.

Taken together, the evidence suggests that NIBS influences a rich and interconnected network of molecular targets—including BDNF, glutamate, GABA, and plasticity-related genes. Although different techniques share some common molecular triggers, they diverge in downstream signaling, leading to distinct functional outcomes. The challenge ahead lies in clarifying these convergences and divergences, and in validating molecular mechanisms across species and clinical populations, in order to fully harness the therapeutic potential of NIBS.

Although several of the pathways discussed—particularly those involving BDNF–TrkB signaling—are hypothesized to contribute to recovery in neurodegenerative and cerebrovascular disorders, current evidence remains largely preclinical, and consistent links between molecular modulation and clinical symptom improvement have yet to be demonstrated. Accordingly, drawing therapeutic conclusions would be premature.

Future Directions and Limitation of Translation in Humans

The translation of NIBS research from preclinical models to human applications remains challenging. A key barrier is that, although NIBS is often interpreted within a simplified LTP/LTD framework, current evidence shows that it engages broader and more dynamic forms of plasticity, including modulation of glutamatergic signaling, BDNF–TrkB pathways, intracellular cascades (ERK, PI3K/Akt, mTOR), epigenetic regulation, neuroinflammatory processes, interneuron-specific GABAergic signaling, neurogenesis, and structural synaptic remodeling. Because these effects extend beyond binary potentiation or depression, future research must employ approaches capable of capturing multi-level plasticity changes rather than focusing solely on excitability shifts. Translational progress is further limited by incomplete mechanistic understanding and inconsistent outcome measures across animal and human studies. The wide range of stimulation parameters available for both tDCS and rTMS further complicates interpretation. Given the emerging evidence for shared molecular and cellular targets across NIBS modalities, preclinical research should prioritize evidence-based frameworks for cortical target selection and the establishment of direct causal links between stimulation, target engagement, and biomarker modulation [118].

Future clinical studies would benefit from integrating multimodal biomarkers, including MEPs, TMS-EEG, EEG, and fMRI, allowing quantification of plasticity-related changes across circuit levels and enabling comparison across protocols [118,119]. Such approaches are necessary because relying only on excitability biomarkers risks overlooking broader adaptations involving neuroimmune, epigenetic, and metabolic pathways.

A persistent obstacle is the complexity of NIBS-induced neural activation: stimulation recruits large neuronal populations, including both inhibitory and excitatory neurons, making it difficult to extrapolate from in vitro models designed for highly controlled synaptic plasticity paradigms [118]. Computational modeling has improved estimation of electric field distributions and interactions with neuronal compartments, [120], yet uncertainty remains regarding whether axons, somas, dendrites, or glial-associated structures are most affected [120,121]. Recognition that NIBS likely influences neuron–glia interactions and neuroinflammatory signaling supports the need for experimental systems that reflect cellular diversity, including co-cultures of neurons, astrocytes, and microglia, which are better suited for studying glial contributions to synaptic and structural plasticity [89,122].

Beyond synaptic efficacy, structural plasticity, including changes in dendritic spine density, synapse formation, dendritic remodeling, and axonal sprouting, is a promising target given its relevance to long-term reorganization of neural networks and recovery of function [15]. Advances in in vivo microscopy and live-cell imaging now permit real-time visualization of plasticity events at subcellular resolution, offering an opportunity to directly link molecular signaling with circuit remodeling over time. Future research should prioritize experimental systems that capture this cellular diversity.

Finally, meaningful translation requires continued parallel progress in preclinical and clinical research. Although convergent molecular pathways across rodents, humans, and induced pluripotent stem cell-derived neural cultures support translational potential, species-specific differences in cortical architecture necessitate iterative cross-validation of mechanisms [118,122]. Integrating mechanistic insights with human neurophysiological and imaging data will be essential to closing the translational gap and defining the therapeutic potential of NIBS across neurological and psychiatric diseases.

5. Conclusions

In this systematic review, we examined how NIBS induces molecular and cellular changes linked to neural plasticity and considered whether these effects may arise from shared or modality-specific pathways. Across studies, several recurring patterns emerged, including consistent influences on key neurotransmitter systems and intracellular signaling cascades. Although the evidence base is heterogeneous and some modalities remain better studied than others, the overall findings indicate that different forms of NIBS tend to engage overlapping molecular processes. Beyond the traditional view of NIBS protocols as producing LTP-like or LTD-like effects, the molecular data point to broader adaptive responses involving epigenetic regulation, modulation of neuroinflammatory pathways, interneuron-specific GABAergic changes, neurogenesis, and synaptic remodeling. These observations suggest that NIBS does not act through a single, binary synaptic mechanism but instead influences multiple levels of plasticity. This molecular perspective complements neurophysiological measures such as MEPs and helps to bridge the gap between observed functional effects and their underlying biological substrates. By integrating findings on neurotransmitter dynamics, gene expression, and intracellular signaling, this review supports a system-level view of how NIBS shapes neuroplasticity while recognizing that further standardized and translational studies—particularly in humans—will be important to refine and extend this emerging mechanistic framework.

Supplementary Materials: The following supporting information can be downloaded at: https://www.mdpi.com/article/10.3390/cells14241996/s1, The completed PRISMA 2020 checklist is provided as Supplementary Table S1.

Author Contributions: Conceptualization, L.R. and A.L.; methodology, L.R.; validation, L.R., A.L., G.L. and M.B.; formal analysis, V.S., M.M. and A.C.; investigation, V.S., M.M. and A.C.; resources, A.L. and L.R.; writing—original draft preparation, V.S., M.M., A.C. and S.A.S.; writing—review and editing, M.B., A.L., L.R., G.L., E.P.C. and V.D.L.; supervision, L.R. and A.L. All authors have read and agreed to the published version of the manuscript.

Funding: This research received no external funding.

Institutional Review Board Statement: Not applicable.

Informed Consent Statement: Not applicable.

Data Availability Statement: No new data were created or analyzed in this study.

Conflicts of Interest: The authors declare no conflicts of interest.

References

1. Polanía, R.; Nitsche, M.A.; Ruff, C.C. Studying and modifying brain function with non-invasive brain stimulation. *Nat. Neurosci.* **2018**, *21*, 174–187. [CrossRef] [PubMed]
2. Bhattacharya, A.; Mrudula, K.; Sreepada, S.S.; Sathyaprabha, T.N.; Pal, P.K.; Chen, R.; Udupa, K. An Overview of Noninvasive Brain Stimulation: Basic Principles and Clinical Applications. *Can. J. Neurol. Sci.* **2022**, *49*, 479–492. [CrossRef] [PubMed]
3. Desarkar, P.; Vicario, C.M.; Soltanlou, M. Non-invasive brain stimulation in research and therapy. *Sci. Rep.* **2024**, *14*, 29334. [CrossRef] [PubMed]
4. Lefaucheur, J.P.; Aleman, A.; Baeken, C.; Benninger, D.H.; Brunelin, J.; Di Lazzaro, V.; Filipović, S.R.; Grefkes, C.; Hasan, A.; Hummel, F.C.; et al. Evidence-based guidelines on the therapeutic use of repetitive transcranial magnetic stimulation (rTMS): An update (2014–2018). *Clin. Neurophysiol.* **2020**, *131*, 474–528. [CrossRef]
5. Trapp, N.T.; Purgianto, A.; Taylor, J.J.; Singh, M.K.; Oberman, L.M.; Mickey, B.J.; Youssef, N.A.; Solzbacher, D.; Zebley, B.; Cabrera, L.Y.; et al. Consensus review and considerations on TMS to treat depression: A comprehensive update endorsed by the National Network of Depression Centers, the Clinical TMS Society, and the International Federation of Clinical Neurophysiology. *Clin. Neurophysiol.* **2025**, *170*, 206–233. [CrossRef]
6. Latorre, A.; Rocchi, L.; Berardelli, A.; Bhatia, K.P.; Rothwell, J.C. The use of transcranial magnetic stimulation as a treatment for movement disorders: A critical review. *Mov. Disord.* **2019**, *34*, 769–782. [CrossRef]
7. Chou, Y.H.; Ton That, V.; Sundman, M. A systematic review and meta-analysis of rTMS effects on cognitive enhancement in mild cognitive impairment and Alzheimer's disease. *Neurobiol. Aging* **2020**, *86*, 1–10. [CrossRef]
8. Koch, G.; Spampinato, D. Alzheimer disease and neuroplasticity. *Handb. Clin. Neurol.* **2022**, *184*, 473–479. [CrossRef]
9. Smith, M.C.; Stinear, C.M. Transcranial magnetic stimulation (TMS) in stroke: Ready for clinical practice? *J. Clin. Neurosci.* **2016**, *31*, 10–14. [CrossRef]
10. Sveva, V.; Cruciani, A.; Mancuso, M.; Santoro, F.; Latorre, A.; Monticone, M.; Rocchi, L. Cerebellar Non-Invasive Brain Stimulation: A Frontier in Chronic Pain Therapy. *J. Pers. Med.* **2024**, *14*, 675. [CrossRef]
11. VanHaerents, S.; Chang, B.S.; Rotenberg, A.; Pascual-Leone, A.; Shafi, M.M. Noninvasive Brain Stimulation in Epilepsy. *J. Clin. Neurophysiol.* **2020**, *37*, 118–130. [CrossRef]
12. Lan, L.; Zhang, X.; Li, X.; Rong, X.; Peng, Y. The efficacy of transcranial magnetic stimulation on migraine: A meta-analysis of randomized controlled trails. *J. Headache Pain* **2017**, *18*, 86. [CrossRef] [PubMed]
13. Magee, J.C.; Grienberger, C. Synaptic Plasticity Forms and Functions. *Annu. Rev. Neurosci.* **2020**, *43*, 95–117. [CrossRef] [PubMed]
14. Citri, A.; Malenka, R.C. Synaptic plasticity: Multiple forms, functions, and mechanisms. *Neuropsychopharmacology* **2008**, *33*, 18–41. [CrossRef] [PubMed]
15. Runge, K.; Cardoso, C.; de Chevigny, A. Dendritic Spine Plasticity: Function and Mechanisms. *Front. Synaptic Neurosci.* **2020**, *12*, 36. [CrossRef]

16. Granger, A.J.; Nicoll, R.A. Expression mechanisms underlying long-term potentiation: A postsynaptic view, 10 years on. *Philos. Trans. R. Soc. Lond. B Biol. Sci.* **2014**, *369*, 20130136. [CrossRef]
17. Sumi, T.; Harada, K. Mechanism underlying hippocampal long-term potentiation and depression based on competition between endocytosis and exocytosis of AMPA receptors. *Sci. Rep.* **2020**, *10*, 14711. [CrossRef]
18. Caporale, N.; Dan, Y. Spike timing-dependent plasticity: A Hebbian learning rule. *Annu. Rev. Neurosci.* **2008**, *31*, 25–46. [CrossRef]
19. Pozo, K.; Goda, Y. Unraveling mechanisms of homeostatic synaptic plasticity. *Neuron* **2010**, *66*, 337–351. [CrossRef]
20. Karabanov, A.; Ziemann, U.; Hamada, M.; George, M.S.; Quartarone, A.; Classen, J.; Massimini, M.; Rothwell, J.; Siebner, H.R. Consensus Paper: Probing Homeostatic Plasticity of Human Cortex with Non-invasive Transcranial Brain Stimulation. *Brain Stimul.* **2015**, *8*, 993–1006. [CrossRef]
21. Georgiev, D.; Rocchi, L.; Tocco, P.; Speekenbrink, M.; Rothwell, J.C.; Jahanshahi, M. Continuous Theta Burst Stimulation over the Dorsolateral Prefrontal Cortex and the Pre-SMA Alter Drift Rate and Response Thresholds Respectively During Perceptual Decision-Making. *Brain Stimul.* **2016**, *9*, 601–608. [CrossRef]
22. Mendez, J.C.; Rocchi, L.; Jahanshahi, M.; Rothwell, J.; Merchant, H. Probing the timing network: A continuous theta burst stimulation study of temporal categorization. *Neuroscience* **2017**, *356*, 167–175. [CrossRef]
23. Dumitru, A.; Rocchi, L.; Saini, F.; Rothwell, J.C.; Roiser, J.P.; David, A.S.; Richieri, R.M.; Lewis, G.; Lewis, G. Influence of theta-burst transcranial magnetic stimulation over the dorsolateral prefrontal cortex on emotion processing in healthy volunteers. *Cogn. Affect. Behav. Neurosci.* **2020**, *20*, 1278–1293. [CrossRef] [PubMed]
24. Bestmann, S.; Krakauer, J.W. The uses and interpretations of the motor-evoked potential for understanding behaviour. *Exp. Brain Res.* **2015**, *233*, 679–689. [CrossRef] [PubMed]
25. Klink, K.; Paßmann, S.; Kasten, F.H.; Peter, J. The Modulation of Cognitive Performance with Transcranial Alternating Current Stimulation: A Systematic Review of Frequency-Specific Effects. *Brain Sci.* **2020**, *10*, 932. [CrossRef] [PubMed]
26. Cruciani, A.; Mancuso, M.; Sveva, V.; Maccarrone, D.; Todisco, A.; Motolese, F.; Santoro, F.; Pilato, F.; Spampinato, D.A.; Rocchi, L.; et al. Using TMS-EEG to assess the effects of neuromodulation techniques: A narrative review. *Front. Hum. Neurosci.* **2023**, *17*, 1247104. [CrossRef]
27. Delvendahl, I.; Jung, N.H.; Kuhnke, N.G.; Ziemann, U.; Mall, V. Plasticity of motor threshold and motor-evoked potential amplitude—A model of intrinsic and synaptic plasticity in human motor cortex? *Brain Stimul.* **2012**, *5*, 586–593. [CrossRef]
28. Rawji, V.; Latorre, A.; Sharma, N.; Rothwell, J.C.; Rocchi, L. On the Use of TMS to Investigate the Pathophysiology of Neurodegenerative Diseases. *Front. Neurol.* **2020**, *11*, 584664. [CrossRef]
29. Zanardi, R.; Magri, L.; Rossini, D.; Malaguti, A.; Giordani, S.; Lorenzi, C.; Pirovano, A.; Smeraldi, E.; Lucca, A. Role of serotonergic gene polymorphisms on response to transcranial magnetic stimulation in depression. *Eur. Neuropsychopharmacol.* **2007**, *17*, 651–657. [CrossRef]
30. Hille, M.; Kühn, S.; Kempermann, G.; Bonhoeffer, T.; Lindenberger, U. From animal models to human individuality: Integrative approaches to the study of brain plasticity. *Neuron* **2024**, *112*, 3522–3541. [CrossRef]
31. Cooke, S.F.; Bliss, T.V. Plasticity in the human central nervous system. *Brain* **2006**, *129*, 1659–1673. [CrossRef] [PubMed]
32. Andrade-Talavera, Y.; Pérez-Rodríguez, M.; Prius-Mengual, J.; Rodríguez-Moreno, A. Neuronal and astrocyte determinants of critical periods of plasticity. *Trends Neurosci.* **2023**, *46*, 566–580. [CrossRef]
33. Heldt, S.A.; Stanek, L.; Chhatwal, J.P.; Ressler, K.J. Hippocampus-specific deletion of BDNF in adult mice impairs spatial memory and extinction of aversive memories. *Mol. Psychiatry* **2007**, *12*, 656–670. [CrossRef] [PubMed]
34. Cheeran, B.; Talelli, P.; Mori, F.; Koch, G.; Suppa, A.; Edwards, M.; Houlden, H.; Bhatia, K.; Greenwood, R.; Rothwell, J.C. A common polymorphism in the brain-derived neurotrophic factor gene (BDNF) modulates human cortical plasticity and the response to rTMS. *J. Physiol.* **2008**, *586*, 5717–5725. [CrossRef] [PubMed]
35. Geng, H.; Chen, H.; Wang, H.; Wang, L. The Histone Modifications of Neuronal Plasticity. *Neural Plast.* **2021**, *2021*, 6690523. [CrossRef]
36. Ullian, E.M.; Sapperstein, S.K.; Christopherson, K.S.; Barres, B.A. Control of synapse number by glia. *Science* **2001**, *291*, 657–661. [CrossRef]
37. Nagai, J.; Yu, X.; Papouin, T.; Cheong, E.; Freeman, M.R.; Monk, K.R.; Hastings, M.H.; Haydon, P.G.; Rowitch, D.; Shaham, S.; et al. Behaviorally consequential astrocytic regulation of neural circuits. *Neuron* **2021**, *109*, 576–596. [CrossRef]
38. Chung, D.W.; Wills, Z.P.; Fish, K.N.; Lewis, D.A. Developmental pruning of excitatory synaptic inputs to parvalbumin interneurons in monkey prefrontal cortex. *Proc. Natl. Acad. Sci. USA* **2017**, *114*, E629–E637. [CrossRef]
39. Lee, J.H.; Kim, J.Y.; Noh, S.; Lee, H.; Lee, S.Y.; Mun, J.Y.; Park, H.; Chung, W.S. Astrocytes phagocytose adult hippocampal synapses for circuit homeostasis. *Nature* **2021**, *590*, 612–617. [CrossRef]
40. Mahmoud, S.; Gharagozloo, M.; Simard, C.; Gris, D. Astrocytes Maintain Glutamate Homeostasis in the CNS by Controlling the Balance between Glutamate Uptake and Release. *Cells* **2019**, *8*, 184. [CrossRef]
41. Lines, J.; Corkrum, M.; Aguilar, J.; Araque, A. The Duality of Astrocyte Neuromodulation: Astrocytes Sense Neuromodulators and Are Neuromodulators. *J. Neurochem.* **2025**, *169*, e70054. [CrossRef] [PubMed]

42. Baek, A.; Kim, J.H.; Pyo, S.; Jung, J.H.; Park, E.J.; Kim, S.H.; Cho, S.R. The Differential Effects of Repetitive Magnetic Stimulation in an In Vitro Neuronal Model of Ischemia/Reperfusion Injury. *Front. Neurol.* **2018**, *9*, 50. [CrossRef] [PubMed]
43. Feng, S.F.; Shi, T.Y.; Fan, Y.; Wang, W.N.; Chen, Y.C.; Tan, Q.R. Long-lasting effects of chronic rTMS to treat chronic rodent model of depression. *Behav. Brain Res.* **2012**, *232*, 245–251. [CrossRef] [PubMed]
44. Gao, F.; Wang, S.; Guo, Y.; Wang, J.; Lou, M.; Wu, J.; Ding, M.; Tian, M.; Zhang, H. Protective effects of repetitive transcranial magnetic stimulation in a rat model of transient cerebral ischaemia: A microPET study. *Eur. J. Nucl. Med. Mol. Imaging* **2010**, *37*, 954–961. [CrossRef]
45. Guo, F.; Lou, J.; Han, X.; Deng, Y.; Huang, X. Repetitive Transcranial Magnetic Stimulation Ameliorates Cognitive Impairment by Enhancing Neurogenesis and Suppressing Apoptosis in the Hippocampus in Rats with Ischemic Stroke. *Front. Physiol.* **2017**, *8*, 559. [CrossRef]
46. Hwang, W.; Choi, J.K.; Bang, M.S.; Park, W.Y.; Oh, B.M. Gene Expression Profile Changes in the Stimulated Rat Brain Cortex After Repetitive Transcranial Magnetic Stimulation. *Brain Neurorehabil.* **2022**, *15*, e27. [CrossRef]
47. Kim, S.K.; Lee, G.Y.; Kim, S.K.; Kwon, Y.J.; Seo, E.B.; Lee, H.; Lee, S.H.; Kim, S.J.; Lee, S.; Ye, S.K. Protective Effects of Repetitive Transcranial Magnetic Stimulation Against Streptozotocin-Induced Alzheimer's Disease. *Mol. Neurobiol.* **2024**, *61*, 1687–1703. [CrossRef]
48. McNerney, M.W.; Heath, A.; Narayanan, S.K.; Yesavage, J. Repetitive Transcranial Magnetic Stimulation Improves Brain-Derived Neurotrophic Factor and Cholinergic Signaling in the 3xTgAD Mouse Model of Alzheimer's Disease. *J. Alzheimers Dis.* **2022**, *86*, 499–507. [CrossRef]
49. Meneses-San Juan, D.; Lamas, M.; Ramírez-Rodríguez, G.B. Repetitive Transcranial Magnetic Stimulation Reduces Depressive-like Behaviors, Modifies Dendritic Plasticity, and Generates Global Epigenetic Changes in the Frontal Cortex and Hippocampus in a Rodent Model of Chronic Stress. *Cells* **2023**, *12*, 2062. [CrossRef]
50. Wang, H.Y.; Crupi, D.; Liu, J.; Stucky, A.; Cruciata, G.; Di Rocco, A.; Friedman, E.; Quartarone, A.; Ghilardi, M.F. Repetitive transcranial magnetic stimulation enhances BDNF-TrkB signaling in both brain and lymphocyte. *J. Neurosci.* **2011**, *31*, 11044–11054. [CrossRef]
51. Wang, H.N.; Wang, L.; Zhang, R.G.; Chen, Y.C.; Liu, L.; Gao, F.; Nie, H.; Hou, W.G.; Peng, Z.W.; Tan, Q. Anti-depressive mechanism of repetitive transcranial magnetic stimulation in rat: The role of the endocannabinoid system. *J. Psychiatr. Res.* **2014**, *51*, 79–87. [CrossRef]
52. Weiler, M.; Stieger, K.C.; Shroff, K.; Klein, J.P.; Wood, W.H., 3rd; Zhang, Y.; Chandrasekaran, P.; Lehrmann, E.; Camandola, S.; Long, J.M.; et al. Transcriptional changes in the rat brain induced by repetitive transcranial magnetic stimulation. *Front. Hum. Neurosci.* **2023**, *17*, 1215291. [CrossRef] [PubMed]
53. Battaglia, F.; Wang, H.Y.; Ghilardi, M.F.; Gashi, E.; Quartarone, A.; Friedman, E.; Nixon, R.A. Cortical plasticity in Alzheimer's disease in humans and rodents. *Biol. Psychiatry* **2007**, *62*, 1405–1412. [CrossRef] [PubMed]
54. Gandolfi, D.; Cerri, S.; Mapelli, J.; Polimeni, M.; Tritto, S.; Fuzzati-Armentero, M.T.; Bigiani, A.; Blandini, F.; Mapelli, L.; D'Angelo, E. Activation of the CREB/c-Fos Pathway during Long-Term Synaptic Plasticity in the Cerebellum Granular Layer. *Front. Cell Neurosci.* **2017**, *11*, 184. [CrossRef] [PubMed]
55. Hu, S.; Wang, X.; Yang, X.; Ouyang, S.; Pan, X.; Fu, Y.; Wu, S. Long-term iTBS Improves Neural Functional Recovery by Reducing the Inflammatory Response and Inhibiting Neuronal Apoptosis Via miR-34c-5p/p53/Bax Signaling Pathway in Cerebral Ischemic Rats. *Neuroscience* **2023**, *527*, 37–51. [CrossRef]
56. Ismail, F.Y.; Krishnan, M.; Jayaraj, R.L.; Bru-Mercier, G.; Pessia, M.; Ljubisavljevic, M.R. Time dependent changes in protein expression induced by intermittent theta burst stimulation in a cell line. *Front. Neurol.* **2024**, *15*, 1396776. [CrossRef]
57. Labedi, A.; Benali, A.; Mix, A.; Neubacher, U.; Funke, K. Modulation of inhibitory activity markers by intermittent theta-burst stimulation in rat cortex is NMDA-receptor dependent. *Brain Stimul.* **2014**, *7*, 394–400. [CrossRef]
58. Ljubisavljevic, M.R.; Javid, A.; Oommen, J.; Parekh, K.; Nagelkerke, N.; Shehab, S.; Adrian, T.E. The Effects of Different Repetitive Transcranial Magnetic Stimulation (rTMS) Protocols on Cortical Gene Expression in a Rat Model of Cerebral Ischemic-Reperfusion Injury. *PLoS ONE* **2015**, *10*, e0139892. [CrossRef]
59. Stekic, A.; Zeljkovic, M.; Zaric Kontic, M.; Mihajlovic, K.; Adzic, M.; Stevanovic, I.; Ninkovic, M.; Grkovic, I.; Ilic, T.V.; Nedeljkovic, N.; et al. Intermittent Theta Burst Stimulation Ameliorates Cognitive Deficit and Attenuates Neuroinflammation via PI3K/Akt/mTOR Signaling Pathway in Alzheimer's-Like Disease Model. *Front. Aging Neurosci.* **2022**, *14*, 889983. [CrossRef]
60. Thomson, A.C.; Kenis, G.; Tielens, S.; de Graaf, T.A.; Schuhmann, T.; Rutten, B.P.F.; Sack, A.T. Transcranial Magnetic Stimulation-Induced Plasticity Mechanisms: TMS-Related Gene Expression and Morphology Changes in a Human Neuron-Like Cell Model. *Front. Mol. Neurosci.* **2020**, *13*, 528396. [CrossRef]
61. Wu, X.; Liu, J.; Hui, Y.; Wu, Z.; Wang, L.; Wang, Y.; Bai, Y.; Li, J.; Zhang, L.; Xi, Y.; et al. Long-term intermittent theta burst stimulation enhanced hippocampus-dependent memory by regulating hippocampal theta oscillation and neurotransmitter levels in healthy rats. *Neurochem. Int.* **2024**, *173*, 105671. [CrossRef] [PubMed]

62. Agrawal, B.; Boulos, S.; Khatib, S.; Feuermann, Y.; Panov, J.; Kaphzan, H. Molecular Insights into Transcranial Direct Current Stimulation Effects: Metabolomics and Transcriptomics Analyses. *Cells* **2024**, *13*, 205. [CrossRef] [PubMed]
63. Cancel, L.M.; Silas, D.; Bikson, M.; Tarbell, J.M. Direct current stimulation modulates gene expression in isolated astrocytes with implications for glia-mediated plasticity. *Sci. Rep.* **2022**, *12*, 17964. [CrossRef] [PubMed]
64. Holmes, B.; Jung, S.H.; Lu, J.; Wagner, J.A.; Rubbi, L.; Pellegrini, M.; Jankord, R. Transcriptomic Modification in the Cerebral Cortex following Noninvasive Brain Stimulation: RNA-Sequencing Approach. *Neural Plast.* **2016**, *2016*, 5942980. [CrossRef]
65. Kim, M.S.; Koo, H.; Han, S.W.; Paulus, W.; Nitsche, M.A.; Kim, Y.-H.; A Yoon, J.; Shin, Y.-I. Repeated anodal transcranial direct current stimulation induces neural plasticity-associated gene expression in the rat cortex and hippocampus. *Restor. Neurol. Neurosci.* **2017**, *35*, 137–146. [CrossRef]
66. Longo, V.; Barbati, S.A.; Re, A.; Paciello, F.; Bolla, M.; Rinaudo, M.; Miraglia, F.; Alù, F.; Di Donna, M.G.; Vecchio, F.; et al. Transcranial Direct Current Stimulation Enhances Neuroplasticity and Accelerates Motor Recovery in a Stroke Mouse Model. *Stroke* **2022**, *53*, 1746–1758. [CrossRef]
67. Magri, C.; Vitali, E.; Cocco, S.; Giacopuzzi, E.; Rinaudo, M.; Martini, P.; Barbon, A.; Grassi, C.; Gennarelli, M. Whole Blood Transcriptome Characterization of 3xTg-AD Mouse and Its Modulation by Transcranial Direct Current Stimulation (tDCS). *Int. J. Mol. Sci.* **2021**, *22*, 7629. [CrossRef]
68. Podda, M.V.; Cocco, S.; Mastrodonato, A.; Fusco, S.; Leone, L.; Barbati, S.A.; Colussi, C.; Ripoli, C.; Grassi, C. Anodal transcranial direct current stimulation boosts synaptic plasticity and memory in mice via epigenetic regulation of Bdnf expression. *Sci. Rep.* **2016**, *6*, 22180. [CrossRef]
69. Sánchez-León, C.A.; Cordones, I.; Ammann, C.; Ausín, J.M.; Gómez-Climent, M.A.; Carretero-Guillén, A.; Campos, G.S.-G.; Gruart, A.; Delgado-García, J.M.; Cheron, G.; et al. Immediate and after effects of transcranial direct-current stimulation in the mouse primary somatosensory cortex. *Sci. Rep.* **2021**, *11*, 3123. [CrossRef]
70. Sun, Y.; Lipton, J.O.; Boyle, L.M.; Madsen, J.R.; Goldenberg, M.C.; Pascual-Leone, A.; Sahin, M.; Rotenberg, A. Direct current stimulation induces mGluR5-dependent neocortical plasticity. *Ann. Neurol.* **2016**, *80*, 233–246. [CrossRef]
71. Walter, H.L.; Pikhovych, A.; Endepols, H.; Rotthues, S.; Bärmann, J.; Backes, H.; Hoehn, M.; Wiedermann, D.; Neumaier, B.; Fink, G.R.; et al. Transcranial-Direct-Current-Stimulation Accelerates Motor Recovery After Cortical Infarction in Mice: The Interplay of Structural Cellular Responses and Functional Recovery. *Neurorehabil. Neural Repair* **2022**, *36*, 701–714. [CrossRef]
72. Hannah, R.; Rocchi, L.; Tremblay, S.; Rothwell, J.C. Controllable Pulse Parameter TMS and TMS-EEG As Novel Approaches to Improve Neural Targeting with rTMS in Human Cerebral Cortex. *Front. Neural Circuits* **2016**, *10*, 97. [CrossRef]
73. Lefaucheur, J.P. Transcranial magnetic stimulation. *Handb. Clin. Neurol.* **2019**, *160*, 559–580. [CrossRef] [PubMed]
74. Huang, Y.Z.; Edwards, M.J.; Rounis, E.; Bhatia, K.P.; Rothwell, J.C. Theta burst stimulation of the human motor cortex. *Neuron* **2005**, *45*, 201–206. [CrossRef] [PubMed]
75. Rossini, P.M.; Burke, D.; Chen, R.; Cohen, L.; Daskalakis, Z.; Di Iorio, R.; Di Lazzaro, V.; Ferreri, F.; Fitzgerald, P.; George, M. Non-invasive electrical and magnetic stimulation of the brain, spinal cord, roots and peripheral nerves: Basic principles and procedures for routine clinical and research application. An updated report from an IFCN Committee. *Clin. Neurophysiol.* **2015**, *126*, 1071–1107. [CrossRef] [PubMed]
76. Di Biasio, F.; Conte, A.; Bologna, M.; Iezzi, E.; Rocchi, L.; Modugno, N.; Berardelli, A. Does the cerebellum intervene in the abnormal somatosensory temporal discrimination in Parkinson's disease? *Park. Relat. Disord.* **2015**, *21*, 789–792. [CrossRef]
77. Rocchi, L.; Spampinato, D.A.; Pezzopane, V.; Orth, M.; Bisiacchi, P.S.; Rothwell, J.C.; Casula, E.P. Cerebellar noninvasive neuromodulation influences the reactivity of the contralateral primary motor cortex and surrounding areas: A TMS-EMG-EEG study. *Cerebellum* **2023**, *22*, 319–331. [CrossRef]
78. Rounis, E.; Huang, Y.Z. Theta burst stimulation in humans: A need for better understanding effects of brain stimulation in health and disease. *Exp. Brain Res.* **2020**, *238*, 1707–1714. [CrossRef]
79. Suppa, A.; Huang, Y.Z.; Funke, K.; Ridding, M.C.; Cheeran, B.; Di Lazzaro, V.; Ziemann, U.; Rothwell, J.C. Ten Years of Theta Burst Stimulation in Humans: Established Knowledge, Unknowns and Prospects. *Brain Stimul.* **2016**, *9*, 323–335. [CrossRef]
80. Stefan, K.; Kunesch, E.; Cohen, L.G.; Benecke, R.; Classen, J. Induction of plasticity in the human motor cortex by paired associative stimulation. *Brain* **2000**, *123*, 572–584. [CrossRef]
81. Suppa, A.; Li Voti, P.; Rocchi, L.; Papazachariadis, O.; Berardelli, A. Early visuomotor integration processes induce LTP/LTD-like plasticity in the human motor cortex. *Cereb. Cortex* **2015**, *25*, 703–712. [CrossRef]
82. Conte, A.; Li Voti, P.; Pontecorvo, S.; Quartuccio, M.E.; Baione, V.; Rocchi, L.; Cortese, A.; Bologna, M.; Francia, A.; Berardelli, A. Attention-related changes in short-term cortical plasticity help to explain fatigue in multiple sclerosis. *Mult. Scler.* **2016**, *22*, 1359–1366. [CrossRef] [PubMed]
83. Batsikadze, G.; Paulus, W.; Kuo, M.F.; Nitsche, M.A. Effect of serotonin on paired associative stimulation-induced plasticity in the human motor cortex. *Neuropsychopharmacology* **2013**, *38*, 2260–2267. [CrossRef] [PubMed]

84. Silverstein, J.; Cortes, M.; Tsagaris, K.Z.; Climent, A.; Gerber, L.M.; Oromendia, C.; Fonzetti, P.; Ratan, R.R.; Kitago, T.; Iacoboni, M.; et al. Paired Associative Stimulation as a Tool to Assess Plasticity Enhancers in Chronic Stroke. *Front. Neurosci.* **2019**, *13*, 792. [CrossRef] [PubMed]
85. Sohn, M.N.; Brown, J.C.; Sharma, P.; Ziemann, U.; McGirr, A. Pharmacological adjuncts and transcranial magnetic stimulation-induced synaptic plasticity: A systematic review. *J. Psychiatry Neurosci.* **2024**, *49*, E59–E76. [CrossRef]
86. Tampi, R.R. Paired associative stimulation (PAS) and Alzheimer's disease (AD). *Int. Psychogeriatr.* **2023**, *35*, 123–125. [CrossRef]
87. Nejati, V.; Vaziri, Z.; Antal, A.; Antonenko, D.; Behroozmand, R.; Bestmann, S.; Brunelin, J.; Brunoni, A.R.; Carvalho, S.; Davis, N.J.; et al. Report Approval for Transcranial Electrical Stimulation (RATES): Expert recommendation based on a Delphi consensus study. *Nat. Protoc.* **2025**. [CrossRef]
88. Nitsche, M.A.; Paulus, W. Excitability changes induced in the human motor cortex by weak transcranial direct current stimulation. *J. Physiol.* **2000**, *527*, 633–639. [CrossRef]
89. Jannati, A.; Oberman, L.M.; Rotenberg, A.; Pascual-Leone, A. Assessing the mechanisms of brain plasticity by transcranial magnetic stimulation. *Neuropsychopharmacology* **2023**, *48*, 191–208. [CrossRef]
90. Latorre, A.; Rocchi, L.; Paparella, G.; Manzo, N.; Bhatia, K.P.; Rothwell, J.C. Changes in cerebellar output abnormally modulate cortical myoclonus sensorimotor hyperexcitability. *Brain* **2024**, *147*, 1412–1422. [CrossRef]
91. Liebetanz, D.; Nitsche, M.A.; Tergau, F.; Paulus, W. Pharmacological approach to the mechanisms of transcranial DC-stimulation-induced after-effects of human motor cortex excitability. *Brain* **2002**, *125*, 2238–2247. [CrossRef] [PubMed]
92. Nitsche, M.A.; Fricke, K.; Henschke, U.; Schlitterlau, A.; Liebetanz, D.; Lang, N.; Henning, S.; Tergau, F.; Paulus, W. Pharmacological modulation of cortical excitability shifts induced by transcranial direct current stimulation in humans. *J. Physiol.* **2003**, *553*, 293–301. [CrossRef] [PubMed]
93. Nitsche, M.A.; Liebetanz, D.; Schlitterlau, A.; Henschke, U.; Fricke, K.; Frommann, K.; Lang, N.; Henning, S.; Paulus, W.; Tergau, F. GABAergic modulation of DC stimulation-induced motor cortex excitability shifts in humans. *Eur. J. Neurosci.* **2004**, *19*, 2720–2726. [CrossRef] [PubMed]
94. Mosayebi-Samani, M.; Melo, L.; Agboada, D.; Nitsche, M.A.; Kuo, M.F. Ca^{2+} channel dynamics explain the nonlinear neuroplasticity induction by cathodal transcranial direct current stimulation over the primary motor cortex. *Eur. Neuropsychopharmacol.* **2020**, *38*, 63–72. [CrossRef]
95. Nitsche, M.A.; Lampe, C.; Antal, A.; Liebetanz, D.; Lang, N.; Tergau, F.; Paulus, W. Dopaminergic modulation of long-lasting direct current-induced cortical excitability changes in the human motor cortex. *Eur. J. Neurosci.* **2006**, *23*, 1651–1657. [CrossRef]
96. Kuo, M.F.; Grosch, J.; Fregni, F.; Paulus, W.; Nitsche, M.A. Focusing effect of acetylcholine on neuroplasticity in the human motor cortex. *J. Neurosci.* **2007**, *27*, 14442–14447. [CrossRef]
97. Akaneya, Y.; Tsumoto, T.; Kinoshita, S.; Hatanaka, H. Brain-derived neurotrophic factor enhances long-term potentiation in rat visual cortex. *J. Neurosci.* **1997**, *17*, 6707–6716. [CrossRef]
98. Balkowiec, A.; Katz, D.M. Cellular mechanisms regulating activity-dependent release of native brain-derived neurotrophic factor from hippocampal neurons. *J. Neurosci.* **2002**, *22*, 10399–10407. [CrossRef]
99. Lu, B. BDNF and activity-dependent synaptic modulation. *Learn. Mem.* **2003**, *10*, 86–98. [CrossRef]
100. Fritsch, B.; Reis, J.; Martinowich, K.; Schambra, H.M.; Ji, Y.; Cohen, L.G.; Lu, B. Direct current stimulation promotes BDNF-dependent synaptic plasticity: Potential implications for motor learning. *Neuron* **2010**, *66*, 198–204. [CrossRef]
101. Ruohonen, J.; Karhu, J. tDCS possibly stimulates glial cells. *Clin. Neurophysiol.* **2012**, *123*, 2006–2009. [CrossRef] [PubMed]
102. Sala, G.; Bocci, T.; Borzì, V.; Parazzini, M.; Priori, A.; Ferrarese, C. Direct current stimulation enhances neuronal alpha-synuclein degradation in vitro. *Sci. Rep.* **2021**, *11*, 2197. [CrossRef] [PubMed]
103. Fridriksson, J.; Elm, J.; Stark, B.C.; Basilakos, A.; Rorden, C.; Sen, S.; George, M.S.; Gottfried, M.; Bonilha, L. BDNF genotype and tDCS interaction in aphasia treatment. *Brain Stimul.* **2018**, *11*, 1276–1281. [CrossRef] [PubMed]
104. Brunoni, A.R.; Carracedo, A.; Amigo, O.M.; Pellicer, A.L.; Talib, L.; Carvalho, A.F.; Lotufo, P.A.; Benseñor, I.M.; Gattaz, W.; Cappi, C. Association of BDNF, HTR2A, TPH1, SLC6A4, and COMT polymorphisms with tDCS and escitalopram efficacy: Ancillary analysis of a double-blind, placebo-controlled trial. *Braz. J. Psychiatry* **2020**, *42*, 128–135. [CrossRef]
105. Dai, C.; Wang, J.; Li, J.; Wang, J.; Zhang, L.; Yin, C.; Li, Y. Repetitive anodal transcranial direct current stimulation improves neurological recovery by preserving the neuroplasticity in an asphyxial rat model of cardiac arrest. *Brain Stimul.* **2021**, *14*, 407–416. [CrossRef]
106. Dumontoy, S.; Ramadan, B.; Risold, P.Y.; Pedron, S.; Houdayer, C.; Etiévant, A.; Cabeza, L.; Haffen, E.; Peterschmitt, Y.; Van Waes, V. Repeated Anodal Transcranial Direct Current Stimulation (RA-tDCS) over the Left Frontal Lobe Increases Bilateral Hippocampal Cell Proliferation in Young Adult but Not Middle-Aged Female Mice. *Int. J. Mol. Sci.* **2023**, *24*, 8750. [CrossRef]
107. Ardolino, G.; Bossi, B.; Barbieri, S.; Priori, A. Non-synaptic mechanisms underlie the after-effects of cathodal transcutaneous direct current stimulation of the human brain. *J. Physiol.* **2005**, *568*, 653–663. [CrossRef]
108. Spampinato, D.; Ibanez, J.; Spanoudakis, M.; Hammond, P.; Rothwell, J.C. Cerebellar transcranial magnetic stimulation: The role of coil type from distinct manufacturers. *Brain Stimul.* **2020**, *13*, 135–156. [CrossRef]

109. Korai, S.A.; Ranieri, F.; Di Lazzaro, V.; Papa, M.; Cirillo, G. Neurobiological After-Effects of Low Intensity Transcranial Electric Stimulation of the Human Nervous System: From Basic Mechanisms to Metaplasticity. *Front. Neurol.* **2021**, *12*, 587771. [CrossRef]
110. Wischnewski, M.; Alekseichuk, I.; Opitz, A. Neurocognitive, physiological, and biophysical effects of transcranial alternating current stimulation. *Trends Cogn. Sci.* **2023**, *27*, 189–205. [CrossRef]
111. Kirsch, D.L.; Nichols, F. Cranial electrotherapy stimulation for treatment of anxiety, depression, and insomnia. *Psychiatr. Clin. North. Am.* **2013**, *36*, 169–176. [CrossRef] [PubMed]
112. Limoge, A.; Robert, C.; Stanley, T.H. Transcutaneous cranial electrical stimulation (TCES): A review 1998. *Neurosci. Biobehav. Rev.* **1999**, *23*, 529–538. [CrossRef] [PubMed]
113. Page, M.J.; McKenzie, J.E.; Bossuyt, P.M.; Boutron, I.; Hoffmann, T.C.; Mulrow, C.D.; Shamseer, L.; Tetzlaff, J.M.; Akl, E.A.; Brennan, S.E.; et al. The PRISMA 2020 statement: An updated guideline for reporting systematic reviews. *BMJ* **2021**, *372*, n71. [CrossRef] [PubMed]
114. Wang, H.; Peng, R.Y. Basic roles of key molecules connected with NMDAR signaling pathway on regulating learning and memory and synaptic plasticity. *Mil. Med. Res.* **2016**, *3*, 26. [CrossRef]
115. Guo, Z.; Jiang, C.H.; Tong, C.; Yang, Y.; Wang, Z.; Lam, S.M.; Wang, D.; Li, R.; Shui, G.; Shi, Y.S.; et al. Activity-dependent PI4P synthesis by PI4KIIIα regulates long-term synaptic potentiation. *Cell Rep.* **2022**, *38*, 110452. [CrossRef]
116. Ivanov, A.; Pellegrino, C.; Rama, S.; Dumalska, I.; Salyha, Y.; Ben-Ari, Y.; Medina, I. Opposing role of synaptic and extrasynaptic NMDA receptors in regulation of the extracellular signal-regulated kinases (ERK) activity in cultured rat hippocampal neurons. *J. Physiol.* **2006**, *572*, 789–798. [CrossRef]
117. Sánchez-León, C.A.; Sánchez-Garrido Campos, G.; Fernández, M.; Sánchez-López, A.; Medina, J.F.; Márquez-Ruiz, J. Somatodendritic orientation determines tDCS-induced neuromodulation of Purkinje cell activity in awake mice. *bioRxiv*, 2023; *in press*. [CrossRef]
118. Edwards, J.D.; Dominguez-Vargas, A.U.; Rosso, C.; Branscheidt, M.; Sheehy, L.; Quandt, F.; Zamora, S.A.; Fleming, M.K.; Azzollini, V.; Mooney, R.A.; et al. A translational roadmap for transcranial magnetic and direct current stimulation in stroke rehabilitation: Consensus-based core recommendations from the third stroke recovery and rehabilitation roundtable. *Neurorehabil. Neural Repair* **2024**, *38*, 19–29. [CrossRef]
119. Xu, S.; Cui, H.; Xiao, X.; Manshaii, F.; Hong, G.; Chen, J. Precision at Deep Brain: Noninvasive Temporal Interference Stimulation. *ACS Nano* **2025**, *19*, 39589–39614. [CrossRef]
120. Molero-Chamizo, A.; Nitsche, M.A.; Gutiérrez Lérida, C.; Salas Sánchez, Á.; Martín Riquel, R.; Andújar Barroso, R.T.; Alameda Bailén, J.R.; García Palomeque, J.C.; Rivera-Urbina, G.N. Standard Non-Personalized Electric Field Modeling of Twenty Typical tDCS Electrode Configurations via the Computational Finite Element Method: Contributions and Limitations of Two Different Approaches. *Biology* **2021**, *10*, 1230. [CrossRef]
121. Callejón-Leblic, M.A.; Miranda, P.C. A comprehensive analysis of the impact of head model extent on electric field predictions in transcranial current stimulation. *J. Neural Eng.* **2021**, *18*, 046024. [CrossRef]
122. Markowska, A.; Tarnacka, B. Molecular Changes in the Ischemic Brain as Non-Invasive Brain Stimulation Targets-TMS and tDCS Mechanisms, Therapeutic Challenges, and Combination Therapies. *Biomedicines* **2024**, *12*, 1560. [CrossRef]

cells MDPI

Review

Epigenetics and Gut Microbiota in the Pathogenesis and Treatment of Bipolar Disorder (BD)

Shabnam Nohesara [1], Hamid Mostafavi Abdolmaleky [1,2,*], Ahmad Pirani [3] and Sam Thiagalingam [1,4,*]

[1] Department of Medicine (Biomedical Genetics), Boston University Chobanian & Avedisian School of Medicine, Boston, MA 02118, USA; snohesar@bu.edu
[2] Department of Medicine, Division of Gastroenterology, Beth Israel Deaconess Medical Center, Harvard Medical School, Boston, MA 02215, USA
[3] Mental Health Research Center, Psychosocial Health Research Institute, Iran University of Medical Sciences, Tehran 1449614535, Iran; a.pirani85@gmail.com
[4] Department of Pathology & Laboratory Medicine, Boston University Chobanian & Avedisian School of Medicine, Boston, MA 02118, USA
* Correspondence: sabdolma@bidmc.harvard.edu (H.M.A.); samthia@bu.edu (S.T.)

Abstract

Bipolar disorder (BD) is a multifactorial mental disease with a prevalence of 1–5% in adults, caused by complex interactions between genetic and environmental factors. Environmental factors contribute to gene expression alterations through epigenetic mechanisms without changing the underlying DNA sequences. Interactions between the gut microbiota (GM) and diverse external factors, such as nutritional composition, may induce epigenetic alterations and increase susceptibility to BD. While epigenetic mechanisms are involved in both the pathogenesis of BD and drug treatment responses, epigenetic marks could be employed as predictors and indicators of drug response. This review highlights recent studies on the potential role of epigenetic aberrations in the development and progression of BD. Next, we focus on drug response-related alterations in the epigenetic landscape, including DNA methylation, histone modifications, and non-coding RNAs. Afterward, we delve into the potential roles of GM-induced epigenetic changes in the pathogenesis of BD and GM-based therapeutic strategies aimed at improving BD outcomes through epigenetic modifications. We also discuss how BD drugs may exert beneficial effects through modulation of the GM and the epigenome. Finally, we consider future research strategies that could address existing challenges.

Keywords: bipolar disorder; epigenetic; microbiome; drug response

1. Introduction

Bipolar disorder (BD), as a multifactorial disease in origin, is caused by interplays between genetic and environmental factors, with a prevalence of 1–5% in adults [1–3]. The main symptoms of BD are the recurrence of manic, hypomanic, and depressive episodes that may be accompanied by intervals of well-being [4]. BD is classified into two types: (i) BD type I (BD I), characterized by a manic episode with or without episodes of major depression, and (ii) BD type II (BD II), characterized by major depressive episodes and at least a hypomanic episode [5]. Given the elevated risk of suicide and disability, BD imposes a significant socioeconomic burden on both the family and society [6,7]. Owing to a remarkable link between heritable traits, epigenetic modifications, and drug response in patients with BD, a growing body of investigations suggests that genetic and epigenetic marks may have predictive value [8]. The explorations of reliable genetic and epigenetic markers of drug response pave the way for a deeper understanding of drug mechanisms of action and improved treatment management for individuals with BD [9]. Multiple lines of evidence over the last decade indicate that, besides genetic factors, epigenetic mechanisms, which regulate gene expression without changing DNA's sequences, can serve as potential biomarkers for assessing both the risk of BD in vulnerable individuals and evaluate their response to various drug treatments [10].

The pathogenesis of BD and responses to medications are also related to diets and other environmental factors, such as gut microbiota (GM) [11]. A variety of environmental factors such as physical activity, diet, and plant extracts contribute to shaping the GM, and thus affect mental health involving the gut–brain axis [12–15].

For instance, in a large cohort study of subjects with psychiatric disorders (N = 1,101), including BD (N = 217) and controls (N = 343), an elevated risk of acute mania has been linked to the consumption of nitrated cured meats [16]. This study also demonstrated that consuming nitrated products in rats is associated with an elevated abundance of some gut bacteria, including *Lachnospiraceae* and *Erysipelotrichales*. Low consumption of polyunsaturated fatty acids (PUFAs), which maintain intestinal epithelium permeability, may give rise to developing BD as well [17,18]. On the other hand, anerobic exercises may support mental health by improving the diversity and abundance of genera from the *Firmcutes* phylum [19].

Therefore, physical activity, manipulating diet, and utilizing antimicrobials and drugs that affect brain function and cognition may provide opportunities to mitigate the development or exacerbation of BD by changing microbial communities and their metabolites, which affect epigenetic mechanisms [20,21]. Microbial communities in the human gastrointestinal tract may also influence responses to pharmacotherapy by affecting drug pharmacokinetics or manipulating the function of hepatic enzymes [22,23]. In fact, the GM, which includes a complex and dynamic population of pathogenic, commensal, and symbiotic microorganisms, possesses a bidirectional communication with the host, influencing their metabolic status [24,25]. Hence, numerous studies have focused on the potential roles of GM composition in the pathogenesis of BD and response to BD drugs. For example, Hu et al. reported that untreated BD patients exhibit a decreased abundance of butyrate-producing bacteria, which are involved in modulating both inflammation and the epigenome [26].

In this narrative review, we first discuss the relationship between epigenetic aberrations and BD pathogenesis. Then, we review the literature on drug response-induced alterations in epigenetic marks to evaluate whether epigenetic markers may be considered as reliable predictors of drug response. Next, we review the key role of the GM in developing BD and GMbased therapeutics for the treatment of BD patients via epigenetic mechanisms. We also discuss how BD drugs may exert their effects by modulating the GM through epigenetic changes.

2. Epigenetic Alterations and the Pathogenesis of BD

2.1. DNA Methylation

The main epigenetic mechanisms include DNA methylation (DNAm), post-translational histone modifications, and non-coding RNAs, whose interplay modulates and fine-tune gene expression levels [27,28]. Distinct DNAm patterns are initiated by the addition of a methyl group to a cytosine–guanine dinucleotide (CpG) across the genome and are generally associated with the suppression of gene expression. However, methylation of some particular CpGs in the gene promoter region or gene body may increase gene expression [29,30]. Genes' transcription fine-tuning can occur mostly by DNAm or demethylation of CpG dinucleotides present in the genes' promoter regions [31].

Aberrant DNAm of several genes' CpG sites has been connected to BD and suicide. For example, Li et al. reported that BD patients exhibit aberrant DNAm in CpG islands (CGIs) flanking regions and CGI sparse promoters in blood cells [32]. Zhang et al. reported 2075 aberrant CGI methylation points in subjects with BD versus normal individuals, which were classified into 24 categories [33]. In their investigation, the authors found DNA hypermethylation of COMT and PPIEL genes in the blood samples of the observation group compared to the control subjects [33]. Gaine et al. found a noticeable hypomethylation of the ARHGEF38 region across four CpG sites in postmortem brain samples from BD patients who died by suicide versus healthy controls [34]. The main findings from other studies regarding the critical roles of DNA methylome alterations in individuals with BD are summarized in Table 1.

Table 1. Altered DNA methylation (DNAm) in BD patients versus control subjects.

Country of Studies/Number of Participants	Sample	Outcomes	Ref.
USA/115	Post-mortem brains (dorsolateral frontal cortex)	Hypomethylation of MB-COMT promoter and increased gene expression in BD, particularly in the left brain vs. controls	[35]
USA/35 patients with schizophrenia, 35 BD vs. 35 matched controls	Post-mortem brains (dorsolateral frontal cortex)	Hypermethylation of HTR2A promoter at and around the −1438A/G, but hypomethylation of its promoter at and around T102C polymorphic sites in BD vs. controls, which altered HTR2A expression	[36]
Canada/34 BD and 35 matched controls	Post-mortem brain, sperm, and blood samples	DNA hypomethylation at an extended HCG9 region (related to major histocompatibility complex and immune function) in BD subjects vs. controls	[37]
Italy/61 BD I and 50 BD II	Peripheral blood mononuclear cells	Greater amounts of BDNF promoter methylation in BD II vs. BD I, which, in general, decrease BDNF expression	[38]
Canada/16 controls, 14 BD with response to lithium, 16 unaffected relatives of BD subjects	Transformed lymphoblasts from subjects with BD	Reduced DNAm in BD patients vs. controls, which, in general, is linked to increased gene expression	[39]
USA and Iran/35 postmortem brains from BD and 35 from controls	Post-mortem brain samples	DTNBP1 promoter DNA hypermethylation decreasing gene expression in psychotic BD patients; DTNBP1 hypoactivity is linked to psychosis	[40]
USA/ 12 BD and 10 controls	Post-mortem prefrontal cortex	Lower KCNQ3 exon 11 DNAm and gene expression in BD vs. controls	[41]
UK/168 controls and 459 BD	Blood samples	DNA hypomethylation of FAM63B (a gene strongly linked to SCZ) in BD vs. controls	[42]
UK/ 40 BD subjects and 38 healthy controls	Blood samples	Complete methylation of four CpG islands (CGIs) across CACNA1C, a top gene linked to BD; hypermethylation of five CpG sites at CGI 3 island in intron 3 in BD vs. controls	[43]
Italy/99 BD 42 and controls	Peripheral blood mononuclear cells	Promoter DNA hypermethylation and reduced PDYN expression in BD IBD II vs. controls	[44]
USA/28 BD and 13 controls	Skeletal muscle samples	Higher global methylation of 5-mC and 5-fC in BD vs. controls	[45]
China/99 BD (N) and 92 controls	Blood samples	Hypermethylation of the AluY A1 and A2 CpG sites and hypo-methylation of A3 CpG site in BD	[46]
USA/166 BD and 162 controls	DNA samples from the Mayo Clinic Bipolar Disorder Biobank	Hypermethylation of CpG site created by Val = G allele of the Val66Met variance of BDNF, which decreases gene expression in BD associated with earlier disease onset	[47]
Japan/34 BD and 35 controls	Post-mortem brains (prefrontal cortex)	Hypermethylation of NTRK2 (encoding a BDNF receptor) and GRIN1 (encoding a subunit of the NMDA receptor) in neuronal cells of subjects with BD	[48]
USA/84 BD with and 79 without a history of suicide attempt vs. 76 controls	Blood samples	Six differentially methylated CpG sites and seven differentially methylated regions (DMRs) in BD with vs. without suicide attempt	[49]
USA/128 BD and 141 controls	Blood samples	Hypermethylation cg19215110 and cg23953820 sites but hypomethylation of cg14279856 and cg03270204 sites of DDR1 (encoding a tyrosine kinase receptor involved in neuronal migration) in BD	[50]

2.2. Histone Modifications

In addition to DNAm, the pathogenesis of BD is associated with changes in histone modifications [51]. Elimination of acetyl groups by histone deacetylases (HDACs) from lysine residues on the tails of histone proteins leads to condensation of the chromatin and a lower permissive state for transcription [52]. On the other hand, HDAC inhibitors (e.g., sodium valproate) provide a more active chromatin structure and subsequently increase genes' transcription [53]. In vivo imaging of HDAC-specific radiotracers demonstrated that changes in the level and activity of HDACs are connected to attention and emotion regulation in subjects with BD [54]. Therefore, inhibitors of HDACs such as valproic acid (VPA) may be considered good candidates for exerting anti-manic effects by influencing specific regions of the brain [55].

2.3. Non-Coding RNAs

Among other epigenetic mechanisms, miRNAs and other non-coding RNAs also act as post-transcriptional regulators of gene expression [56]. In this field, an animal study reported that the development of manic-like behaviors after 72 h REM sleep deprivation (SD) in rats is correlated with reduced expression levels of miRNA-325-3p, miR-326-3p, and miR-330-5p in the prefrontal cortex (PFC) [57]. In BD, individuals with a higher genetic risk of developing the disease state exhibited increased expression levels of miR-15b, miR-132, and miR-652 [58]. Furthermore, it has been shown that depressive episodes of the BD subjects are connected to downregulation of miR-499, miR-708, and miR-1908 [59]. A study by Bame et al. identified 58 differentially expressed miRNAs between control- and BD patient-derived neurons, some of which are involved in axon guidance, cell migration, dendrite and synapse function, and brain development [60]. Considering these findings, it has been proposed that miRNA may be used as a potential biomarker for BD in the clinical settings and for developing new therapeutic agents and approaches in the near future [61].

In addition to miRNA, dysregulation of long non-coding RNAs (lncRNAs) is linked to BD pathogenesis. While a previous study reported reduced levels of the lncRNA MALAT1 in the blood cells of patients with BD [62], a more recent study has shown increased expression of two other lncRNAs, HOXA-AS2, and MEG3 in the blood samples of patients with BD [63]. Other findings related to the potential roles of miRNA dysregulation in BD pathogenesis are summarized in Table 2.

Table 2. Dysregulated miRNAs in BD patients versus controls.

Country of Study/Number of Participants	Sample Source	Outcomes	Ref.
Italy/15 BD and 9 controls	Peripheral blood samples	Elevated expression of miR-150-5p, miR-25-3p, miR-451a, and miR-144-3p involved in metabolic processes but reduced expression of miR-363-3p, miR-4454 + miR-7975, miR-873-3p, miR-548al, miR-598-3p, miR-4443, miR-551a, and miR-6721-5p involved in neurogenesis and neurodevelopment in BD	[64]
Turkey/19 manic, 39 euthymic, and 51 controls	Blood samples	Up-regulation of miR-125b-5p, miR-106b-5p, miR-9-5p, miR-107, miR-29a-3p, miR-106a-5p, and miR-125a-3p in manic BD vs. controls and their association with manic episodes	[65]
Turkey/69 BD including 15 depressed, 27 manic, 27 euthymic and 41 controls	Plasma-derived exosomal miRNA	Reduced levels of miR-484, miR-652-3p, miR-142-3p, and an increased level of miR-185-5p in BD vs. controls, which are involved in the regulation of PI3K/Akt signaling, fatty acid biosynthesis/metabolism, extracellular matrix and adhesion pathways	[66]
Taiwan/79 BD IBD II and 95 controls	Blood samples	Up-regulation of miR-7-5p, miR-23b-3p, miR-142-3p, miR-221-5p, and miR-370-3p in BD IBD II vs. controls and their potential role as useful tools for the diagnosis of BD IBD II.	[67]
China/27 BD and 32 unipolar depressive disorder	Blood samples	Higher levels of miR-19b-3p in BD vs. unipolar depressive disorder that is involved in the pathway of inflammatory dysregulation associated with experiencing early childhood trauma	[68]
Germany/960 BD and 960 controls	Blood samples	Association between miR-499a dysregulation and BD progression and probable BD susceptibility	[69]
Lithuania/26 BD and 74 controls	Blood samples	Overexpression of let-7e-5p and miR-125a-5p in BD subjects vs. controls and their potential role as peripheral biomarkers for BD	[70]
Iran/240 controls and 260 BD	Blood samples	Increased expression of miR-23b-3p and reduced expression of miR-19b-3p in BD vs. controls and their potential role as biomarkers for BD	[71]
United Kingdom/112 independent samples (51 female, 61 male)	Human fetal brain tissue	Correlation between the elevated prenatal expression of miR-1908-5p and susceptibility to BD	[72]

3. Epigenetic Alterations in Response to Mood Stabilizers and Antipsychotic Drugs

3.1. Valproate

VPA is a short-chain fatty acid, which is served as an FDA-approved mood stabilizer for the treatment of BD and an anticonvulsant drug for the epilepsy treatment [73,74]. VPA has demonstrated a capacity for changing the gene expression of cell cycle regulators via modifying promoter DNAm, which may influence cell proliferation in the hippocampus region [75]. VPA is also capable of inducing chromatin remodeling and attenuating DNA hypermethylation of GABAergic genes' promoters that are responsible for disturbances in GABAergic neurotransmission in psychiatric diseases [76]. Dong et al. found that the protective impact of valproate in decreasing anxiety, cognitive dysfunction, and symptoms of BD is linked to elevated expression levels of reelin (RELN) gene and glutamic acid decarboxylase 67 (Gad-67) via decreasing DNAm of their promoters in nuclear extracts from the brain of adult mice [77].

Regarding DNAm, Aizawa et al. found that VPA administration could elevate the expression of p21 (both its mRNA and protein levels) in the mouse hippocampus, accompanied by changes in DNAm at the distal CGI of the p21 gene, without changing the expression of DNA methyltransferase (Dnmt) 1 or Dnmt3a [78]. Backlund et al. found that combination therapy with lithium (Li) and VPA in BD patients resulted in a hypermethylated pattern in leukocytes compared to Li monotherapy [79].

As another example, Chen et al. reported that prolonged (3 days) VPA treatment of mouse fibroblast 3T3-L1 cells could influence mitochondrial epigenetic marks by reducing 5-hydroxy methylation (5hmC) of mitochondrial DNA [80]. D'Addario et al. found that the increased DNAm of the BDNF promoter in subjects with BD II was reversed by various pharmacological therapies, especially VPA and Li [81]. D'Addario et al. also reported DNA hypermethylation at the PDYN gene promoter in BD IBD II subjects compared to healthy controls, which was reversed by both VPA and Li [44].

In addition to its effects on DNAm, VPA acts as a potent inhibitor of HDACs at therapeutic concentrations (0.4–0.8 mM), contributing to the regulation of gene expression through both HDAC-dependent and -independent mechanisms [82–85]. VPA has also been shown to activate the human FGF1 gene promoter by suppressing HDAC and GSK-3 activities [86]. Moreover, VPA is capable of inducing histone H3 methylation and acetylation in the leptin receptor gene in the hippocampus of Brown Norway rats [87]. In BD patients, treatment with VPA for four weeks increases the levels of acetylated Histone 3 (H3ac) and acetylated Histone 4 proteins (H4ac), as observed in lymphocyte nuclear protein extracts [88].

In animal models of acute mania, hyperactivity is linked to elevated HDAC activity in the PFC, and VPA has been shown to improve BD-like symptoms [89]. Valvassori et al. also developed an animal model of mania induced by d-AMPH and found elevated activity of HDACs in the PFC [90]. In their study, treatment of animals with VPA could reduce the activity of HDACs in the PFC and striatum of rats. In a rat model of mania induced by ouabain, Varela et al. found that VPA was able to reverse hyperactivity by reducing the activity of DNA methyltransferase and HDACs [91]. Logan et al. found that normalizing the manic-like behavioral profile in the ClockΔ19 model of bipolar mania in mice, treated with VPA, is mediated by HDAC2 inhibition in the ventral tegmental area (VTA) [92].

Besides DNAm and histone modifications, VPA has been shown to influence neuronal function by altering the expression of non-coding RNAs such as microRNAs [93,94]. Kim et al. found that chronic treatment with VPA is capable of influencing the expression of a BD-associated miRNA, miR-1908-5p, and its target genes such as *STX1A*, *CLSTN1*, *DLGAP4*, *GRIN1*, and *GRM4* [95]. In their studies, VPA reduced the expression of miR-1908-5p in neural progenitor cells derived from dermal fibroblasts of a BD patient while increasing its expression in control human neural progenitor cells. Additionally, it has been shown that lncRNAs may contribute to the mechanisms of action of BD drugs. For example, VPA could increase the expression of the hub lncRNA of module 5, GAS6-AS1, in human neuronal-like (NT2-N) cells [96]. Figure 1 illustrates a summary of findings related to the effects of VPA on different epigenetic mechanisms.

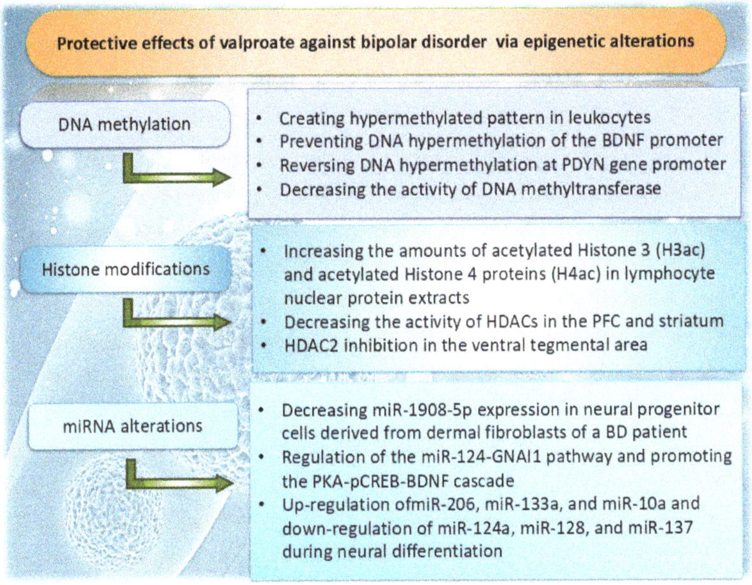

Figure 1. Protective effects of valproate mediated by epigenetic mechanisms in BD. Valproate is capable of exerting beneficial impact in BD patients via alleviating abnormal changes in DNA methylation, histone modifications, and miRNAs. Figure was created using the Microsoft PowerPoint software 2010.

3.2. *Lithium (Li)*

Li is one of the gold standard drugs and first-line treatments for BD [97]. Approximately 20–30% of patients with BD are excellent responders to Li, while more than 40% show little to no clinical response [98]. Differential DNAm patterns have been identified between responders and non-responders to Li treatment. Claire et al. reported 111 differentially methylated regions (DMRs) in the DNA of blood cells between the responders and non-responders to long-term Li treatment [99]. Zafrilla-López et al. reported 130 differentially methylated positions (DMPs) and 16 DMRs between responders and non-responders to Li treatment, affecting 122 genes, including *CACNA1B*, *ANK3*, *TENM2*, *SORCS2*, *EEF2K*, *CYP1A1*, *HOXB6*, *HOXB3*, and *HOXB-AS3*, as well as genes involved in the GSK3β signaling pathway [100]. Furthermore, Asai et al. reported that CpG sites within the *SLC6A4* gene, which were hypermethylated in subjects with BD, exhibited significant hypomethylation in the neuroblastoma cells following treatment with Li, VPA, and carbamazepine [101].

Similarly to VPA, Li has also demonstrated the ability to reverse increased locomotor activity induced by d-AMPH in animal models of acute mania by inhibiting HDAC activity in the PFC [89]. In another animal model of mania caused by REM sleep deprivation, Andrabi et al. found that treatment with Li hampered derangements in histone acetyl transferase/histone deacetylase (HAT/HDAC) activity and suppressed inflammation, oxidative stress, mitochondrial impairment, and apoptotic cell death [102].

In humans, the beneficial effects of Li in BD have been shown to be mediated, in part, by the normalization of miRNAs' expression. An earlier study demonstrated that gene-to-gene interplay between miRNA-206 and BDNF affects susceptibility to BD I and treatment response to mood stabilizers [103]. In a later study, Pisanu et al. found that miR-320a, miR-155-3p, and their targeted genes, CAPNS1 (Calpain Small Subunit 1) and RGS16 (Regulator of G Protein Signaling 16) for miR-320), and SP4 (Sp4 Transcription Factor) for miR-155-3p, may have a potential role in clinical response to Li [104]. In their study, increased levels of miR-320a and decreased expression of its target genes were observed in excellent responders versus non-responders [104]. In a large multinational GWAS study, miR-499a was also linked to Li response in BD [105]. In a more recent study, Cattane et al. reported 77 differentially expressed miRNAs in Li responders compared to non-responders. The affected miRNAs are involved in neurodevelopment, intracellular signal transduction, and immune system response [106]. Figure 2 provides a summary of the influence of Li on diverse epigenetic mechanisms and their corresponding genes in BD.

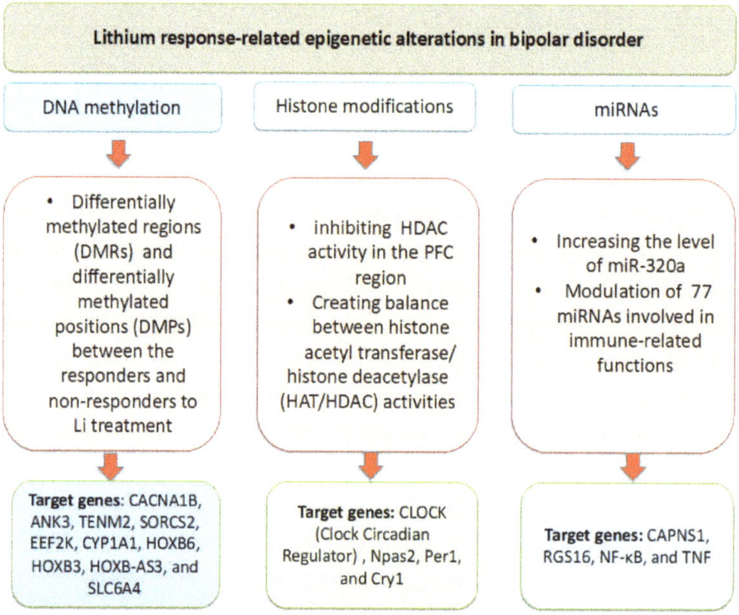

Figure 2. Lithium effects on different epigenetic mechanisms and their corresponding affected genes. Lithium has demonstrated the ability to exert its protective effects via normalizing aberrations in DNA methylation, histone modifications, and miRNAs. Figure was created using Microsoft PowerPoint software 2010.

3.3. *Lamotrigine (LTG)*

Lamotrigine (LTG) is a phenyltriazine-derived anti-seizure medication and a mood stabilizer that is used for the treatment of depressive episodes in BD [107]. LTG, in combination with Li and VPA, is capable of alleviating the symptoms relevant to cyclic episodes of mania and hypomania, as well as depression [108]. LTG may exert its neuroprotective effects via epigenetic mechanisms. For example, Kidnapillai et al. found that LTG in combination with other BD drugs could elevate the expression of miR-128 and miR-378, which regulate the expression of genes involved in neurite outgrowth and neural plasticity [109]. Moreover, the protective effects of LTG may be associated with histone modifications. For example, the neuroprotective effects of LTG against glutamate excitotoxicity are linked to elevated levels of acetylated histone H3 and H4 and to increasing the histone acetylation and the Bcl-2 promoter activity [110].

3.4. Atypical Antipsychotics

Atypical antipsychotic drugs have been found to influence epigenetic mechanisms, especially DNA and histone methylation in neuropsychiatric disorders [111,112]. For example, quetiapine, as an atypical antipsychotic drug, is capable of reducing DNAm level of the promoter region of *SLC6A4* (serotonin transporter gene) in human neuroblastoma cells [113]. As another example, Houtepen et al. found that treatment of BD patients with both quetiapine and VPA alter their DNAm signatures in specific blood cells. In their study, the beneficial effects of quetiapine were related to immune and neurogenesis-related DNAm networks [114].

Non-coding RNAs are also considered as master modulators of the action of antipsychotic drugs in the treatment of BD patients. For instance, BD patients treated with a range of antipsychotics and mood stabilizers (Li, VPA, risperidone, olanzapine, clozapine, and quetiapine) for four weeks demonstrated higher plasma levels of miRNA-134 versus unmedicated BD patients [115]. Quetiapine and lamotrigine also increased the expression of miR-128 and miR-378 in cultured human neuronal cells, and their combination could reduce the expression of their target genes (*GRIN3A*, *VIM*, and *NOVA1*), which are involved in suppressing neurite outgrowth and neurogenesis [109]. Additionally, quetiapine is capable of elevating the expression of hub lncRNA LINC02381 in NT2-N cells [96].

The protective effects of asenapine and risperidone in subjects with BD are related to changes in miRNAs levels as well. Notably, 12 weeks of treatment with asenapine resulted in the overexpression of 14 miRNAs (miR-15a-5p, miR-17-3p, miR-17-5p, miR-18a-5p, miR-19b-3p, miR-20a-5p, miR-27a-3p, miR-30b-5p, miR-145-5p, miR-148b-3p, miR-106a-5p, miR-106b-5p, miR-210-3p, and miR-339-5p) and reduced expression of 2 miRNAs (miR-92b-5p and miR-1343-5p) in individuals with BD [116]. Furthermore, patients with BD treated with risperidone for 12 weeks exhibited downregulation of three miRNAs, including miR-146b-5p, miR-664b-5p, and miR-6778-5p [116].

4. The Gut Microbiota, BD, and the Disease Severity

Emerging evidence suggests that the GM is a key player in modulating mental health, while its composition changes are associated with the development of neuropsychiatric disorders such as BD [117,118]. Owing to the increased inflammation and disrupting intestinal barrier integrity, individuals with BD are affected by the translocation of bacteria and their toxins like lipopolysaccharide from the intestinal lumen to blood circulation and subsequently into the brain [119]. Disruption of the brain–gut-microbial axis in BD heavily affects host endocrine functions relevant to neuronal functions (i.e., controlling the production of neurotransmitters such as serotonin or mediating the activation of the hypothalamic–pituitary–adrenal axis) and host immune system in the intestine and brain [21].

The potential role of the GM in the pathogenesis of BD is related to epigenetic aberrations, since GM-derived metabolites, particularly SCFAs, can function as HDAC inhibitors. For example, a decreased abundance of *Faecalibacterium*, a butyrate-producing gut bacterium, was observed in subjects with BD, and its level was directly associated with self-reported health status [120]. Moreover, changes in the GM composition involved in the pathogenesis of BD are associated with altered DNA methylation patterns. For example, GM diversity in participants with BD was negatively correlated with methylation of the clock gene *ARNTL*, a main regulator of monoamine oxidase A transcription and circadian rhythms [121]. Also, the GM–miRNA interactions may play a key role in the pathogenesis of psychiatric disorders [122,123]. The GM influences host health by regulating host miRNAs. As an example, the inhibition of miR-34a-5p by an antagomir could improve total abdominal irradiation-mediated cognitive impairment through restoring GM composition [124]. The imbalances in gut bacteria and epigenetic metabolites, such as butyrate, in BD patients may also increase oxidative stress and neuroinflammation, which in turn negatively affect mood and cognitive function. To this end, Huang et al., investigated the alterations in the GM composition and its association with inflammation in depressed BD patients and reported decreased abundance of some bacteria (*Butyricicoccus, Lachnospiraceae incertae sedis*, and *Dorea*) involved in indirect release of anti-inflammatory cytokines such as IL-4, IL-10, and IL-11 [125]. In another study, untreated BD patients exhibited lower levels of the anti-inflammatory butyrate-producing bacteria versus healthy controls [26]. Painold et al. reported that the GM composition is associated with distinct changes in inflammatory and metabolic factors in BD patients. In their study, BD patients exhibited higher abundance of pro-inflammatory bacteria, like *Streptococcus*, which was linked to greater serum IL-6 levels [126]. Additionally, Coello et al. found that the pathogenesis of BD may be related to *Flavonifractor*, a bacterial genus that may contribute to inducing oxidative stress and inflammation in the host [127]. In a study conducted by Guo et al., unmedicated depressive BD II patients showed increased levels of pro-inflammatory bacteria (e.g., *Proteobacteria, Enterobacteriaceae, Porphyromonadaceae, Pseudescherichia*), higher levels of inflammatory cytokines levels, and changes in glutamate (Glu) and gamma-aminobutyric acid (GABA) metabolism-related bacteria (*Bifidobacterium, Escherichia, Bacteroides, Parabacteroides, Eggerthella*) [128].

5. Microbiota-Based Therapeutic Approaches for the Treatment of BD

Several approaches can be employed to modulate the GM and its associated epigenetic effects in managing BD.

5.1. Probiotics

Probiotics are microorganisms with the capability for creating health benefits to the host when they are consumed in proper concentrations [129–131]. Some probiotics, known as brain probiotics, can promote the production of neurotransmitters such as GABA and serotonin and improve brain functions [132,133]. Specific SCFA-producing probiotics known as "psychobiotics" have the potential to improve disease outcomes by mitigating epigenetic dysregulations, as many SCFAs are epigenetically active metabolites [134]. SCFA-producing probiotics like *Lactobacillus, Faecalibacterium*, and *Ruminococcus* can alleviate depressive-like behaviors and cognitive impairments in BD patients [135]. Similarly, probiotic organisms such as *Lactobacillus rhamnosus* strain GG and *Bifidobacterium animalis* subsp. *lactis* strain Bb12 have been shown to improve symptoms in patients with mania and BD I, possibly through enhancing butyrate production, maintaining the mucosal barrier, and reducing oxidative stress and systemic inflammation [136–139]. Several other studies have shown that probiotic supplementation may improve manic symptoms and cognitive function in BD subjects [139–141]. As an interesting example, Zhang et al. found that oral administration of a probiotic containing *Bifidobacterium animals subsp. lactis BAMA-B06/BAu-B0111* could alleviate depressive mood in subjects with depressive BD type I by reducing the abundance of phylum *Firmicutes* and improving metabolic disarrays [137]. In another study, Borkent et al. found that beneficial effects of a probiotic, containing nine *Bifidobacterium* and *Lactococcus* bacterial strains in BD subjects, reduce intestinal permeability and suppress inflammation [142].

5.2. Postbiotics

SCFAs, well-known postbiotics and metabolic byproducts generated by the microbial community in the gut during the fermentation of dietary fibers, play an important role in host health, since they act as histone deacetylase inhibitors [143,144]. In addition to histone modifications, the beneficial effects of SCFAs, such as butyrate, in the treatment of neuropsychiatric diseases may also be attributed to their key role in modulating DNA methylation [145,146]. Several studies have demonstrated that SCFAs may be used to alleviate symptoms of BD in human and animal models [147,148]. For example, Moretti et al. found that rats exposed to an animal model of mania induced by d-AMPH exhibited hyperactivity and reduced activity of mitochondrial respiratory chain complexes. Treatment with sodium butyrate was able to counteract these abnormal changes [149]. Valvassori et al. suggested that sodium butyrate may be considered a potential new mood stabilizer, as it restores proteins from the neurotrophin family—including BDNF, NGF, and GDNF—and exerts neuroprotective effects against oxidative damage in the frontal cortex and hippocampus of rats exposed to an ouabain-induced animal model of mania [150,151].

5.3. Specific Diets (e.g., Ketogenic Diet)

Diet also has a great influence on improving the symptoms of subjects with BD through modulation of the GM composition and epigenetic modifications. For instance, a healthier diet containing fish, beans, fruits, and nuts triggers a greater diversity and richness of GM in BD patients [152].

Ketogenic diet (KD) is also a promising diet capable of mimicking the mechanisms of mood stabilizers for repairing or bypassing metabolism dysregulations, reducing neuroinflammation, and improving manic symptoms via two mechanisms [153,154]. First, KD influences the synthesis of neurotransmitters such as GABA and glutamate via increasing the production of acetyl-CoA from ketone bodies, particularly β-hyroxybutyrate (βOHB) [155]. Second, βOHB acts as class I HDAC inhibitor that elevates histone acetylation and thereby enhances the expression of several genes involved in mitochondrial function, antioxidant defense, and the reduction in inflammation [156]. βOHB also decreases HDAC2 levels, which in turn elevates the expression of genes involved in oxidative stress and neuroinflammation. Polito et al. found that βOHB produced by KD could increase microglial polarization towards an M2 anti-inflammatory phenotype, which in turn decreases the pro-inflammatory cytokine IL-17 and increases the anti-inflammatory cytokine IL-10 [157]. Sethi et al. reported that a ketogenic diet may contribute to improved metabolic health with a greater than 1-point improvement in 69% of subjects with BD [158].

5.4. Fecal Microbiota Transplantation (FMT)

Fecal microbiota transplantation (FMT) is an approach in which fecal bacteria or other microbes are transferred from a donor into the gastrointestinal tract of a recipient [159–161]. FMT may be considered a potential therapeutic procedure for investigating the role of gut microbiota in the pathogenesis of BD and its treatment in animal models [162,163]. For example, FMT from BD subjects to mice is associated with delivering behavioral and metabolic characteristics relevant to BD. In fact, mice receiving FMT from BD patients exhibited anxiety-like behavior and reduced sociability, which was associated with an elevated abundance of *Bacteroidota* and reduced abundances of *Parabacteroides merdae* and *Akkermansia muciniphila* and reductions in the levels of epigenetic metabolites, including acetic acid and butyric acid [164]. As another example, Hinton examined the effect of FMT in a BD I subject unresponsive to different types of psychotropic medications and found that this patient exhibited drug-free remission of psychiatric symptoms during the follow-up years [165].

5.5. Antipsychotic Medications

It is also worth noting that the atypical antipsychotics (AAPs) and non-AAPs strongly affect microbial communities and GM diversity in subjects with BD [166]. A recent study has demonstrated that BD patients treated with antipsychotics and Li exhibit an elevated abundance of 23 genera and 18 bacterial species, such as butyrate-producing bacteria (*Ruminococcus obeum*, *Eubacterium biforme*, and *Eubacterium rectale*) [167]. Kamath et al. found that lurasidone is also capable of modulating the GM via elevating the mean operational taxonomical units (OTUs) and alpha diversity in rats [168]. Treatment of BD patients with quetiapine (300 mg/d) for four weeks led to a rebound in the abundance of *Bifidobacteria* and *Eubacterium rectale*, bacteria involved in reducing inflammation by producing butyrate and other SCFAs [169].

Another recent study showed a positive relationship between lurasidone bioavailability, microbial diversity, and SCFA levels, which is mediated by adjusting luminal pH [170]. High concentrations of SCFA also provide a favorable environment for lurasidone solubility via reducing intestinal pH, whereas GM depletion leads to reduced solubility of lurasidone [170]. Notably, while lurasidone has not been associated with weight gain [171], there is evidence that adolescents treated with risperidone (versus psychiatric controls) exhibited a lower ratio of *Bacteroidetes–Firmicutes* bacteria, which contributed to weight gain [172]. Figure 3 illustrates how interactions between the GM and BD medications, mediated through epigenetic mechanisms, which contribute to therapeutic benefits for patients with BD.

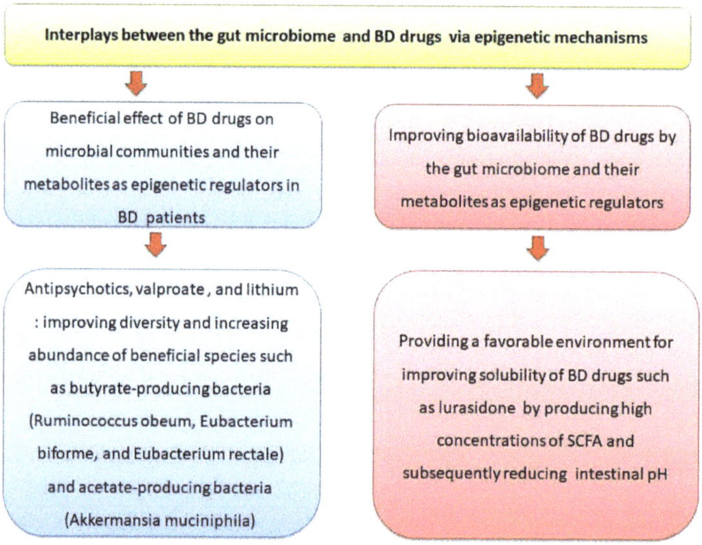

Figure 3. Interactions between the GM and bipolar disorder medications, which are mediated through epigenetic mechanisms and contribute to therapeutic benefits for patients with BD. Figure was created using the Microsoft PowerPoint software 2010.

6. Conclusions, Challenges, and Prospectives

The severity of BD and the outcome of therapy can be related to alterations in epigenetic mechanisms, including DNAm, post-translational histone modifications, and non-coding RNAs. Understanding epigenetic aberrations involved in the pathogenesis of BD and the mechanisms of action of BD drugs may help identify individuals susceptible to BD. This knowledge could also serve as a basis for developing biomarkers that enable more accurate diagnosis, treatment, and the exploration of novel personalized therapies. It could also pave the way for improving treatment outcomes in subjects with treatment-resistant BD who do not respond to available mood stabilizers and antipsychotic agents. In this review, we first examined the existing literature on the potential roles of epigenetic modifications in determining the therapeutic efficacy of mood stabilizers and antipsychotic drugs in BD patients.

One of the major limitations of these studies is that the vast body of current studies has not focused on the identification of epigenetic biomarkers in long-term therapies. Another challenge for pharmaco-epigenetic studies in BD is the small sample sizes and inadequate statistical power that should be addressed by using larger sample sizes and fostering collaborative cohort studies. Although multiple studies have shown that drugs such as VPA and Lithiumare are capable of targeting various epigenetic mechanisms, almost all published studies only examined one epigenetic mechanism at a time. Therefore, future studies should focus on the investigation of the impact of disease or a drug on a combination of several epigenetic mechanisms. Moreover, while GM research shows promise for developing more effective therapies to improve outcomes in BD patients, several significant gaps need to be addressed in future studies. For example, small sample sizes and the inconsistencies across studies, as well as variations in dietary habits and microbial composition across different populations, may lead to more controversial findings in the field. Additionally, most GM data are compositional, focusing on relative bacterial abundance, and therefore do not provide information on their specific absolute abundance or potential pathological impacts.

Replicability of GM research in BD could be more difficult since the disease has diverse features in different phases of BD. Additional difficulties come from technical differences (e.g., taxonomy databases, sequencing platforms, DNA extraction approaches, PCR primers) and investigating various patient cohorts (e.g., type1 or 2 BD, cyclothymia, rapid-cycling BD, BD with mixed features, BD with seasonal pattern, medicated vs. unmedicated, first-episode, chronic) that may cause systematic biases, which in turn hamper obtaining precise information about compositional differences in BD patients. Larger sample sizes and stronger demographic and phenotypic characterization of clinical samples in future studies could help to obtain greater insights into the functional roles of the GM in mental health and the pathogenesis of BD via epigenetic mechanisms. Likewise, other biological factors relevant to the gut–brain axis should be considered in investigations of the GM to facilitate better understanding of the mechanisms by which the gut ecosystem influences the cerebral tissue in BD patients.

Owing to the potential role of an accelerated biological aging and changes in biological clocks as well as epigenetic age dysregulation in BD, future studies should also focus on clarifying the correlation between age and gut microbial composition effects on epigenetic alterations. In order to determine whether alterations in the GM and gut-derived metabolites as epigenetic modifiers are trait- or state-related, more longitudinal sampling and studies should be implemented during different periods of treatment (e.g., acute vs. long-term), various stages and episodes (e.g., manic vs. depressive episodes) of BD. Despite current challenges, the prospects for novel discoveries in the GM–epigenome relationship and its influence on the brain through the "gut–brain axis" are incredibly promising. This excitement is largely due to recent technological advancements such as single cell sequencing, whole-genome epigenetic profiling, multi-omics approaches, large-scale multinational research initiatives, and advanced bioinformatics and machine learning.

Author Contributions: Conceptualization, H.M.A. and S.T.; Writing—original draft preparation, S.N.; Writing—review & editing, A.P.; Methodology, H.M.A.; Software, A.P.; Validation, S.T.; Formal analysis and investigation, S.N. and H.M.A.; Visualization, S.T. All authors have read and agreed to the published version of the manuscript.

Funding: This research received no external funding.

Data Availability Statement: Not applicable.

Acknowledgments: We acknowledge Faria Ashrafi for her assistance with the illustrations in this manuscript.

Conflicts of Interest: The authors declare no conflicts of interest.

References

1. Robinson, N.; Bergen, S.E. Environmental risk factors for schizophrenia and bipolar disorder and their relationship to genetic risk: Current knowledge and future directions. *Front. Genet.* **2021**, *12*, 686666. [CrossRef] [PubMed]
2. Gordovez, F.J.A.; McMahon, F.J. The genetics of bipolar disorder. *Mol. Psychiatry* **2020**, *25*, 544–559. [CrossRef] [PubMed]
3. Menculini, G.; Balducci, P.M.; Attademo, L.; Bernardini, F.; Moretti, P.; Tortorella, A. Environmental risk factors for bipolar disorders and high-risk states in adolescence: A systematic review. *Medicina* **2020**, *56*, 689. [CrossRef] [PubMed]
4. McIntyre, R.S.; Berk, M.; Brietzke, E.; Goldstein, B.I.; López-Jaramillo, C.; Kessing, L.V.; Malhi, G.S.; Nierenberg, A.A.; Rosenblat, J.D.; Majeed, A. Bipolar disorders. *Lancet* **2020**, *396*, 1841–1856. [CrossRef] [PubMed]
5. Guzman-Parra, J.; Streit, F.; Forstner, A.J.; Strohmaier, J.; González, M.J.; Gil Flores, S.; Cabaleiro Fabeiro, F.J.; del Río Noriega, F.; Perez Perez, F.; Haro González, J. Clinical and genetic differences between bipolar disorder type 1 and 2 in multiplex families. *Transl. Psychiatry* **2021**, *11*, 31. [CrossRef] [PubMed]
6. Cloutier, M.; Greene, M.; Guerin, A.; Touya, M.; Wu, E. The economic burden of bipolar I disorder in the United States in 2015. *J. Affect. Disord.* **2018**, *226*, 45–51. [CrossRef] [PubMed]
7. Wisłowska-Stanek, A.; Kołosowska, K.; Maciejak, P. Neurobiological basis of increased risk for suicidal behaviour. *Cells* **2021**, *10*, 2519. [CrossRef] [PubMed]
8. Rybakowski, J.K. Genetic influences on response to mood stabilizers in bipolar disorder: Current status of knowledge. *CNS Drugs* **2013**, *27*, 165–173. [CrossRef] [PubMed]
9. Lockwood, L.E.; Youssef, N.A. Systematic review of epigenetic effects of pharmacological agents for bipolar disorders. *Brain Sci.* **2017**, *7*, 154. [CrossRef] [PubMed]
10. Fries, G.; Walss-Bass, C.; Soares, J.; Quevedo, J. Non-genetic transgenerational transmission of bipolar disorder: Targeting DNA methyltransferases. *Mol. Psychiatry* **2016**, *21*, 1653–1654. [CrossRef] [PubMed]
11. Verma, A.; Inslicht, S.S.; Bhargava, A. Gut-brain axis: Role of microbiome, metabolomics, hormones, and stress in mental health disorders. *Cells* **2024**, *13*, 1436. [CrossRef] [PubMed]
12. Nohesara, S.; Mostafavi-Abdolmaleky, H.; Dickerson, F.; Pinto-Tomas, A.A.; Jeste, D.V.; Thiagalingam, S. Associations of microbiome pathophysiology with social activity and behavior are mediated by epigenetic modulations: Avenues for designing innovative therapeutic strategies. *Neurosci. Biobehav. Rev.* **2025**, *174*, 106208. [CrossRef] [PubMed]
13. Cataldi, S.; Poli, L.; Şahin, F.N.; Patti, A.; Santacroce, L.; Bianco, A.; Greco, G.; Ghinassi, B.; Di Baldassarre, A.; Fischetti, F. The effects of physical activity on the gut microbiota and the gut–brain Axis in preclinical and human models: A narrative review. *Nutrients* **2022**, *14*, 3293. [CrossRef] [PubMed]
14. Sanborn, V.; Gunstad, J. The potential mediation of the effects of physical activity on cognitive function by the gut microbiome. *Geriatrics* **2020**, *5*, 63. [CrossRef] [PubMed]
15. Codoñer-Franch, P.; Gombert, M.; Martínez-Raga, J.; Cenit, M.C. Circadian disruption and mental health: The chronotherapeutic potential of microbiome-based and dietary strategies. *Int. J. Mol. Sci.* **2023**, *24*, 7579. [CrossRef] [PubMed]
16. Khambadkone, S.G.; Cordner, Z.A.; Dickerson, F.; Severance, E.G.; Prandovszky, E.; Pletnikov, M.; Xiao, J.; Li, Y.; Boersma, G.J.; Talbot, C.C., Jr. Nitrated meat products are associated with mania in humans and altered behavior and brain gene expression in rats. *Mol. Psychiatry* **2020**, *25*, 560–571. [CrossRef] [PubMed]
17. Evans, S.J.; Ringrose, R.N.; Harrington, G.J.; Mancuso, P.; Burant, C.F.; McInnis, M.G. Dietary intake and plasma metabolomic analysis of polyunsaturated fatty acids in bipolar subjects reveal dysregulation of linoleic acid metabolism. *J. Psychiatr. Res.* **2014**, *57*, 58–64. [CrossRef] [PubMed]
18. Kankaanpää, P.E.; Salminen, S.J.; Isolauri, E.; Lee, Y.K. The influence of polyunsaturated fatty acids on probiotic growth and adhesion. *FEMS Microbiol. Lett.* **2001**, *194*, 149–153. [CrossRef] [PubMed]
19. Dalton, A.; Mermier, C.; Zuhl, M. Exercise influence on the microbiome–gut–brain axis. *Gut Microbes* **2019**, *10*, 555–568. [CrossRef] [PubMed]

20. Singh, S.; Sharma, P.; Pal, N.; Kumawat, M.; Shubham, S.; Sarma, D.K.; Tiwari, R.R.; Kumar, M.; Nagpal, R. Impact of environmental pollutants on gut microbiome and mental health via the gut–brain axis. *Microorganisms* **2022**, *10*, 1457. [CrossRef] [PubMed]
21. Ortega, M.A.; Álvarez-Mon, M.A.; García-Montero, C.; Fraile-Martínez, Ó.; Monserrat, J.; Martinez-Rozas, L.; Rodríguez-Jiménez, R.; Álvarez-Mon, M.; Lahera, G. Microbiota–gut–brain axis mechanisms in the complex network of bipolar disorders: Potential clinical implications and translational opportunities. *Mol. Psychiatry* **2023**, *28*, 2645–2673. [CrossRef] [PubMed]
22. ElRakaiby, M.; Dutilh, B.E.; Rizkallah, M.R.; Boleij, A.; Cole, J.N.; Aziz, R.K. Pharmacomicrobiomics: The impact of human microbiome variations on systems pharmacology and personalized therapeutics. *Omics A J. Integr. Biol.* **2014**, *18*, 402–414. [CrossRef]
23. Selwyn, F.P.; Cheng, S.L.; Klaassen, C.D.; Cui, J.Y. Regulation of hepatic drug-metabolizing enzymes in germ-free mice by conventionalization and probiotics. *Drug Metab. Dispos.* **2016**, *44*, 262–274. [CrossRef] [PubMed]
24. Kuziel, G.A.; Rakoff-Nahoum, S. The gut microbiome. *Curr. Biol.* **2022**, *32*, R257–R264. [CrossRef] [PubMed]
25. Ross, F.C.; Patangia, D.; Grimaud, G.; Lavelle, A.; Dempsey, E.M.; Ross, R.P.; Stanton, C. The interplay between diet and the gut microbiome: Implications for health and disease. *Nat. Rev. Microbiol.* **2024**, *22*, 671–686. [CrossRef] [PubMed]
26. Hu, S.; Li, A.; Huang, T.; Lai, J.; Li, J.; Sublette, M.E.; Lu, H.; Lu, Q.; Du, Y.; Hu, Z. Gut microbiota changes in patients with bipolar depression. *Adv. Sci.* **2019**, *6*, 1900752. [CrossRef] [PubMed]
27. Radford, E.J. An introduction to epigenetic mechanisms. *Prog. Mol. Biol. Transl. Sci.* **2018**, *158*, 29–48. [PubMed]
28. Smigielski, L.; Jagannath, V.; Rössler, W.; Walitza, S.; Grünblatt, E. Epigenetic mechanisms in schizophrenia and other psychotic disorders: A systematic review of empirical human findings. *Mol. Psychiatry* **2020**, *25*, 1718–1748. [CrossRef] [PubMed]
29. Smith, Z.D.; Hetzel, S.; Meissner, A. DNA methylation in mammalian development and disease. *Nat. Rev. Genet.* **2024**, *26*, 7–30. [CrossRef] [PubMed]
30. Tian, W.; Zhou, J.; Bartlett, A.; Zeng, Q.; Liu, H.; Castanon, R.G.; Kenworthy, M.; Altshul, J.; Valadon, C.; Aldridge, A. Single-cell DNA methylation and 3D genome architecture in the human brain. *Science* **2023**, *382*, eadf5357. [CrossRef] [PubMed]
31. Shi, J.; Xu, J.; Chen, Y.E.; Li, J.S.; Cui, Y.; Shen, L.; Li, J.J.; Li, W. The concurrence of DNA methylation and demethylation is associated with transcription regulation. *Nat. Commun.* **2021**, *12*, 5285. [CrossRef] [PubMed]
32. Li, Y.; Camarillo, C.; Xu, J.; Arana, T.B.; Xiao, Y.; Zhao, Z.; Chen, H.; Ramirez, M.; Zavala, J.; Escamilla, M.A. Genome-Wide Methylome Analyses Reveal Novel Epigenetic Regulation Patterns in Schizophrenia and Bipolar Disorder. *BioMed Res. Int.* **2015**, *2015*, 201587. [CrossRef] [PubMed]
33. Zhang, H.; Ke, X.; Hu, L.; Wang, J.; Gao, L.; Xie, J. Study on the epigenetic methylation modification of bipolar disorder major genes. *Eur. Rev. Med. Pharmacol. Sci.* **2018**, *22*, 1421–1425. [PubMed]
34. Gaine, M.E.; Seifuddin, F.; Sabunciyan, S.; Lee, R.S.; Benke, K.S.; Monson, E.T.; Zandi, P.P.; Potash, J.B.; Willour, V.L. Differentially methylated regions in bipolar disorder and suicide. *Am. J. Med. Genet. Part B Neuropsychiatr. Genet.* **2019**, *180*, 496–507. [CrossRef] [PubMed]
35. Abdolmaleky, H.M.; Cheng, K.-H.; Faraone, S.V.; Wilcox, M.; Glatt, S.J.; Gao, F.; Smith, C.L.; Shafa, R.; Aeali, B.; Carnevale, J. Hypomethylation of MB-COMT promoter is a major risk factor for schizophrenia and bipolar disorder. *Hum. Mol. Genet.* **2006**, *15*, 3132–3145. [CrossRef] [PubMed]
36. Abdolmaleky, H.M.; Yaqubi, S.; Papageorgis, P.; Lambert, A.W.; Ozturk, S.; Sivaraman, V.; Thiagalingam, S. Epigenetic dysregulation of HTR2A in the brain of patients with schizophrenia and bipolar disorder. *Schizophr. Res.* **2011**, *129*, 183–190. [CrossRef] [PubMed]
37. Kaminsky, Z.; Tochigi, M.; Jia, P.; Pal, M.; Mill, J.; Kwan, A.; Ioshikhes, I.; Vincent, J.; Kennedy, J.; Strauss, J. A multi-tissue analysis identifies HLA complex group 9 gene methylation differences in bipolar disorder. *Mol. Psychiatry* **2012**, *17*, 728–740. [CrossRef] [PubMed]
38. Dell, B.; Palazzo, M.C.; Benatti, B.; Camuri, G.; Galimberti, D.; Fenoglio, C.; Scarpini, E.; Di Francesco, A.; Maccarrone, M.; Altamura, A.C. Epigenetic modulation of BDNF gene: Differences in DNA methylation between unipolar and bipolar patients. *J. Affect. Disord.* **2014**, *166*, 330–333. [CrossRef] [PubMed]
39. Huzayyin, A.A.; Andreazza, A.C.; Turecki, G.; Cruceanu, C.; Rouleau, G.A.; Alda, M.; Young, L.T. Decreased global methylation in patients with bipolar disorder who respond to lithium. *Int. J. Neuropsychopharmacol.* **2014**, *17*, 561–569. [CrossRef] [PubMed]
40. Abdolmaleky, H.M.; Pajouhanfar, S.; Faghankhani, M.; Joghataei, M.T.; Mostafavi, A.; Thiagalingam, S. Antipsychotic drugs attenuate aberrant DNA methylation of DTNBP1 (dysbindin) promoter in saliva and post-mortem brain of patients with schizophrenia and psychotic bipolar disorder. *Am. J. Med. Genet. Part B Neuropsychiatr. Genet.* **2015**, *168*, 687–696. [CrossRef] [PubMed]
41. Kaminsky, Z.; Jones, I.; Verma, R.; Saleh, L.; Trivedi, H.; Guintivano, J.; Akman, R.; Zandi, P.; Lee, R.S.; Potash, J.B. DNA methylation and expression of KCNQ 3 in bipolar disorder. *Bipolar Disord.* **2015**, *17*, 150–159. [CrossRef] [PubMed]

42. Starnawska, A.; Demontis, D.; McQuillin, A.; O'Brien, N.L.; Staunstrup, N.H.; Mors, O.; Nielsen, A.L.; Børglum, A.D.; Nyegaard, M. Hypomethylation of FAM63B in bipolar disorder patients. *Clin. Epigenetics* **2016**, *8*, 52. [CrossRef] [PubMed]
43. Starnawska, A.; Demontis, D.; Pen, A.; Hedemand, A.; Nielsen, A.; Staunstrup, N.; Grove, J.; Als, T.; Jarram, A.; O'brien, N. CACNA1C hypermethylation is associated with bipolar disorder. *Transl. Psychiatry* **2016**, *6*, e831. [CrossRef] [PubMed]
44. D'Addario, C.; Palazzo, M.C.; Benatti, B.; Grancini, B.; Pucci, M.; Di Francesco, A.; Camuri, G.; Galimberti, D.; Fenoglio, C.; Scarpini, E. Regulation of gene transcription in bipolar disorders: Role of DNA methylation in the relationship between prodynorphin and brain derived neurotrophic factor. *Prog. Neuro Psychopharmacol. Biol. Psychiatry* **2018**, *82*, 314–321. [CrossRef] [PubMed]
45. Burghardt, K.J.; Howlett, B.H.; Sanders, E.; Dass, S.E.; Msallaty, Z.; Mallisho, A.; Seyoum, B.; Yi, Z. Skeletal muscle DNA methylation modifications and psychopharmacologic treatment in bipolar disorder. *Eur. Neuropsychopharmacol.* **2019**, *29*, 1365–1373. [CrossRef] [PubMed]
46. Li, S.; Zong, L.; Hou, Y.; Zhang, W.; Zhou, L.; Yang, Q.; Wang, L.; Jiang, W.; Li, Q.; Huang, X. Altered DNA methylation of the Alu y subfamily in schizophrenia and bipolar disorder. *Epigenomics* **2019**, *11*, 581–586. [CrossRef] [PubMed]
47. Nassan, M.; Veldic, M.; Winham, S.; Frye, M.A.; Larrabee, B.; Colby, C.; Biernacka, J.; Bellia, F.; Pucci, M.; Terenius, L. Methylation of Brain Derived Neurotrophic Factor (BDNF) Val66Met CpG site is associated with early onset bipolar disorder. *J. Affect. Disord.* **2020**, *267*, 96–102. [CrossRef] [PubMed]
48. Bundo, M.; Ueda, J.; Nakachi, Y.; Kasai, K.; Kato, T.; Iwamoto, K. Decreased DNA methylation at promoters and gene-specific neuronal hypermethylation in the prefrontal cortex of patients with bipolar disorder. *Mol. Psychiatry* **2021**, *26*, 3407–3418. [CrossRef] [PubMed]
49. Mirza, S.; Lima, C.N.; Del Favero-Campbell, A.; Rubinstein, A.; Topolski, N.; Cabrera-Mendoza, B.; Kovács, E.H.; Blumberg, H.P.; Richards, J.G.; Williams, A.J. Blood epigenome-wide association studies of suicide attempt in adults with bipolar disorder. *Transl. Psychiatry* **2024**, *14*, 70. [CrossRef] [PubMed]
50. Garcia-Ruiz, B.; Jiménez, E.; Aranda, S.; Verdolini, N.; Gutiérrez-Zotes, A.; Sáez, C.; Losantos, E.; Alonso-Lana, S.; Fatjó-Vilas, M.; Sarró, S. Associations of altered leukocyte DDR1 promoter methylation and childhood trauma with bipolar disorder and suicidal behavior in euthymic patients. *Mol. Psychiatry* **2024**, *29*, 2478–2486. [CrossRef] [PubMed]
51. Chakravarty, S.; Bhat, U.A.; Reddy, R.G.; Gupta, P.; Kumar, A. Histone deacetylase inhibitors and psychiatric disorders. In *Epigenetics in Psychiatry*; Elsevier: Amsterdam, The Netherlands, 2021; pp. 657–699.
52. Nestler, E.J.; Peña, C.J.; Kundakovic, M.; Mitchell, A.; Akbarian, S. Epigenetic basis of mental illness. *Neuroscientist* **2016**, *22*, 447–463. [CrossRef] [PubMed]
53. Benes, F.M.; Lim, B.; Matzilevich, D.; Walsh, J.P.; Subburaju, S.; Minns, M. Regulation of the GABA cell phenotype in hippocampus of schizophrenics and bipolars. *Proc. Natl. Acad. Sci. USA* **2007**, *104*, 10164–10169. [CrossRef] [PubMed]
54. Tseng, C.-E.J.; Gilbert, T.M.; Catanese, M.C.; Hightower, B.G.; Peters, A.T.; Parmar, A.J.; Kim, M.; Wang, C.; Roffman, J.L.; Brown, H.E. In vivo human brain expression of histone deacetylases in bipolar disorder. *Transl. Psychiatry* **2020**, *10*, 224. [CrossRef] [PubMed]
55. Arent, C.O.; Valvassori, S.S.; Fries, G.R.; Stertz, L.; Ferreira, C.L.; Lopes-Borges, J.; Mariot, E.; Varela, R.B.; Ornell, F.; Kapczinski, F. Neuroanatomical profile of antimaniac effects of histone deacetylases inhibitors. *Mol. Neurobiol.* **2011**, *43*, 207–214. [CrossRef] [PubMed]
56. Ghafouri-Fard, S.; Badrlou, E.; Taheri, M.; Dürsteler, K.M.; Beatrix Brühl, A.; Sadeghi-Bahmani, D.; Brand, S. A comprehensive review on the role of non-coding RNAs in the pathophysiology of bipolar disorder. *Int. J. Mol. Sci.* **2021**, *22*, 5156. [CrossRef] [PubMed]
57. Leem, K.H.; Kim, S.; Kim, H.W.; Park, H.J. Downregulation of microRNA-330-5p induces manic-like behaviors in REM sleep-deprived rats by enhancing tyrosine hydroxylase expression. *CNS Neurosci. Ther.* **2023**, *29*, 1525–1536. [CrossRef] [PubMed]
58. Walker, R.M.; Rybka, J.; Anderson, S.M.; Torrance, H.S.; Boxall, R.; Sussmann, J.E.; Porteous, D.J.; McIntosh, A.M.; Evans, K.L. Preliminary investigation of miRNA expression in individuals at high familial risk of bipolar disorder. *J. Psychiatr. Res.* **2015**, *62*, 48–55. [CrossRef] [PubMed]
59. Banach, E.; Dmitrzak-Weglarz, M.; Pawlak, J.; Kapelski, P.; Szczepankiewicz, A.; Rajewska-Rager, A.; Slopien, A.; Skibinska, M.; Czerski, P.; Hauser, J. Dysregulation of miR-499, miR-708 and miR-1908 during a depression episode in bipolar disorders. *Neurosci. Lett.* **2017**, *654*, 117–119. [CrossRef] [PubMed]
60. Bame, M.; McInnis, M.G.; O'Shea, K.S. MicroRNA alterations in induced pluripotent stem cell-derived neurons from bipolar disorder patients: Pathways involved in neuronal differentiation, axon guidance, and plasticity. *Stem Cells Dev.* **2020**, *29*, 1145–1159. [CrossRef] [PubMed]
61. Martinez, B.; Peplow, P.V. MicroRNAs as potential diagnostic biomarkers for bipolar disorder. *Neural Regen. Res.* **2025**, *20*, 1681–1695. [CrossRef] [PubMed]
62. Shirvani Farsani, Z.; Zahirodin, A.; Ghaderian, S.M.H.; Shams, J.; Naghavi Gargari, B. The role of long non-coding RNA MALAT1 in patients with bipolar disorder. *Metab. Brain Dis.* **2020**, *35*, 1077–1083. [CrossRef] [PubMed]

63. Hosseini, M.; Mokhtari, M.J. Up-regulation of HOXA-AS2 and MEG3 long non-coding RNAs acts as a potential peripheral biomarker for bipolar disorder. *J. Cell. Mol. Med.* **2024**, *28*, e70150. [CrossRef] [PubMed]
64. Tabano, S.; Caldiroli, A.; Terrasi, A.; Colapietro, P.; Grassi, S.; Carnevali, G.S.; Fontana, L.; Serati, M.; Vaira, V.; Altamura, A.C. A miRNome analysis of drug-free manic psychotic bipolar patients versus healthy controls. *Eur. Arch. Psychiatry Clin. Neurosci.* **2020**, *270*, 893–900. [CrossRef] [PubMed]
65. Camkurt, M.A.; Karababa, İ.F.; Erdal, M.E.; Kandemir, S.B.; Fries, G.R.; Bayazıt, H.; Ay, M.E.; Kandemir, H.; Ay, Ö.I.; Coşkun, S. MicroRNA dysregulation in manic and euthymic patients with bipolar disorder. *J. Affect. Disord.* **2020**, *261*, 84–90. [CrossRef] [PubMed]
66. Ceylan, D.; Tufekci, K.U.; Keskinoglu, P.; Genc, S.; Özerdem, A. Circulating exosomal microRNAs in bipolar disorder. *J. Affect. Disord.* **2020**, *262*, 99–107. [CrossRef] [PubMed]
67. Lee, S.-Y.; Lu, R.-B.; Wang, L.-J.; Chang, C.-H.; Lu, T.; Wang, T.-Y.; Tsai, K.-W. Serum miRNA as a possible biomarker in the diagnosis of bipolar II disorder. *Sci. Rep.* **2020**, *10*, 1131. [CrossRef] [PubMed]
68. Chen, Y.; Shi, J.; Liu, H.; Wang, Q.; Chen, X.; Tang, H.; Yan, R.; Yao, Z.; Lu, Q. Plasma microRNA array analysis identifies overexpressed miR-19b-3p as a biomarker of bipolar depression distinguishing from unipolar depression. *Front. Psychiatry* **2020**, *11*, 757. [CrossRef] [PubMed]
69. Tielke, A.; Martins, H.; Pelzl, M.A.; Maaser-Hecker, A.; David, F.S.; Reinbold, C.S.; Streit, F.; Sirignano, L.; Schwarz, M.; Vedder, H. Genetic and functional analyses implicate microRNA 499A in bipolar disorder development. *Transl. Psychiatry* **2022**, *12*, 437. [CrossRef] [PubMed]
70. Gecys, D.; Dambrauskiene, K.; Simonyte, S.; Patamsyte, V.; Vilkeviciute, A.; Musneckis, A.; Butkute-Sliuoziene, K.; Lesauskaite, V.; Zemaitis, L.; Usaite, D. Circulating hsa-let-7e-5p and hsa-miR-125a-5p as Possible Biomarkers in the Diagnosis of Major Depression and Bipolar Disorders. *Dis. Markers* **2022**, *2022*, 3004338. [CrossRef] [PubMed]
71. Davarinejad, O.; Foruzandeh, Z.; Mohammadi, P.; Golmohammadi, F.; Ghavi, D.; Alivand, M.; Chogan, M.; Zhaleh, H.; Rahmati, Y. Hsa-miR-19b-3p and Hsa-miR-23b-3P are the candidate biomarkers for bipolar disorder. *Inform. Med. Unlocked* **2022**, *30*, 100959. [CrossRef]
72. Toste, C.C.; O'Donovan, M.C.; Bray, N.J. Mapping microRNA expression quantitative trait loci in the prenatal human brain implicates miR-1908-5p expression in bipolar disorder and other brain-related traits. *Hum. Mol. Genet.* **2023**, *32*, 2941–2949. [CrossRef] [PubMed]
73. Schäfer, M.; Brandl, E.J. Mood stabilizers: Valproate. In *NeuroPsychopharmacotherapy*; Springer: Cham, Switzerland, 2020; pp. 1–9.
74. Mari, J.; Dieckmann, L.H.J.; Prates-Baldez, D.; Haddad, M.; da Silva, N.R.; Kapczinski, F. The efficacy of valproate in acute mania, bipolar depression and maintenance therapy for bipolar disorder: An overview of systematic reviews with meta-analyses. *BMJ Open* **2024**, *14*, e087999. [CrossRef] [PubMed]
75. Haghshenas, S.; Putoux, A.; Reilly, J.; Levy, M.A.; Relator, R.; Ghosh, S.; Kerkhof, J.; McConkey, H.; Edery, P.; Lesca, G. Discovery of DNA methylation signature in the peripheral blood of individuals with history of antenatal exposure to valproic acid. *Genet. Med.* **2024**, *26*, 101226. [CrossRef] [PubMed]
76. Guidotti, A.; Dong, E.; Kundakovic, M.; Satta, R.; Grayson, D.R.; Costa, E. Characterization of the action of antipsychotic subtypes on valproate-induced chromatin remodeling. *Trends Pharmacol. Sci.* **2009**, *30*, 55–60. [CrossRef] [PubMed]
77. Dong, E.; Chen, Y.; Gavin, D.P.; Grayson, D.R.; Guidotti, A. Valproate induces DNA demethylation in nuclear extracts from adult mouse brain. *Epigenetics* **2010**, *5*, 730–735. [CrossRef] [PubMed]
78. Aizawa, S.; Yamamuro, Y. Valproate administration to mice increases hippocampal p21 expression by altering genomic DNA methylation. *Neuroreport* **2015**, *26*, 915–920. [CrossRef] [PubMed]
79. Backlund, L.; Wei, Y.B.; Martinsson, L.; Melas, P.A.; Liu, J.J.; Mu, N.; Östenson, C.-G.; Ekström, T.J.; Schalling, M.; Lavebratt, C. Mood stabilizers and the influence on global leukocyte DNA methylation in bipolar disorder. *Mol. Neuropsychiatry* **2015**, *1*, 76–81. [CrossRef] [PubMed]
80. Chen, H.; Dzitoyeva, S.; Manev, H. Effect of valproic acid on mitochondrial epigenetics. *Eur. J. Pharmacol.* **2012**, *690*, 51–59. [CrossRef] [PubMed]
81. D'addario, C.; Dell'Osso, B.; Palazzo, M.C.; Benatti, B.; Lietti, L.; Cattaneo, E.; Galimberti, D.; Fenoglio, C.; Cortini, F.; Scarpini, E. Selective DNA methylation of BDNF promoter in bipolar disorder: Differences among patients with BDI and BDII. *Neuropsychopharmacology* **2012**, *37*, 1647–1655. [CrossRef] [PubMed]
82. Phiel, C.J.; Zhang, F.; Huang, E.Y.; Guenther, M.G.; Lazar, M.A.; Klein, P.S. Histone deacetylase is a direct target of valproic acid, a potent anticonvulsant, mood stabilizer, and teratogen. *J. Biol. Chem.* **2001**, *276*, 36734–36741. [CrossRef] [PubMed]
83. Chiu, C.-T.; Wang, Z.; Hunsberger, J.G.; Chuang, D.-M. Therapeutic potential of mood stabilizers lithium and valproic acid: Beyond bipolar disorder. *Pharmacol. Rev.* **2013**, *65*, 105–142. [CrossRef] [PubMed]
84. Sinha, P.; Cree, S.L.; Miller, A.L.; Pearson, J.F.; Kennedy, M.A. Transcriptional analysis of sodium valproate in a serotonergic cell line reveals gene regulation through both HDAC inhibition-dependent and independent mechanisms. *Pharmacogenomics J.* **2021**, *21*, 359–375. [CrossRef] [PubMed]

85. Lopes-Borges, J.; Valvassori, S.S.; Varela, R.B.; Tonin, P.T.; Vieira, J.S.; Goncalves, C.L.; Streck, E.L.; Quevedo, J. Histone deacetylase inhibitors reverse manic-like behaviors and protect the rat brain from energetic metabolic alterations induced by ouabain. *Pharmacol. Biochem. Behav.* **2015**, *128*, 89–95. [CrossRef] [PubMed]
86. Kao, C.Y.; Hsu, Y.C.; Liu, J.W.; Lee, D.C.; Chung, Y.F.; Chiu, I.M. The mood stabilizer valproate activates human FGF 1 gene promoter through inhibiting HDAC and GSK-3 activities. *J. Neurochem.* **2013**, *126*, 4–18. [CrossRef] [PubMed]
87. Lee, R.; Pirooznia, M.; Guintivano, J.; Ly, M.; Ewald, E.; Tamashiro, K.; Gould, T.; Moran, T.; Potash, J. Search for common targets of lithium and valproic acid identifies novel epigenetic effects of lithium on the rat leptin receptor gene. *Transl. Psychiatry* **2015**, *5*, e600. [CrossRef] [PubMed]
88. Sharma, R.P.; Rosen, C.; Kartan, S.; Guidotti, A.; Costa, E.; Grayson, D.R.; Chase, K. Valproic acid and chromatin remodeling in schizophrenia and bipolar disorder: Preliminary results from a clinical population. *Schizophr. Res.* **2006**, *88*, 227–231. [CrossRef] [PubMed]
89. Stertz, L.; Fries, G.R.; Aguiar, B.W.D.; Pfaffenseller, B.; Valvassori, S.S.; Gubert, C.; Ferreira, C.L.; Moretti, M.; Ceresér, K.M.; Kauer-Sant'Anna, M. Histone deacetylase activity and brain-derived neurotrophic factor (BDNF) levels in a pharmacological model of mania. *Braz. J. Psychiatry* **2013**, *36*, 39–46. [CrossRef] [PubMed]
90. Valvassori, S.S.; Resende, W.R.; Varela, R.B.; Arent, C.O.; Gava, F.F.; Peterle, B.R.; Dal-Pont, G.C.; Carvalho, A.F.; Andersen, M.L.; Quevedo, J. The effects of histone deacetylase inhibition on the levels of cerebral cytokines in an animal model of mania induced by dextroamphetamine. *Mol. Neurobiol.* **2018**, *55*, 1430–1439. [CrossRef] [PubMed]
91. Varela, R.B.; Resende, W.R.; Dal-Pont, G.C.; Gava, F.F.; Tye, S.J.; Quevedo, J.; Valvassori, S.S. HDAC inhibitors reverse mania-like behavior and modulate epigenetic regulatory enzymes in an animal model of mania induced by Ouabain. *Pharmacol. Biochem. Behav.* **2020**, *193*, 172917. [CrossRef] [PubMed]
92. Logan, R.W.; Ozburn, A.R.; Arey, R.N.; Ketchesin, K.D.; Winquist, A.; Crain, A.; Tobe, B.T.; Becker-Krail, D.; Jarpe, M.B.; Xue, X. Valproate reverses mania-like behaviors in mice via preferential targeting of HDAC2. *Mol. Psychiatry* **2021**, *26*, 4066–4084. [CrossRef] [PubMed]
93. Aluru, N.; Deak, K.L.; Jenny, M.J.; Hahn, M.E. Developmental exposure to valproic acid alters the expression of microRNAs involved in neurodevelopment in zebrafish. *Neurotoxicol. Teratol.* **2013**, *40*, 46–58. [CrossRef] [PubMed]
94. Wang, X.; Luo, Y.; Liu, S.; Tan, L.; Wang, S.; Man, R. MicroRNA-134 plasma levels before and after treatment with valproic acid for epilepsy patients. *Oncotarget* **2017**, *8*, 72748. [CrossRef] [PubMed]
95. Kim, Y.; Zhang, Y.; Pang, K.; Kang, H.; Park, H.; Lee, Y.; Lee, B.; Lee, H.-J.; Kim, W.-K.; Geum, D. Bipolar disorder associated microRNA, miR-1908-5p, regulates the expression of genes functioning in neuronal glutamatergic synapses. *Exp. Neurobiol.* **2016**, *25*, 296. [CrossRef] [PubMed]
96. Truong, T.T.; Bortolasci, C.C.; Spolding, B.; Panizzutti, B.; Liu, Z.S.; Kidnapillai, S.; Richardson, M.; Gray, L.; Smith, C.M.; Dean, O.M. Co-expression networks unveiled long non-coding RNAs as molecular targets of drugs used to treat bipolar disorder. *Front. Pharmacol.* **2022**, *13*, 873271. [CrossRef] [PubMed]
97. Ochoa, E.L. Lithium as a neuroprotective agent for bipolar disorder: An overview. *Cell. Mol. Neurobiol.* **2022**, *42*, 85–97. [CrossRef] [PubMed]
98. Papiol, S.; Schulze, T.G.; Heilbronner, U. Lithium response in bipolar disorder: Genetics, genomics, and beyond. *Neurosci. Lett.* **2022**, *785*, 136786. [CrossRef] [PubMed]
99. Marie-Claire, C.; Lejeune, F.-X.; Mundwiller, E.; Ulveling, D.; Moszer, I.; Bellivier, F.; Etain, B. A DNA methylation signature discriminates between excellent and non-response to lithium in patients with bipolar disorder type 1. *Sci. Rep.* **2020**, *10*, 12239. [CrossRef] [PubMed]
100. Zafrilla-López, M.; Acosta-Díez, M.; Mitjans, M.; Giménez-Palomo, A.; Saiz, P.A.; Barrot-Feixat, C.; Jiménez, E.; Papiol, S.; Ruiz, V.; Gavín, P. Lithium response in bipolar disorder: Epigenome-wide DNA methylation signatures and epigenetic aging. *Eur. Neuropsychopharmacol.* **2024**, *85*, 23–31. [CrossRef] [PubMed]
101. Asai, T.; Bundo, M.; Sugawara, H.; Sunaga, F.; Ueda, J.; Tanaka, G.; Ishigooka, J.; Kasai, K.; Kato, T.; Iwamoto, K. Effect of mood stabilizers on DNA methylation in human neuroblastoma cells. *Int. J. Neuropsychopharmacol.* **2013**, *16*, 2285–2294. [CrossRef] [PubMed]
102. Andrabi, M.; Andrabi, M.M.; Kunjunni, R.; Sriwastva, M.K.; Bose, S.; Sagar, R.; Srivastava, A.K.; Mathur, R.; Jain, S.; Subbiah, V. Lithium acts to modulate abnormalities at behavioral, cellular, and molecular levels in sleep deprivation-induced mania-like behavior. *Bipolar Disord.* **2020**, *22*, 266–280. [CrossRef] [PubMed]
103. Wang, Z.; Zhang, C.; Huang, J.; Yuan, C.; Hong, W.; Chen, J.; Yu, S.; Xu, L.; Gao, K.; Fang, Y. MiRNA-206 and BDNF genes interacted in bipolar I disorder. *J. Affect. Disord.* **2014**, *162*, 116–119. [CrossRef] [PubMed]
104. Pisanu, C.; Merkouri Papadima, E.; Melis, C.; Congiu, D.; Loizedda, A.; Orrù, N.; Calza, S.; Orrù, S.; Carcassi, C.; Severino, G. Whole genome expression analyses of miRNAs and mRNAs suggest the involvement of miR-320a and miR-155-3p and their targeted genes in lithium response in bipolar disorder. *Int. J. Mol. Sci.* **2019**, *20*, 6040. [CrossRef] [PubMed]

105. Reinbold, C.S.; Forstner, A.J.; Hecker, J.; Fullerton, J.M.; Hoffmann, P.; Hou, L.; Heilbronner, U.; Degenhardt, F.; Adli, M.; Akiyama, K. Analysis of the influence of microRNAs in lithium response in bipolar disorder. *Front. Psychiatry* **2018**, *9*, 207. [CrossRef] [PubMed]
106. Cattane, N.; Courtin, C.; Mombelli, E.; Maj, C.; Mora, C.; Etain, B.; Bellivier, F.; Marie-Claire, C.; Cattaneo, A. Transcriptomics and miRNomics data integration in lymphoblastoid cells highlights the key role of immune-related functions in lithium treatment response in Bipolar disorder. *BMC Psychiatry* **2022**, *22*, 665. [CrossRef] [PubMed]
107. Costa, B.; Vale, N. Understanding lamotrigine's role in the CNS and possible future evolution. *Int. J. Mol. Sci.* **2023**, *24*, 6050. [CrossRef] [PubMed]
108. Żełabowski, K.; Wojtysiak, K.; Ratka, Z.; Biedka, K.; Chłopaś-Konowałek, A. Lamotrigine Therapy: Relation Between Treatment of Bipolar Affective Disorder and Incidence of Stevens–Johnson Syndrome—A Narrative Review of the Existing Literature. *J. Clin. Med.* **2025**, *14*, 4103. [CrossRef] [PubMed]
109. Kidnapillai, S.; Wade, B.; Bortolasci, C.C.; Panizzutti, B.; Spolding, B.; Connor, T.; Crowley, T.; Jamain, S.; Gray, L.; Leboyer, M. Drugs used to treat bipolar disorder act via microRNAs to regulate expression of genes involved in neurite outgrowth. *J. Psychopharmacol.* **2020**, *34*, 370–379. [CrossRef] [PubMed]
110. Leng, Y.; Fessler, E.B.; Chuang, D.-M. Neuroprotective effects of the mood stabilizer lamotrigine against glutamate excitotoxicity: Roles of chromatin remodelling and Bcl-2 induction. *Int. J. Neuropsychopharmacol.* **2013**, *16*, 607–620. [CrossRef] [PubMed]
111. Ignácio, Z.M.; Réus, G.Z.; Abelaira, H.M.; Maciel, A.L.; de Moura, A.B.; Matos, D.; Demo, J.P.; da Silva, J.B.; Gava, F.F.; Valvassori, S.S. Quetiapine treatment reverses depressive-like behavior and reduces DNA methyltransferase activity induced by maternal deprivation. *Behav. Brain Res.* **2017**, *320*, 225–232. [CrossRef] [PubMed]
112. Chen, X.; Liu, H.; Gan, J.; Wang, X.; Yu, G.; Li, T.; Liang, X.; Yu, B.; Xiao, L. Quetiapine modulates histone methylation status in oligodendroglia and rescues adolescent behavioral alterations of socially isolated mice. *Front. Psychiatry* **2020**, *10*, 984. [CrossRef] [PubMed]
113. Sugawara, H.; Bundo, M.; Asai, T.; Sunaga, F.; Ueda, J.; Ishigooka, J.; Kasai, K.; Kato, T.; Iwamoto, K. Effects of quetiapine on DNA methylation in neuroblastoma cells. *Prog. Neuro Psychopharmacol. Biol. Psychiatry* **2015**, *56*, 117–121. [CrossRef] [PubMed]
114. Houtepen, L.C.; Van Bergen, A.H.; Vinkers, C.H.; Boks, M.P. DNA methylation signatures of mood stabilizers and antipsychotics in bipolar disorder. *Epigenomics* **2016**, *8*, 197–208. [CrossRef] [PubMed]
115. Rong, H.; Liu, T.B.; Yang, K.J.; Yang, H.C.; Wu, D.H.; Liao, C.P.; Hong, F.; Yang, H.Z.; Wan, F.; Ye, X.Y. MicroRNA-134 plasma levels before and after treatment for bipolar mania. *J. Psychiatr. Res.* **2011**, *45*, 92–95. [CrossRef] [PubMed]
116. Lim, C.H.; Zainal, N.Z.; Kanagasundram, S.; Zain, S.M.; Mohamed, Z. Preliminary examination of microRNA expression profiling in bipolar disorder I patients during antipsychotic treatment. *Am. J. Med. Genet. Part B Neuropsychiatr. Genet.* **2016**, *171*, 867–874. [CrossRef] [PubMed]
117. Chen, Y.H.; Zhou, C.H.; Yu, H.; Wu, W.J.; Wang, Y.W.; Liu, L.; Hu, G.T.; Li, B.J.; Peng, Z.W.; Wang, H.N. Gut microbial signatures and differences in bipolar disorder and schizophrenia of emerging adulthood. *CNS Neurosci. Ther.* **2023**, *29*, 5–17. [CrossRef] [PubMed]
118. Zhang, P.; Kong, L.; Huang, H.; Pan, Y.; Zhang, D.; Jiang, J.; Shen, Y.; Xi, C.; Lai, J.; Ng, C.H. Gut microbiota–A potential contributor in the pathogenesis of bipolar disorder. *Front. Neurosci.* **2022**, *16*, 830748. [CrossRef] [PubMed]
119. Gondalia, S.; Parkinson, L.; Stough, C.; Scholey, A. Gut microbiota and bipolar disorder: A review of mechanisms and potential targets for adjunctive therapy. *Psychopharmacology* **2019**, *236*, 1433–1443. [CrossRef] [PubMed]
120. Evans, S.J.; Bassis, C.M.; Hein, R.; Assari, S.; Flowers, S.A.; Kelly, M.B.; Young, V.B.; Ellingrod, V.E.; McInnis, M.G. The gut microbiome composition associates with bipolar disorder and illness severity. *J. Psychiatr. Res.* **2017**, *87*, 23–29. [CrossRef] [PubMed]
121. Bengesser, S.; Mörkl, S.; Painold, A.; Dalkner, N.; Birner, A.; Fellendorf, F.; Platzer, M.; Queissner, R.; Hamm, C.; Maget, A. Epigenetics of the molecular clock and bacterial diversity in bipolar disorder. *Psychoneuroendocrinology* **2019**, *101*, 160–166. [CrossRef] [PubMed]
122. Tong, C.-Q.; Li, M.-J.; Liu, Y.; Zhou, Q.; Sun, W.-Q.; Chen, J.-Y.; Wang, D.; Li, F.; Chen, Z.-J.; Song, Y.-H. Regulation of hippocampal miRNA expression by intestinal flora in anxiety-like mice. *Eur. J. Pharmacol.* **2024**, *984*, 177016. [CrossRef] [PubMed]
123. Behrouzi, A.; Ashrafian, F.; Mazaheri, H.; Lari, A.; Nouri, M.; Rad, F.R.; Tavassol, Z.H.; Siadat, S.D. The importance of interaction between MicroRNAs and gut microbiota in several pathways. *Microb. Pathog.* **2020**, *144*, 104200. [CrossRef] [PubMed]
124. Cui, M.; Xiao, H.; Li, Y.; Dong, J.; Luo, D.; Li, H.; Feng, G.; Wang, H.; Fan, S. Total abdominal irradiation exposure impairs cognitive function involving miR-34a-5p/BDNF axis. *Biochim. Biophys. Acta BBA-Mol. Basis Dis.* **2017**, *1863*, 2333–2341. [CrossRef] [PubMed]
125. Huang, T.; Shang, Y.; Dai, C.; Zhang, Q.; Hu, S.; Xie, J. Gut microbiota and its relation to inflammation in patients with bipolar depression: A cross-sectional study. *Ann. Gen. Psychiatry* **2023**, *22*, 21. [CrossRef] [PubMed]

126. Painold, A.; Mörkl, S.; Kashofer, K.; Halwachs, B.; Dalkner, N.; Bengesser, S.; Birner, A.; Fellendorf, F.; Platzer, M.; Queissner, R. A step ahead: Exploring the gut microbiota in inpatients with bipolar disorder during a depressive episode. *Bipolar Disord.* **2019**, *21*, 40–49. [CrossRef] [PubMed]
127. Coello, K.; Hansen, T.H.; Sørensen, N.; Munkholm, K.; Kessing, L.V.; Pedersen, O.; Vinberg, M. Gut microbiota composition in patients with newly diagnosed bipolar disorder and their unaffected first-degree relatives. *Brain Behav. Immun.* **2019**, *75*, 112–118. [CrossRef] [PubMed]
128. Guo, Z.; Xiao, S.; Chen, G.; Zhong, S.; Zhong, H.; Sun, S.; Chen, P.; Tang, X.; Yang, H.; Jia, Y. Disruption of the gut microbiota-inflammation-brain axis in unmedicated bipolar disorder II depression. *Transl. Psychiatry* **2024**, *14*, 495. [CrossRef] [PubMed]
129. Mazziotta, C.; Tognon, M.; Martini, F.; Torreggiani, E.; Rotondo, J.C. Probiotics mechanism of action on immune cells and beneficial effects on human health. *Cells* **2023**, *12*, 184. [CrossRef] [PubMed]
130. Nohesara, S.; Mostafavi Abdolmaleky, H.; Pirani, A.; Pettinato, G.; Thiagalingam, S. The Obesity–Epigenetics–Microbiome Axis: Strategies for Therapeutic Intervention. *Nutrients* **2025**, *17*, 1564. [CrossRef] [PubMed]
131. Ķimse, L.; Reinis, A.; Miķelsone-Jansone, L.; Gintere, S.; Krūmiņa, A. A narrative review of psychobiotics: Probiotics that influence the gut–brain axis. *Medicina* **2024**, *60*, 601. [CrossRef] [PubMed]
132. Cheon, M.-J.; Lim, S.-M.; Lee, N.-K.; Paik, H.-D. Probiotic properties and neuroprotective effects of Lactobacillus buchneri KU200793 isolated from Korean fermented foods. *Int. J. Mol. Sci.* **2020**, *21*, 1227. [CrossRef] [PubMed]
133. Bermúdez-Humarán, L.G.; Salinas, E.; Ortiz, G.G.; Ramirez-Jirano, L.J.; Morales, J.A.; Bitzer-Quintero, O.K. From probiotics to psychobiotics: Live beneficial bacteria which act on the brain-gut axis. *Nutrients* **2019**, *11*, 890. [CrossRef] [PubMed]
134. Cheng, Y.; Liu, J.; Ling, Z. Short-chain fatty acids-producing probiotics: A novel source of psychobiotics. *Crit. Rev. Food Sci. Nutr.* **2022**, *62*, 7929–7959. [CrossRef] [PubMed]
135. Obi-Azuike, C.; Ebiai, R.; Gibson, T.; Hernandez, A.; Khan, A.; Anugwom, G.; Urhi, A.; Prasad, S.; Souabni, S.A.; Oladunjoye, F. A systematic review on gut-brain axis aberrations in bipolar disorder and methods of balancing the gut microbiota. *Brain Behav.* **2023**, *13*, e3037. [CrossRef] [PubMed]
136. Dickerson, F.; Adamos, M.; Katsafanas, E.; Khushalani, S.; Origoni, A.; Savage, C.; Schweinfurth, L.; Stallings, C.; Sweeney, K.; Goga, J. Adjunctive probiotic microorganisms to prevent rehospitalization in patients with acute mania: A randomized controlled trial. *Bipolar Disord.* **2018**, *20*, 614–621. [CrossRef] [PubMed]
137. Zhang, J.-Z.; Zhu, F.-Y.; Feng, E.-X.; Wu, F.; Fang, X.; Tu, H.-J.; Wei, J.; Wan, A.-L.; Chen, T.-T. Adjunct therapy with probiotics for depressive episodes of bipolar disorder type I: A randomized placebo-controlled trial. *J. Funct. Foods* **2023**, *105*, 105553. [CrossRef]
138. Reininghaus, E.Z.; Wetzlmair, L.-C.; Fellendorf, F.T.; Platzer, M.; Queissner, R.; Birner, A.; Pilz, R.; Hamm, C.; Maget, A.; Rieger, A. Probiotic treatment in individuals with euthymic bipolar disorder: A pilot-study on clinical changes and compliance. *Neuropsychobiology* **2020**, *79*, 71–79. [CrossRef] [PubMed]
139. Zeng, C.; Qiu, Y.; Li, S.; Teng, Z.; Xiang, H.; Chen, J.; Wu, X.; Cao, T.; Zhang, S.; Chen, Q. Effect of probiotic supplements on oxidative stress biomarkers in first-episode bipolar disorder patients: A randomized, placebo-controlled trial. *Front. Pharmacol.* **2022**, *13*, 829815. [CrossRef] [PubMed]
140. Shahrbabaki, M.E.; Sabouri, S.; Sabahi, A.; Barfeh, D.; Divsalar, P.; Esmailzadeh, M.; Ahmadi, A. The efficacy of probiotics for treatment of bipolar disorder-type 1: A randomized, double-blind, placebo controlled trial. *Iran. J. Psychiatry* **2020**, *15*, 10. [CrossRef]
141. Reininghaus, E.Z.; Wetzlmair, L.-C.; Fellendorf, F.T.; Platzer, M.; Queissner, R.; Birner, A.; Pilz, R.; Hamm, C.; Maget, A.; Koidl, C. The impact of probiotic supplements on cognitive parameters in euthymic individuals with bipolar disorder: A pilot study. *Neuropsychobiology* **2020**, *79*, 63–70. [CrossRef] [PubMed]
142. Borkent, J.; Ioannou, M.; Neijzen, D.; Haarman, B.C.; Sommer, I.E. Probiotic Formulation for Patients With Bipolar or Schizophrenia Spectrum Disorder: A Double-Blind, Randomized Placebo-Controlled Trial. *Schizophr. Bull.* **2024**, sbae188. [CrossRef] [PubMed]
143. Watt, R.; Parkin, K.; Martino, D. The potential effects of short-chain fatty acids on the epigenetic regulation of innate immune memory. *Challenges* **2020**, *11*, 25. [CrossRef]
144. Du, Y.; He, C.; An, Y.; Huang, Y.; Zhang, H.; Fu, W.; Wang, M.; Shan, Z.; Xie, J.; Yang, Y. The role of short chain fatty acids in inflammation and body health. *Int. J. Mol. Sci.* **2024**, *25*, 7379. [CrossRef] [PubMed]
145. Xie, A.; Ensink, E.; Li, P.; Gordevičius, J.; Marshall, L.L.; George, S.; Pospisilik, J.A.; Aho, V.T.; Houser, M.C.; Pereira, P.A. Bacterial butyrate in Parkinson's disease is linked to epigenetic changes and depressive symptoms. *Mov. Disord.* **2022**, *37*, 1644–1653. [CrossRef] [PubMed]
146. Wei, Y.B.; Melas, P.A.; Wegener, G.; Mathé, A.A.; Lavebratt, C. Antidepressant-like effect of sodium butyrate is associated with an increase in TET1 and in 5-hydroxymethylation levels in the Bdnf gene. *Int. J. Neuropsychopharmacol.* **2015**, *18*, pyu032. [CrossRef] [PubMed]
147. Resende, W.R.; Valvassori, S.S.; Reus, G.Z.; Varela, R.B.; Arent, C.O.; Ribeiro, K.F.; Bavaresco, D.V.; Andersen, M.L.; Zugno, A.I.; Quevedo, J. Effects of sodium butyrate in animal models of mania and depression: Implications as a new mood stabilizer. *Behav. Pharmacol.* **2013**, *24*, 569–579. [CrossRef] [PubMed]

148. Steckert, A.V.; Valvassori, S.S.; Varela, R.B.; Mina, F.; Resende, W.R.; Bavaresco, D.V.; Ornell, F.; Dal-Pizzol, F.; Quevedo, J. Effects of sodium butyrate on oxidative stress and behavioral changes induced by administration of D-AMPH. *Neurochem. Int.* **2013**, *62*, 425–432. [CrossRef] [PubMed]
149. Moretti, M.; Valvassori, S.S.; Varela, R.B.; Ferreira, C.L.; Rochi, N.; Benedet, J.; Scaini, G.; Kapczinski, F.; Streck, E.L.; Zugno, A.I. Behavioral and neurochemical effects of sodium butyrate in an animal model of mania. *Behav. Pharmacol.* **2011**, *22*, 766–772. [CrossRef] [PubMed]
150. Valvassori, S.S.; Dal-Pont, G.C.; Steckert, A.V.; Varela, R.B.; Lopes-Borges, J.; Mariot, E.; Resende, W.R.; Arent, C.O.; Carvalho, A.F.; Quevedo, J. Sodium butyrate has an antimanic effect and protects the brain against oxidative stress in an animal model of mania induced by ouabain. *Psychiatry Res.* **2016**, *235*, 154–159. [CrossRef] [PubMed]
151. Varela, R.B.; Valvassori, S.S.; Lopes-Borges, J.; Mariot, E.; Dal-Pont, G.C.; Amboni, R.T.; Bianchini, G.; Quevedo, J. Sodium butyrate and mood stabilizers block ouabain-induced hyperlocomotion and increase BDNF, NGF and GDNF levels in brain of Wistar rats. *J. Psychiatr. Res.* **2015**, *61*, 114–121. [CrossRef] [PubMed]
152. Riedinger, M.; Mesbah, R.; Koenders, M.; Henderickx, J.; Smits, W.; El Filali, E.; Geleijnse, J.; Van der Wee, N.; de Leeuw, M.; Giltay, E. A healthy dietary pattern is associated with microbiota diversity in recently diagnosed bipolar patients: The Bipolar Netherlands Cohort (BINCO) study. *J. Affect. Disord.* **2024**, *355*, 157–166. [CrossRef] [PubMed]
153. Brenda, J.Y.; Oz, R.S.; Sethi, S. Ketogenic diet as a metabolic therapy for bipolar disorder: Clinical developments. *J. Affect. Disord. Rep.* **2023**, *11*, 100457.
154. Campbell, I.; Campbell, H. Mechanisms of insulin resistance, mitochondrial dysfunction and the action of the ketogenic diet in bipolar disorder. Focus on the PI3K/AKT/HIF1-a pathway. *Med. Hypotheses* **2020**, *145*, 110299. [CrossRef] [PubMed]
155. Bough, K.J.; Rho, J.M. Anticonvulsant mechanisms of the ketogenic diet. *Epilepsia* **2007**, *48*, 43–58. [CrossRef] [PubMed]
156. Freyberg, Z.; Andreazza, A.C.; McClung, C.A.; Phillips, M.L. Linking mitochondrial dysfunction, neurotransmitter, neural network abnormalities and mania: Elucidating neurobiological mechanisms of the therapeutic effect of the ketogenic diet in Bipolar Disorder. *Biol. Psychiatry Cogn. Neurosci. Neuroimaging* **2025**, *10*, 267–277. [CrossRef] [PubMed]
157. Polito, R.; La Torre, M.E.; Moscatelli, F.; Cibelli, G.; Valenzano, A.; Panaro, M.A.; Monda, M.; Messina, A.; Monda, V.; Pisanelli, D. The ketogenic diet and neuroinflammation: The action of beta-hydroxybutyrate in a microglial cell line. *Int. J. Mol. Sci.* **2023**, *24*, 3102. [CrossRef] [PubMed]
158. Sethi, S.; Wakeham, D.; Ketter, T.; Hooshmand, F.; Bjornstad, J.; Richards, B.; Westman, E.; Krauss, R.M.; Saslow, L. Ketogenic diet intervention on metabolic and psychiatric health in bipolar and schizophrenia: A pilot trial. *Psychiatry Res.* **2024**, *335*, 115866. [CrossRef] [PubMed]
159. Nohesara, S.; Abdolmaleky, H.M.; Zhou, J.-R.; Thiagalingam, S. Microbiota-induced epigenetic alterations in depressive disorders are targets for nutritional and probiotic therapies. *Genes* **2023**, *14*, 2217. [CrossRef] [PubMed]
160. Zikou, E.; Koliaki, C.; Makrilakis, K. The role of fecal microbiota transplantation (FMT) in the management of metabolic diseases in humans: A narrative review. *Biomedicines* **2024**, *12*, 1871. [CrossRef] [PubMed]
161. Quaranta, G.; Guarnaccia, A.; Fancello, G.; Agrillo, C.; Iannarelli, F.; Sanguinetti, M.; Masucci, L. Fecal microbiota transplantation and other gut microbiota manipulation strategies. *Microorganisms* **2022**, *10*, 2424. [CrossRef] [PubMed]
162. Anand, N.; Gorantla, V.R.; Chidambaram, S.B. The role of gut dysbiosis in the pathophysiology of neuropsychiatric disorders. *Cells* **2022**, *12*, 54. [CrossRef] [PubMed]
163. Biazzo, M.; Deidda, G. Fecal microbiota transplantation as new therapeutic avenue for human diseases. *J. Clin. Med.* **2022**, *11*, 4119. [CrossRef] [PubMed]
164. Bukowski-Thall, G.; Fellendorf, F.T.; Gorkiewicz, S.; Ip, K.C.K.; Schmidt, L.; Durdevic, M.; Habisch, H.; Mörkl, S.; Wagner-Skacel, J.; Bengesser, S.A. Fecal microbiota transplantation from individual with bipolar disorder and healthy control elicits distinct behaviors and metabolite profiles in mice. *bioRxiv* **2023**. bioRxiv:2023.2011. 2016.566698.
165. Hinton, R. A case report looking at the effects of faecal microbiota transplantation in a patient with bipolar disorder. *Aust. N. Z. J. Psychiatry* **2020**, *54*, 649–650. [CrossRef] [PubMed]
166. Flowers, S.A.; Evans, S.J.; Ward, K.M.; McInnis, M.G.; Ellingrod, V.L. Interaction between atypical antipsychotics and the gut microbiome in a bipolar disease cohort. *Pharmacother. J. Hum. Pharmacol. Drug Ther.* **2017**, *37*, 261–267. [CrossRef] [PubMed]
167. Bui, T.A.; O'Croinin, B.R.; Dennett, L.; Winship, I.R.; Greenshaw, A. Pharmaco-psychiatry and gut microbiome: A systematic review of effects of psychotropic drugs for bipolar disorder. *Microbiology* **2025**, *171*, 001568. [CrossRef] [PubMed]
168. Kamath, S.; Hunter, A.; Collins, K.; Wignall, A.; Joyce, P. The atypical antipsychotics lurasidone and olanzapine exert contrasting effects on the gut microbiome and metabolic function of rats. *Br. J. Pharmacol.* **2024**, *181*, 4531–4545. [CrossRef] [PubMed]
169. Lu, Q.; Lai, J.; Lu, H.; Ng, C.; Huang, T.; Zhang, H.; Ding, K.; Wang, Z.; Jiang, J.; Hu, J. Gut microbiota in bipolar depression and its relationship to brain function: An advanced exploration. *Front. Psychiatry* **2019**, *10*, 784. [CrossRef] [PubMed]
170. Collins, K.; Kamath, S.; Meola, T.R.; Wignall, A.; Joyce, P. The Oral Bioavailability of Lurasidone is Impacted by Changes to the Gut Microbiome: Implications for Antipsychotic Therapy. *bioRxiv* **2024**. bioRxiv:2024.2007. 2017.604016.

171. Pochiero, I.; Calisti, F.; Comandini, A.; Del Vecchio, A.; Costamagna, I.; Rosignoli, M.T.; Cattaneo, A.; Nunna, S.; Peduto, I.; Heiman, F. Impact of lurasidone and other antipsychotics on body weight: Real-world, retrospective, comparative study of 15,323 adults with schizophrenia. *Int. J. Gen. Med.* **2021**, *14*, 4081–4094. [CrossRef] [PubMed]
172. Bahr, S.; Tyler, B.; Wooldridge, N.; Butcher, B.; Burns, T.; Teesch, L.; Oltman, C.; Azcarate-Peril, M.; Kirby, J.; Calarge, C. Use of the second-generation antipsychotic, risperidone, and secondary weight gain are associated with an altered gut microbiota in children. *Transl. Psychiatry* **2015**, *5*, e652. [CrossRef] [PubMed]

cells

MDPI

Review

Autophagy and Alzheimer's Disease: Mechanisms and Impact Beyond the Brain

Zaw Myo Hein [1], Thirupathirao Vishnumukkala [2], Barani Karikalan [3], Aisyah Alkatiri [2], Farida Hussan [2], Saravanan Jagadeesan [4], Mohd Amir Kamaruzzaman [5], Muhammad Danial Che Ramli [6], Che Mohd Nasril Che Mohd Nassir [7,*] and Prarthana Kalerammana Gopalakrishna [8,*]

1 Department of Basic Medical Sciences, College of Medicine, Ajman University, Ajman P.O. Box 346, United Arab Emirates; z.hein@ajman.ac.ae
2 Anatomy Discipline, Human Biology Department, School of Medicine, IMU University, Bukit Jalil, Kuala Lumpur 57000, Malaysia; thirupathirao@imu.edu.my (T.V.); 00000038358@student.imu.edu.my (A.A.); faridahussan@imu.edu.my (F.H.)
3 Department of Pathology, Faculty of Medicine, MAHSA University, Bandar Saujana Putra, Petaling Jaya 42610, Selangor, Malaysia; barani@mahsa.edu.my
4 Department of Anatomy, School of Medicine, Lakeside Campus, Taylor's University, Subang Jaya 47500, Selangor, Malaysia; saravanan.jagadeesan@taylors.edu.my
5 Department of Anatomy, Faculty of Medicine, University Kebangsaan Malaysia, Jalan Yaacob Latif, Bandar Tun Razak, Cheras, Kuala Lumpur 56000, Malaysia; mohdamir@ukm.edu.my
6 Faculty of Health and Life Sciences, Management and Science University, Seksyen 13, Shah Alam 40100, Selangor, Malaysia; muhddanial_cheramli@msu.edu.my
7 Department of Anatomy and Physiology, Faculty of Medicine, University Sultan Zainal Abidin, Kuala Terengganu 20400, Terengganu, Malaysia
8 Physiology Discipline, Human Biology Department, School of Medicine, IMU University, Bukit Jalil, Kuala Lumpur 57000, Malaysia
* Correspondence: nasrilnassir@unisza.edu.my (C.M.N.C.M.N.); prarthana@imu.edu.my (P.K.G.)

Abstract: Alzheimer's disease (AD) is a progressive neurodegenerative disorder marked by neuronal loss, cognitive decline, and pathological hallmarks such as amyloid-beta (Aβ) plaques and tau neurofibrillary tangles. Recent evidence highlights autophagy as a pivotal mechanism in cellular homeostasis, mediating the clearance of misfolded proteins and damaged organelles. However, impaired autophagy contributes significantly to AD pathogenesis by disrupting proteostasis, exacerbating neuroinflammation, and promoting synaptic dysfunction. This review aims to scrutinize the intricate relationship between autophagy dysfunction and AD progression, explaining key pathways including macroautophagy, chaperone-mediated autophagy (CMA), and selective autophagy processes such as mitophagy and aggrephagy. This further extends the discussion beyond the central nervous system, evaluating the role of hepatic autophagy in Aβ clearance and systemic metabolic regulation. An understanding of autophagy's involvement in AD pathology via various mechanisms could give rise to a novel therapeutic strategy targeting autophagic modulation to mitigate disease progression in the future.

Keywords: Alzheimer's disease; autophagy; amyloid-beta clearance; tau pathology; neurodegeneration

1. Introduction

Alzheimer's disease (AD) is a progressive and irreversible neurodegenerative disorder primarily characterized by the destruction of neurons, leading to cognitive decline, memory loss, and changes in behavior and personality. It is the most common cause of dementia worldwide and represents a growing public health concern as global life expectancy increases. As of 2020, dementia, including Alzheimer-related forms, affects over 50 million people globally, with projections suggesting this number will rise to more than 152 million by 2050 [1]. In the United States alone, AD ranks as the sixth leading cause of death, affecting over 5 million individuals and incurring an estimated annual healthcare cost of 305 billion USD, which is expected to escalate to 1.1 trillion USD by 2050 [2].

In Asia, a region experiencing rapid demographic aging, the burden of AD is similarly mounting. For instance, in Malaysia, the prevalence of AD was estimated at 0.126% in 2020 and is projected to increase to 0.454% by 2050, amounting to over 260,000 affected individuals [3]. Comparable trends are observed in other Asian countries such as China [4, 5] and Japan [6], highlighting the urgent need for regional as well as global strategies in AD management and research.

The hallmark neuropathological features of AD include extracellular neuritic plaques and intracellular neurofibrillary tangles. These are primarily composed of aggregated amyloid-beta (Aβ) peptides and hyperphosphorylated tau proteins, respectively [7]. The accumulation of Aβ is believed to trigger a cascade of pathological events such as tau aggregation, synaptic dysfunction, oxidative stress, mitochondrial impairment, and neuroinflammation, which, ultimately, lead to neuronal death and brain atrophy [8]. Despite advancements in understanding these mechanisms, the precise etiology and sequence of pathological changes in AD remain only partially understood. Current therapeutic options are mainly symptomatic and do not prevent disease progression, highlighting the need for further exploration of underlying cellular processes.

One such process is autophagy, a cellular degradation and recycling pathway essential for maintaining protein homeostasis and cellular health. Emerging evidence implies dysfunctional autophagy in the pathogenesis of various neurodegenerative disorders, including AD [9]. In neurons, impaired autophagy may contribute to the accumulation of toxic proteins such as Aβ and hyperphosphorylated tau, thereby exacerbating synaptic loss and neuronal degeneration [10]. Moreover, autophagy intersects key pathological features of AD, including mitochondrial dysfunction, oxidative stress, and inflammation [8].

This review aims to explore the intricate relationship between autophagy and AD pathogenesis. By examining recent advances in our understanding of autophagic dysfunction in AD, we seek to highlight potential molecular targets and therapeutic strategies. Furthermore, we consider whether other interrelated cellular systems may modulate these pathways, offering novel insights into the complex biology of AD and new directions for intervention.

Methodology

This narrative review was conducted by systematically identifying and synthesizing relevant peer-reviewed literature on autophagy and Alzheimer's disease from multiple electronic databases, including PubMed, Scopus, and Web of Science. Searches were performed using keywords such as "autophagy," "Alzheimer's disease," "amyloid-beta," "tau pathology," "selective autophagy," and "neurodegeneration". Studies published in English for the last 10 years were included (with the exception of certain relevant literature from longer than 10 years), with emphasis on recent high-impact publications and original research, mechanistic studies, and comprehensive reviews. Articles were selected based on their relevance to the scope of this review, methodological rigor, and novelty in elucidating autophagy-related mechanisms in AD pathogenesis and therapeutic targeting. Additional sources were identified through manual screening of reference lists from key articles.

2. Overview of Autophagy

Autophagy is a crucial cellular process that plays a vital role in maintaining cellular homeostasis. It is a tightly controlled mechanism that eliminates damaged organelles and misfolded proteins, aiding in cellular survival and adaptation to stressful conditions. It involves autophagosomes, which are a formation of double-membraned vesicles engulfing cytoplasmic material for degradation, promoting energy balance and renewal [11]. Christian de Duve first used the term "autophagy" in the 1960s, and it has since grown to be a well-researched phenomenon with important applications in a range of physiological and pathological conditions [12]. By ensuring the elimination of harmful proteins, organelles, and intracellular pathogens, autophagy acts as a quality control system to preserve cellular integrity and stop the buildup of toxic substances [13]. However, the dysregulation of autophagy has been implicated in a variety of pathological conditions, including cancer, metabolic syndromes, and neurodegenerative diseases such as AD. Thus, it is a crucial protective mechanism in cells that particularly targets and destroys unwanted substances and also microorganisms [14]. The dysregulation of autophagy can result in several illnesses, including cancer, metabolic syndromes, and AD [15]. This study reviewed the intricate relationship between autophagy malfunction and neuronal death in the etiology of AD.

Moreover, autophagy is classified into three primary types based on the contents of the cell that are integrated into the lysosome for degradation, for example, chaperone-mediated autophagy (CMA), macroautophagy, and microautophagy [16]. In CMA, the breakdown of cytosolic proteins involves transporting them directly into lysosomes and the lysosomal lumen [17]. In microautophagy, cytoplasmic material is internalized into the lysosome through direct invagination of the lysosomal membrane [18]. In macroautophagy, degradable cytoplasmic contents are enclosed within subcellular double-membrane structures known as "autophagosomes" that convey cellular "waste" to lysosomes for breakdown [19].

2.1. Macroautophagy

Macroautophagy is the primary and most thoroughly researched form of autophagy, often abbreviated as "autophagy." The basal level of autophagy is low under physiological conditions; however, it can be rapidly induced by various stimuli, including energy deprivation, nutrient starvation, misfolded proteins, damaged organelles, infection, inflammation, and other stressors [20]. The primary event in autophagy is the formation of the autophagosome, which is facilitated by various proteins known as autophagy-related genes (ATGs). The autophagy process, referred to as autophagic flux, consists of four stages categorized by the status of autophagosomes: initiation, nucleation, elongation, closure, and fusion. Each step is precisely controlled by various molecules [21] (Figure 1).

2.1.1. Initiation

The ULK complex, which includes ULK1, ULK2, ATG13, FIP200, and ATG101 in mammals, is primarily responsible for the initiation of autophagy. ULK1 is like ATG1 in yeast, which is a serine/threonine kinase that phosphorylates downstream proteins to signal autophagosome formation [22,23]. Moreover, the mechanistic target of rapamycin complex 1 (mTORC1) acts as a sensor for nutrients and energy, inhibiting the ULK complex in nutrient-rich situations. Whereby, in times of starvation or stress, the AMP-activated protein kinase (AMPK) phosphorylates ULK1 while suppressing mTORC1, enabling the start of autophagy. ATG101 stabilizes the ULK1/2-ATG13-FIP200 complex and boosts its kinase activity [24] (Figure 1).

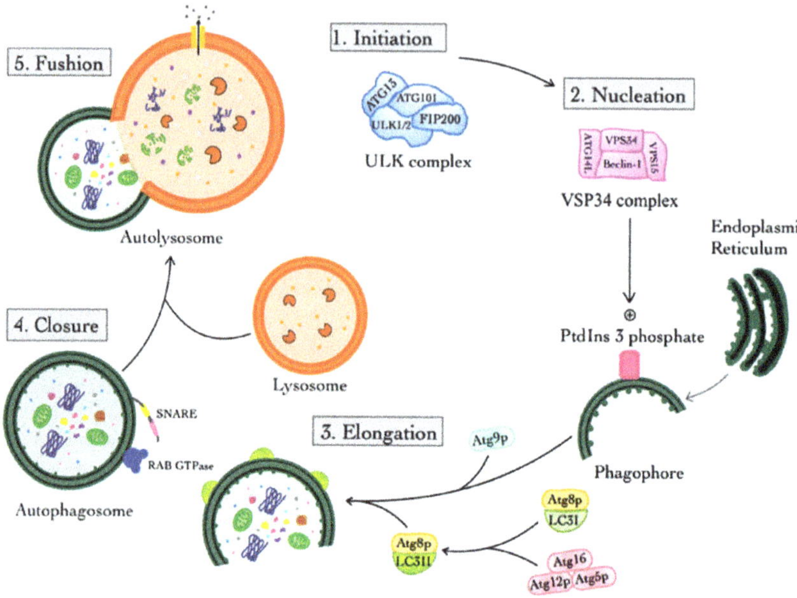

Figure 1. Schematic illustration of the five major stages of macroautophagy: (1) initiation begins with activation of the Unc-51-like autophagy-activating kinase (ULK) complex (ULK1/2, ATG13, ATG101, and FAK family kinase-interacting protein of 200 kDa or FIP200), triggering downstream events; (2) nucleation involves the vacuolar protein sorting 34 (VPS34) complex (including VPS34, VPS15, Beclin-1, and Atg14L) generating phosphatidylinositol 3-phosphate (PtdIns3P) at the phagophore formation site near the endoplasmic reticulum; (3) elongation is mediated by the conjugation systems involving Atg9p, Atg8p/LC3I/LC3II, Atg12–Atg5–Atg16L complex, expanding the phagophore membrane; (4) closure leads to the formation of a double-membraned autophagosome, facilitated by soluble N-ethylmaleimide-sensitive-factor attachment protein receptor (SNARE) proteins and RAB GTPases; and (5) fusion merges the autophagosome with a lysosome to form an autolysosome, enabling degradation of the enclosed cytoplasmic material. This pathway is essential for cellular homeostasis and the clearance of damaged organelles and proteins.

2.1.2. Nucleation

Nucleation is the formation of the phagophore assembly site (PAS) or isolation membrane, which signals the start of autophagic membrane creation (Figure 1). VPS34, Beclin 1 (BECN1), ATG14L, and p150/VPS15 are all necessary components of the Class III Phosphoinositide 3-Kinase (PI3K) complex [20]. In this case, VPS34 produces phosphatidylinositol 3-phosphate (PtdIns3P) at the PAS, which attracts PtdIns3P-binding proteins such as the WD repeat domain, phosphoinositide interacting 1 and 2 (WIPI1/2), and double FYVE-containing protein 1 (DFCP1), hence facilitating membrane nucleation. On the other hand, Beclin 1 is controlled by Bcl-2 family proteins, and Bcl-2 binding inhibits autophagy under normal conditions. Moreover, under stress, Bcl-2 dissociates to allow complex activation [20].

2.1.3. Elongation

Elongation converts the phagophore into a double membrane autophagosome. Two significant ubiquitin-like conjugation pathways exist: firstly, the ATG12-ATG5-ATG16L1 complex that acts as an E3 ligase, promoting the lipidation of microtubule-associated protein 1A/1B light chain 3 (LC3). Secondly, the LC3/Atg8 conjugation pathway involves ATG4 cleaving LC3 to produce LC3-I, which is then conjugated to phosphatidyl-ethanolamine (PE) to form LC3-II and embedded in the autophagosomal membrane. Moreover, LC3-II is a marker for autophagosomes that help with membrane tethering, expansion, and cargo recognition. Moreover, ATG9/ATG9A vesicles provide additional membrane sources and are controlled by scramblase activity, which aids in lipid redistribution and membrane expansion (see Figure 1). WIPI interacts with ATG16L1, facilitating the recruitment of the LC3 lipidation machinery to the phagophore [25].

2.1.4. Closure

The closure is the process by which the open autophagosome membrane seals, forming a complete vesicle. The endosomal sorting complex required for transport (ESCRT) process has been linked to the closure of phagophores. Moreover, the ATG2-WIPI complex is involved in both lipid transport and membrane closure. Additionally, syntaxin 17 (STX17) is found on mature autophagosomes and is required for the subsequent fusion process [26,27].

2.1.5. Fusion

Finally, the acidic lysosomal environment activates hydrolases, which degrade the sequestered cytoplasmic contents, including damaged organelles and misfolded proteins. Lysosomal permeases release nutrients into the cytosol for recycling, including amino acids and fatty acids [28].

2.2. *Selective Features and Regulation of Macroautophagy*

Despite being characterized as a non-selective bulk degradation system, macroautophagy has highly regulated and selective mechanisms that allow for the targeted removal of certain cellular components via receptor-mediated processes. ATG proteins simplify cargo recognition, lipid conjugation, and membrane elongation. To ensure cargo capture and autophagy completion, autophagosome production is closely regulated [29–31].

The autophagy pathway breaks down certain cellular elements, or "cargoes," called selective autophagy [32]. Selective autophagy involves designation, targeting, sequestration, and degradation. Designating distinct payloads with a molecular tag helps the autophagy machinery recognize them. A receptor directs payloads onto growing autophagosomes, which elongate and trap them. This is targeting and sequestration. Finally, the autophagosome merges with a vacuole/lysosome to break down cargo [32].

Moreover, autophagy cargo is assigned by multiple processes. In mammalian cells, ubiquitination sends a "eat me" signal to the central autophagic machinery [33]. These regulatory components can also remove cargo identification signals, like ubiquitination, to avoid autophagy [34]. Many mammals contain autophagy receptors, including p62 (also known as Sequestosome 1 or SQSTM1), NBR1, optineurin (OPTN), nuclear dot protein 52 kDa (NDP52), Tax-binding protein 1 (TAX1BP), and NIP3-like protein X (NIX) (also known as BCL2/adenovirus E1B 19 kDa protein-interacting protein 3-like or BNIP3L) [34]. Selectivity results from high-affinity connections between these receptors and cargo.

Additionally, the receptors use ubiquitin-binding domains and motifs to recognize and bind cargo, connecting with Atg8p/LC3 on autophagosomes through evolutionarily conserved patterns: the LIR in humans and the AIM in yeast to promote cargo sequestration [35]. Fusion with the lytic compartment degrades autophagy receptors. Recently, nuclear Dot Protein 52 kDa (NDP52) and OPTN have been found to directly engage with autophagy initiation machinery, attracting the ULK1-initiation complex to damaged mitochondria and cytosol-invading bacteria to degrade them [33,36,37].

3. Selective Autophagy

Selective autophagy targets specific cargos for degradation and involves several pathways that are classified according to their target organelles: mitophagy (mitochondria), lysophagy (lysosomes), pexophagy (peroxisomes), aggrephagy (protein and RNA aggregates), and ER-phagy (endoplasmic reticulum) [26] (Figure 2).

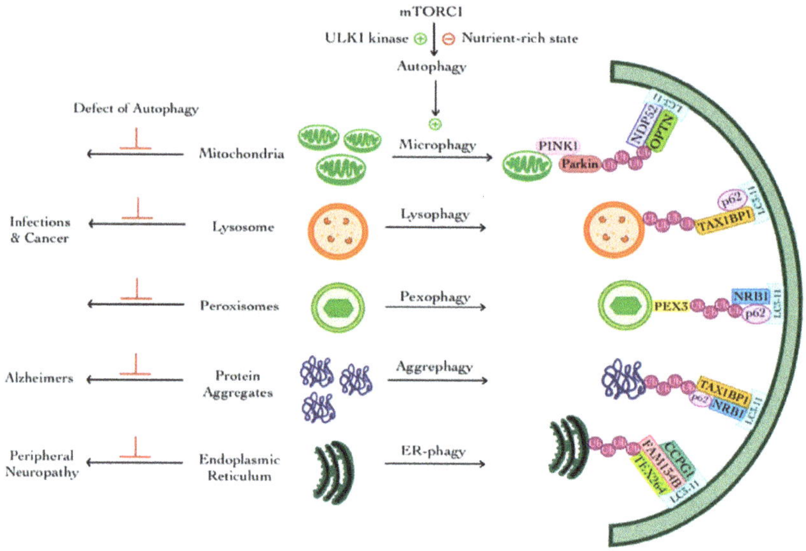

Figure 2. Schematic illustration of selective autophagy pathways and their implications in human diseases. Selective autophagy is activated under nutrient-deprived conditions through the inhibition of mechanistic target of rapamycin complex 1 (mTORC1) and activation of ULK1 kinase, leading to targeted degradation of specific organelles or protein aggregates. Various forms include the following: Mitophagy (mitochondrial clearance), regulated by PTEN-induced kinase 1/parkin (PINK1/Parkin), associated with neurodegeneration in autophagy-deficient states. Lysophagy (lysosomal turnover), critical in infections and cancer, is mediated by ubiquitination and autophagy receptors such as Tax1 (Human T-cell Leukemia Virus Type I) binding protein 1 (or TAX1BP1) and p62. Pexophagy (peroxisome degradation), involving peroxisomal biogenesis factor 3 (PEX3) and NRB1, affects metabolic disorders. Aggrephagy (clearance of protein aggregates), relevant to Alzheimer's disease pathology, employing TAX1BP1, NRB1, and p62. ER-phagy (endoplasmic reticulum turnover), regulated by receptors like family with sequence similarity 134 member B (FAM134B) and cell cycle progression gene 1 (CCPG1), is implicated in peripheral neuropathy. Each process recruits LC3-interacting autophagy adaptors to facilitate engulfment by the autophagosome. Dysfunction in these selective autophagy pathways contributes to various human diseases.

3.1. *Mitophagy*

The selective removal of damaged mitochondria through the process of mitophagy is necessary to maintain the integrity of the cell. The most well-studied example of selective autophagy involves the PTEN-induced kinase 1 (PINK1)-Parkin mitophagy pathway, where functional mitochondria import PINK1 into the inner membrane, while dysfunctional ones accumulate PINK1 on the outer membrane [38,39]. Phospho-ubiquitin recruits mitophagy factors like OPTN and NDP52, while Parkin amplifies the signal by ubiquitinating mitochondrial outer membrane proteins. Designated mitochondria are targeted and sequestered within autophagosomes, allowing selective targeting of dysfunctional ones [40–42].

3.2. *Lysophagy*

Lysosomal membrane permeabilization, a common stress condition, is linked to degenerative diseases, infections, and cancer. Cells repair lysosomes using ESCRT machinery.

When the repair is a failure, ubiquitin-tagged lysosomes begin the process of selective macroautophagy, or lysophagy, which leads to their removal [43]. This suggests that ESCRT-mediated repair and lysophagy are independent processes, clearing only unrepaired lysosomes. Selective autophagy involves autophagy receptors linking ubiquitinated cargo to the phagophore.

In a study involving L-leucyl-L-leucine methyl ester (LLOMe)-induced lysosomal damage, ubiquitination was shown to facilitate efficient lysophagy by recruiting autophagy receptors such as SQSTM1/p62 and TAX1BP1 [44]. Furthermore, optineurin is recruited to ubiquitin-tagged damaged lysosomes in the case of lysosomal damage by α-synuclein fibrils in neurodegenerative disorders [44]. Other studies also indicate that tank-binding kinase 1 (TBK1) collaborates with NDP52, an autophagy receptor, to prevent Salmonella-induced endomembrane damage, which recruits ULK1 and coordinates anti-bacterial autophagy [45]. This suggests that cells employ distinct autophagy receptors and regulation mechanisms for various types of endo-lysosomal membranes and damaging agents [45]. Moreover, according to recent literature, there are several new ubiquitination factors that attract autophagy machinery along with further layers of regulation, such as the processing of ubiquitinated proteins by the ATP-driven chaperone valosin-containing protein (AAA-ATPase VCP/p9) [43].

3.3. *Pexophagy*

Pexophagy is the selective autophagic degradation of damaged or excess peroxisomes, playing a crucial role in maintaining peroxisomal homeostasis. The mechanisms of pexophagy differ between yeast and mammalian cells. In yeast, the loss of peroxisomal AAA-type ATPases, specifically the Pex1p–Pex6p–Pex15p complex, renders peroxisomes import-incompetent, thereby promoting their degradation via pexophagy [46,47]. When key components of the peroxisomal protein import machinery are absent, ubiquitinated Pex5p accumulates on the peroxisomal membrane, signaling the organelle for selective degradation through specific autophagy receptors [46,47].

On the other hand, in mammalian cells, pexophagy is triggered by signals such as oxidative stress or excess peroxisome numbers [48]. The peroxisomal membrane protein PEX3 can promote pexophagy by recruiting autophagy receptors such as NBR1 and p62, which recognize ubiquitinated cargo and facilitate binding to LC3-II on the autophagosomal membrane. This interaction enables the sequestration and subsequent degradation of peroxisomes within the autolysosome [48].

3.4. *Aggrephagy*

Aggrephagy, a specialized form of selective autophagy, is responsible for the degradation of misfolded or aggregation-prone proteins. These protein aggregates are recognized and targeted by autophagy receptors such as TAX1BP1, NBR1, Tollip (Cue5 in yeast), OPTN, and p62. These receptors typically bind to ubiquitinated aggregates and facilitate their recruitment to the autophagic machinery for degradation [49]. mTORC1 plays a central role in regulating aggrephagy through its control over transcription factor EB (TFEB), a master regulator of lysosomal biogenesis and autophagy. Under nutrient-rich conditions, mTORC1 phosphorylates TFEB, retaining it on the lysosomal membrane and preventing its nuclear translocation [50]. Upon mTORC1 inhibition, either through nutrient deprivation or pharmacological intervention, TFEB becomes dephosphorylated and translocates to the nucleus. This activates the transcription of a network of genes involved in autophagy and lysosome formation, thereby promoting aggregate clearance through enhanced autophagic flux and lysosomal biogenesis [51].

3.5. ER-Phagy (Reticulophagy)

Autophagy plays a vital role in maintaining cellular homeostasis, and one of its specialized forms, ER-phagy (also known as reticulophagy), is essential for the selective degradation and remodeling of the endoplasmic reticulum (ER). ER-phagy is mediated by specific ER-resident receptors that recognize and target portions of the ER for encapsulation into autophagosomes. One key ER-phagy receptor is FAM134B, which localizes to the ER membrane and interacts with autophagy-related proteins such as LC3 and GABARAP. It also associates with reticulon 3, enabling the recognition and sequestration of specific ER regions destined for degradation. Additional ER-phagy receptors include CCPG1, Sec62, Atlastin-3, and testis-expressed 264 (TEX264), each contributing to the regulation of ER turnover in response to stress or cellular demands [52,53]. Mutations or dysregulation of these ER-phagy receptors have been implicated in various diseases, including hereditary sensory and autonomic neuropathies, underscoring the importance of ER-phagy in neuronal health and protein quality control [54].

The degradation of organelles like peroxisomes, in conjunction with selective autophagy, further exemplifies autophagy's function in maintaining cellular homeostasis under various conditions. In addition to their role in lipid metabolism, peroxisomes are essential regulators of oxidative metabolism and cellular responses to hypoxia. Key transcription factors such as hypoxia-inducible factor-1α (HIF-1α) and HIF-2α mediate these adaptive responses [55–58]. Under hypoxic conditions, the inhibition of HIF-2α by the von Hippel–Lindau protein (VHL) is alleviated, leading to increased ubiquitination of PEX5, a peroxisomal import receptor, which, in turn, can initiate selective peroxisomal degradation [59]. Recent studies have also identified Tankyrase 1 (TNKS1) and Tankyrase 2 (TNKS2), members of the poly (ADP-ribose) polymerase (PARP) family, as novel regulators of non-canonical pexophagy, functioning through interaction with PEX14 to modulate peroxisome turnover. These mechanisms are also implicated in mitochondrial clearance, proinsulin granule degradation, and cellular iron homeostasis [60,61]. Consequently, like how ER-phagy focuses on the endoplasmic reticulum for quality assurance, pexophagy facilitates the degradation of peroxisomes, thus connecting autophagy to more extensive cellular reactions to stress and energy demands.

The multifunctional roles of autophagy-related (ATG) proteins extend beyond organelle-specific degradation, encompassing non-autophagic functions that enhance their influence on cellular homeostasis. For example, LC3-associated phagocytosis (LAP) plays a vital role in immune regulation and inflammatory responses in phagocytic cells. LAP is activated via Toll-like receptors (TLRs) and immunoglobulin receptors, facilitating LC3-associated endocytosis (LANDO), a pathway involved in the clearance of Aβ and the amelioration of AD pathology [62,63]. Additionally, components of the canonical autophagy machinery, including the ATG conjugation system, ULK1 complex, and PI3K–Beclin 1–VPS34 complex, are involved in various non-canonical autophagy pathways in a cell-type-specific manner [64,65]. The ubiquitin-proteasome system (UPS) further complements autophagic degradation by regulating protein turnover, with ubiquitin serving as a key signal for both proteasomal and autophagic clearance. Notably, p62 acts as a bridge between the UPS and autophagy. Proteasome inhibition has been shown to enhance nuclear translocation of autophagy regulators and activate compensatory autophagic pathways [66–68]. The varying actions of these ATG proteins and their interaction with other degradation systems underline the complex significance of autophagy in cellular health, linking organelle-specific processes such as ER-phagy and pexophagy to broader regulatory mechanisms.

While selective macroautophagy encapsulates certain cellular components in double-membrane vesicles, autophagy also exists in mechanistically separate forms that do not need autophagosome production. These encompass chaperone-mediated autophagy (CMA) and microautophagy, both of which directly carry substrates to the lysosome, highlighting the adaptability of autophagy in preserving cellular homeostasis in various environments.

4. Chaperone-Mediated Autophagy (CMA)

CMA is a selective form of autophagy in which specific substrate proteins are recognized and translocated directly into the lysosome for degradation. This process is initiated when substrate proteins containing a pentapeptide sequence [five amino acids: K (Lysine), F (Phenylalanine), E (Glutamic acid), R (Arginine), and Q (Glutamine)] or KFERQ-like motif are recognized by the cytosolic chaperone heat shock cognate 71-kDa protein (HSC70). The resulting HSC70–substrate complex binds to the cytosolic tail of lysosome-associated membrane protein type 2A (LAMP-2A) on the lysosomal membrane, triggering the multimerization of LAMP-2A into a functional translocation complex [69].

Once the substrate is engaged, it is unfolded and translocated into the lysosomal lumen, a process assisted by luminal HSC70. Following translocation, the LAMP-2A complex is disassembled to allow recycling of its components. Importantly, the rate-limiting step of CMA is the expression level and dynamic assembly/disassembly of LAMP-2A at the lysosomal membrane. Regulation of the LAMP-2A complex is mediated by a set of interacting factors, including lysosomal AKT, monomeric glial fibrillary acidic protein (GFAP), and elongation factor-1α (EF1α), which modulate complex stability in a GTP-dependent manner across all studied cell types [70].

Phosphorylation of AKT by mTORC2 suppresses CMA activity, whereas dephosphorylation by PHLPP1 enhances CMA, leading to its maximal activation. CMA selectively targets a variety of proteins implicated in key cellular processes, including GFAP, PI3K, nuclear factor of activated T-cells (NFATs), nuclear factor erythroid 2-related factor 2 (NRF2), and tumor protein D52 (TPD52). Conversely, retinoic acid receptor alpha (RARα) and growth hormone signaling act as negative regulators of CMA [71,72] (Figure 3).

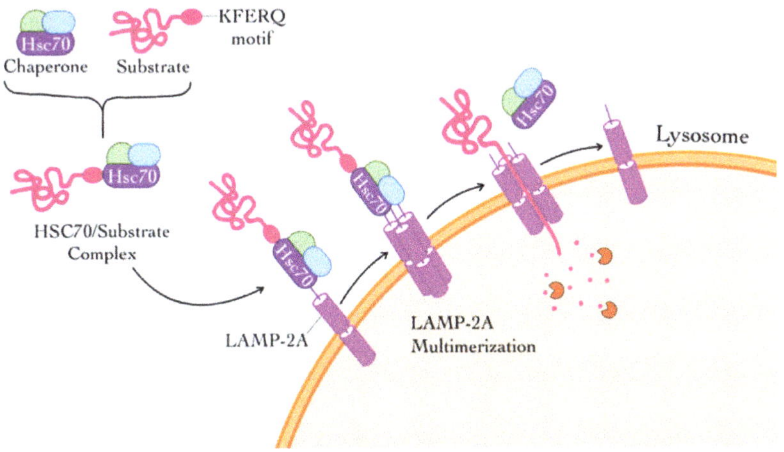

Figure 3. Schematic illustration of the steps involved in chaperone-mediated autophagy (CMA). Substrate proteins containing a KFERQ-like motif are recognized by the cytosolic chaperone heat shock cognate 71-kDa protein (HSC70), forming a substrate–HSC70 complex. This complex is then targeted to the lysosomal membrane, where it binds to the lysosome-associated membrane protein type 2A (LAMP-2A). Upon binding, LAMP-2A undergoes multimerization to form a translocation complex, facilitating the unfolding and transport of the substrate protein into the lysosomal lumen for degradation.

5. Microautophagy and Endosomal Microautophagy

Microautophagy is a form of autophagy where cytoplasmic components are directly engulfed by lysosomes or endosomes through membrane invagination or protrusion, facilitating their subsequent degradation [73]. A specialized subtype of this process in mammals, known as endosomal microautophagy, occurs within late endosomes/multivesicular bodies (LE/MVBs). The endosomal microautophagy targets soluble cytosolic proteins and can be further classified based on cargo selectivity and the molecular machinery involved [74].

Similar to CMA, one endosomal microautophagy pathway also recognizes proteins containing a KFERQ-like motif, which are bound by HSC70. These HSC70–substrate complexes are internalized into LE/MVBs via an endosomal sorting complexes required for transport (ESCRT)-dependent mechanism [75]. After internalization, lysosomal fusion with LE/MVBs enables the degradation of cargo proteins.

A distinct, HSC70-independent endosomal microautophagy pathway has also been identified in mammalian cells. This pathway still relies on ESCRT components but functions to degrade macroautophagy receptors, potentially playing a regulatory role in shifting the balance between selective and bulk macroautophagy [76]. These findings highlight the versatility and functional importance of endosomal microautophagy in maintaining proteostasis and fine-tuning autophagic responses under varying cellular conditions.

6. Autophagy in AD and Organ-Specific Impacts

6.1. *Autophagic Impairment in Protein Aggregate Clearance*

Mounting evidence implicates autophagy dysregulation as a critical contributor to the pathogenesis of AD. As early as 1967, Suzuki observed numerous aberrant subcellular vesicles and tau protein aggregates accumulating in dystrophic neurites in the brains of AD patients [77,78]. At the time, the identity of these vesicles was unknown. In 2005, Nixon and colleagues revealed that these vesicles were immature autophagic vacuoles (AVs), indicating a disruption in the autophagic pathway in AD [79,80].

Subsequent studies using AD mouse models and age-matched controls demonstrated that AVs begin to accumulate in neuronal dendrites and somata even before the appearance of extracellular Aβ plaques [81]. Notably, the aberrant accumulation of immature AVs in axons, particularly within hippocampal neurons, was observed well before detectable synaptic and neuronal loss, suggesting that impaired autophagic flux is an early event in AD pathogenesis [82]. Similar findings have been replicated in various animal models of AD, reinforcing the link between AVs accumulation and disease progression [83].

Moreover, autophagy plays a pivotal role in the clearance of pathological protein aggregates, including hyperphosphorylated tau. Studies have demonstrated that autophagic dysfunction, particularly autophagic gridlock, where AVs fail to mature and fuse with lysosomes, can exacerbate tau accumulation and contribute to the development of AD-like tauopathy [84,85]. Indeed, the abundance of autophagic vacuoles in AD brains, which is rarely observed in healthy controls, supports the hypothesis that impaired autophagy–lysosome function underlies the buildup of pathogenic proteins such as tau and Aβ [86].

Recent research has also highlighted the importance of mitophagy, a selective form of autophagy responsible for mitochondrial quality control, in AD. Impaired mitophagy contributes to the accumulation of dysfunctional mitochondria, leading to increased oxidative stress, energy deficits, and heightened neuronal vulnerability factors that further exacerbate AD pathology [86].

6.2. *Lysosomal Dysfunction and Genetic Regulators*

Amyloid precursor protein (APP), presenilin-1 (PS1), and presenilin-2 (PS2) are the key genes implicated in familial AD [87]. While mutations in PS1 and PS2 are known to increase Aβ production via their role in the γ-secretase complex, studies have shown that wild-type PS1 but not its mutated forms are also essential for lysosomal acidification, a function independent of its γ-secretase activity. This process is critical for maintaining the autophagy–lysosome degradation system [88].

In the case of sporadic AD, apolipoprotein E4 (ApoE4) represents the most significant genetic risk factor. ApoE4 has been found to disrupt autophagic function, contributing to disease pathology [89]. In transgenic mouse models expressing ApoE4, elevated levels of Aβ42 have been observed within lysosomes, leading to hippocampal neuronal death [90]. Additionally, in Neuro-2a cells, apoE4 exacerbates lysosomal membrane leakage and increases apoptosis triggered by Aβ peptides [91].

As discussed, a hallmark of AD is the progressive accumulation of AVs and lysosomal dysfunction in neurons. However, it remains debated whether this autophagic failure is a cause or consequence of AD pathogenesis [92,93]. Emerging evidence suggests that sex differences may also influence autophagy–lysosome system dysfunction in AD, although the mechanisms remain unclear [94]. Intriguingly, autophagy appears to play a neuroprotective role during the early stages of AD; however, in later stages, excessive or dysregulated autophagy may contribute to neuronal degeneration [95].

The earliest morphological evidence of autophagic impairment in AD is the abnormal accumulation of autophagosomes in neurons, particularly in hippocampal regions [81]. Despite decades of research, the exact mechanisms underlying autophagy failure in AD remain unresolved [96]. This failure may arise from one or more of the following: increased autophagy induction, impaired autophagic flux at the degradation stage, defective lysosomal function, or impaired fusion of autophagosomes with lysosomes [97]. Furthermore, lysosomal biogenesis is transcriptionally regulated by TFEB, a master regulator of autophagy and lysosomal gene expression. In AD, TFEB nuclear translocation is impaired, which hinders lysosomal renewal and functional degradation capacity [95]. Notably, enhancing TFEB activity in preclinical models has been shown to restore autophagic flux and reduce Aβ accumulation, offering a promising therapeutic avenue.

6.3. *Autophagic Flux*

Although both human AD brains and animal models exhibit significant accumulation of autophagosomes, this does not necessarily indicate enhanced autophagy activation [96]. In fact, the presence of accumulated autophagosomes may reflect impaired autophagic flux, where the degradation phase, particularly lysosomal fusion or clearance, is dysfunctional.

A key protein in autophagy initiation, Beclin-1, is found at reduced levels in neurons of AD patients when compared to healthy controls [84]. This reduction is believed to result from excessive caspase-3 activity, which cleaves Beclin-1, thereby attenuating autophagy initiation [98]. In transgenic APP mouse models with Beclin-1 deletion, intracellular Aβ accumulation increases significantly, and basal autophagic activity is disrupted, supporting the importance of Beclin-1 in AD pathology [99].

Additionally, the autophagic cargo receptor p62 (also known as SQSTM1), which binds to LC3 to facilitate cargo delivery to autophagosomes, has been evaluated in AD models. In a triple transgenic mouse model of AD, p62 expression was reported to be significantly reduced, potentially reflecting impaired autophagic cargo recognition or turnover [100]. However, other studies suggest an upregulation of autophagy in AD. For instance, a genome-wide transcriptomic analysis revealed that many positive regulators of autophagy are transcriptionally upregulated in AD, potentially as a compensatory response [101]. This may be driven by the reactive oxygen species (ROS)-mediated activation of class III PI3K and other autophagy-initiating kinases [102].

Interestingly, an increased autophagic flux has been reported in early-stage AD, particularly in CA1 hippocampal neurons [103]. In these stages, enhanced expression of ATGs, increased autophagosome formation, and elevated lysosomal biogenesis have been observed [104]. These findings suggest a biphasic pattern of autophagy regulation in AD, with initial upregulation followed by late-stage impairment. This concept is further supported by advanced imaging studies using tandem fluorescent-tagged LC3 reporters, which allow real-time visualization of autophagic flux in vivo [104]. These experiments demonstrate that early-stage AD is characterized by increased autophagosome production, whereas late-stage disease features autophagosome accumulation due to failed autophagosome-lysosome fusion and subsequent degradation deficits [104].

Given these conflicting findings, it is critical to develop stage-specific assessments of autophagy dynamics in AD. Autophagy may be subject to distinct regulatory mechanisms in different phases of disease progression [105]. Thus, future research should prioritize the elucidation of these stage-dependent autophagic alterations and invest in the development of reliable in vivo tools to measure autophagy flux. Such insights are essential for designing therapeutic strategies that target autophagy appropriately at different stages of AD pathogenesis.

6.4. Axonal Transport and Autophagic Vacuole Accumulation

In healthy neurons, AVs are retrogradely transported along axons to the soma, where they undergo degradation [106]. However, in AD, this transport is disrupted, leading to the pathological accumulation of AVs in axons and dendrites [107]. Tau protein, which is primarily localized to axons, plays a pivotal role in stabilizing microtubules and facilitating AV trafficking [108]. However, in AD, tau becomes hyperphosphorylated, disrupting microtubule dynamics and contributing to both defective AV transport and broader autophagic dysfunction. Hyperphosphorylated tau can also impede autophagosome-lysosome fusion and promote the accumulation of toxic protein aggregates [109].

Tau is efficiently degraded through the autophagy pathway and is also known to modulate autophagy itself [97]. Disruption of the autophagy–lysosome system can lead to tau oligomerization and aggregation, processes which are alleviated by enhanced autophagy [110]. Furthermore, tau pathology may induce lysosomal abnormalities, suggesting a reciprocal relationship between tau hyperphosphorylation and autophagy failure [111]. Tau also plays a role in autophagosome maturation and retrograde trafficking by promoting microtubule stability and facilitating fusion with lysosomes [110]. While tau deficiency appears to confer protection against Aβ toxicity, the presence of tau is essential for mediating Aβ-induced neurotoxicity in Alzheimer's disease models [112].

Conversely, other evidence suggests that lysosomal protease dysfunction, rather than tau abnormalities, may be the primary driver of axonal transport deficits [113]. Further studies are needed to delineate the precise molecular defects contributing to AV trafficking failure in AD. Notably, PS1 is essential for lysosomal acidification and autophagosome-lysosome fusion [114]. Fibroblasts from individuals with familial AD-associated PS1 mutations show AV accumulation and impaired turnover of aggregated proteins [115]. Some data suggest that the endolysosomal dysfunction in PS1-deficient cells arises from calcium homeostasis disruption, rather than proton pump defects [116].

6.5. Autophagy–Lysosome Pathway Disruption: Mechanistic Insights

Mechanistic studies have shown that autophagy–lysosome dysfunction in AD is linked to impaired proteolysis within autolysosomes [117]. Both human and mouse AD brains exhibit autolysosomes filled with incompletely degraded cargo, such as LC3-II, ubiquitinated proteins, and Aβ. Inhibition of lysosomal cathepsins, key proteases for cargo degradation, results in similar AVs accumulation, underscoring the importance of these enzymes [118].

Mutations in PS1 may further impair lysosomal calcium homeostasis, thereby compromising autophagosome-lysosome fusion [93]. Additionally, defects in lysosomal acidification have been associated with cholesterol accumulation in AD neurons, which inhibits hydrolase activity and impairs autophagosome clearance [119]. These converging defects result in a collapse of proteostasis, driving neurotoxicity and disease progression. Therapeutic strategies aimed at restoring lysosomal pH and modifying lipid composition may represent promising avenues for reversing autophagic failure in AD. Additionally, beyond classical molecular frameworks, alternative perspectives such as the biostructural theory proposed by Eugen Macovschi conceptualize living matter as a structurally organized, non-equilibrium system whose functional decline manifests through alterations in structural integrity and energy dissipation [120]. Although less explored in mainstream biomedical literature, this outlook provides a broader view of cellular degeneration. Furthermore, even subtle decreases in intracellular pH can significantly influence lysosomal activity, enhancing protease function and modulating the dynamics of autophagic degradation [121,122]. Such physicochemical parameters may alter the threshold for proteostasis and neurodegeneration, adding an important layer of complexity to our current mechanistic understanding.

Moreover, while hyperphosphorylated tau has been mechanistically linked to impaired autophagosome-lysosome fusion and aggregate accumulation, alternative frameworks suggest that tau hyperphosphorylation may also reflect deeper disruptions in cellular structure and energy regulation [123]. According to Eugen Macovschi's biostructural theory, degeneration involves a breakdown of the living biostructure, which destabilizes intracellular organization and may lead to uncontrolled ATP-dependent reactions such as aberrant phosphorylation [120]. This perspective suggests that tau pathology might not only be a molecular driver but also a secondary manifestation of broader biostructural failure within neurons.

6.6. *Aβ Secretion and Autophagy Crosstalk*

APP-derived metabolites can accumulate intracellularly by inhibiting autophagy before lysosomal fusion. The genetic deletion of autophagy-related genes such as ATG5 or ATG7 recapitulates AD-like pathology, demonstrating the importance of intact autophagy in preventing Aβ accumulation [120]. Autophagy plays a dual role in AD: it degrades Aβ and may also contribute to its production and secretion under pathological conditions [81,124].

While autophagy typically promotes Aβ clearance, emerging evidence indicates that it can also facilitate extracellular Aβ release, contributing to plaque formation [125]. The genetic deletion of autophagy components exacerbates intracellular Aβ buildup, highlighting the critical balance between degradation and secretion. Aβ itself may regulate autophagy in a feedback loop, inducing autophagic responses while simultaneously disrupting autolysosomal membrane integrity, leading to degradation failure [125].

Autophagy activation by Aβ42 appears to be initially protective; however, prolonged activation can shift to a pro-death role, contributing to neuronal degeneration [126]. Furthermore, autophagy-derived exosomes may serve as vehicles for intercellular transmission of Aβ and tau, thereby facilitating the spatiotemporal spread of pathology across brain regions [106]. Targeting the crosstalk between autophagy, Aβ metabolism, and intercellular signaling may offer novel therapeutic strategies.

6.7. *The Liver's Role in Autophagy and AD Pathogenesis*

Emerging evidence suggests that the liver contributes to cerebral Aβ burden, acting both as a site of peripheral Aβ clearance and a potential source of circulating Aβ [127].

Mutations in APP, PS1, and PS2, key regulators of Aβ production, are associated with increased Aβ42 formation and deposition in AD mouse models [127]. Surprisingly, PS2 activity is heritable and liver-specific in mice, suggesting that hepatic dysfunction may contribute to cerebral Aβ accumulation [128].

Aβ can reach the brain through several pathways, including blood–brain barrier (BBB) transport and the glymphatic system, which is mediated by brain low-density lipoprotein receptor-related protein 1 (LRP1) [129]. Conversely, the receptor for advanced glycation end-products (RAGE) facilitates Aβ influx across the BBB, promoting cerebral accumulation [127] (Figure 4).

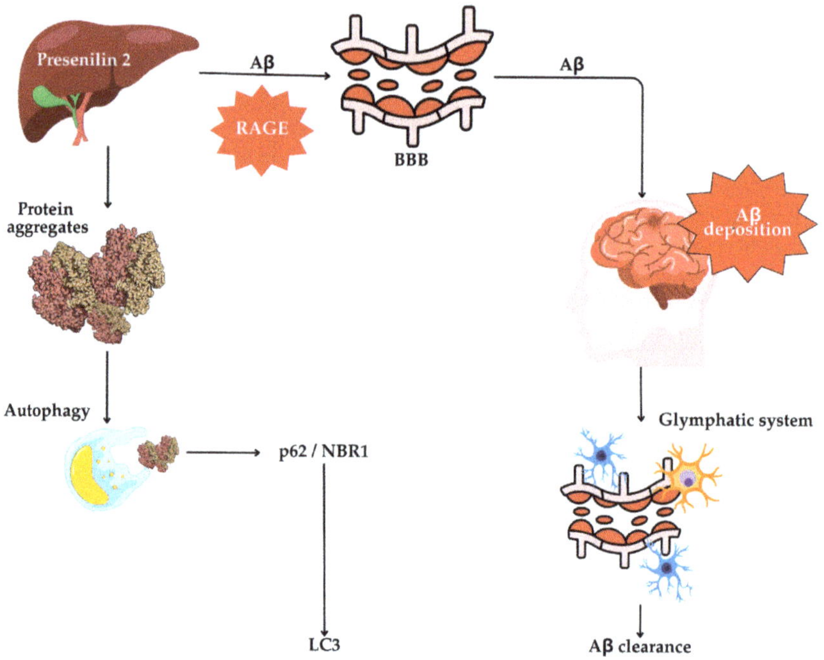

Figure 4. Presenilin 2 (PS2)-mediated amyloid beta (Aβ) synthesis in the liver and its contribution to cerebral amyloid burden. PS2 activity in the liver promotes the synthesis of Aβ, which can enter the brain via the blood–brain barrier (BBB) through the receptor for advanced glycation end-products (RAGE), contributing to cerebral Aβ accumulation. Clearance of Aβ from the brain occurs via the glymphatic system, the brain's waste removal pathway. Intracellularly, misfolded protein aggregates, including Aβ, are targeted for degradation by autophagy. The cargo adaptor proteins p62 (also known as Sequestosome 1 or SQSTM1) and NBR1 bind ubiquitinated protein aggregates and facilitate their delivery to autophagosomal membrane proteins such as microtubule-associated protein 1A/1B light chain 3 (LC3), enabling lysosomal degradation and maintaining proteostasis.

Additionally, autophagy plays a central role in hepatic protein quality control, clearing misfolded and aggregated proteins that accumulate during stress or metabolic dysfunction [129,130]. In Atg7 knockout mice, the loss of hepatic autophagy leads to the accumulation of ubiquitinated protein aggregates, peroxisomes, and defective mitochondria, ultimately causing hepatomegaly and liver damage [131]. The autophagy receptors p62/SQSTM1 and NBR1 mediate the degradation of ubiquitinated proteins, and their co-accumulation upon autophagy inhibition underscores their importance in selective autophagy [132–134].

In the context of alcoholic liver disease, p62 is found in Mallory bodies, suggesting a role in inclusion body formation [116]. Liver-specific deletion of Atg7 results in p62 accumulation, which alters gene expression, enhances nuclear factor kappa B (NF-κB) activity, increases oxidative stress, and promotes tumorigenesis [135]. Moreover, p62 can also facilitate caspase-8 activation via aggregation, further linking autophagy dysfunction to apoptotic signaling [136].

Thus, hepatic autophagy is a crucial physiological process for maintaining proteostasis and systemic homeostasis. Its impairment may influence both peripheral and central Aβ dynamics, and restoring hepatic autophagy could represent an innovative therapeutic approach in AD. However, further research is warranted to clarify the mechanistic links and evaluate interventional strategies.

Importantly, clinical evidence supports a link between hepatic dysfunction and cognitive decline. A UK-based cohort (over 400,000 participants) found that markers of liver dysfunction including elevated aspartate aminotransferase (AST), gamma-glutamyl transpeptidase (GGT), AST/ALT ratio, and hypoproteinemia were significantly associated with the development of dementia, which includes cognitive decline and volumetric loss in hippocampal and subcortical structures [137]. Similarly, a smaller retrospective cohort of 4582 individuals with metabolic-associated steatotic liver disease (MASLD) showed a nearly threefold increased risk of AD compared to controls (RR 2.80; 95% CI 1.79–4.38) [138]. Large-scale studies also found that severe liver fibrosis/cirrhosis was associated with a 9–18% increased risk of dementia [139]. Finally, up to 10% of patients initially diagnosed with dementia may have reversible cognitive decline due to hepatic encephalopathy from undiagnosed cirrhosis [140]. Together, these data support a potential causal and bidirectional relationship between liver health and AD risk, reinforcing the systemic nature of proteostatic and metabolic dysfunction in neurodegeneration.

6.8. Relative Contributions of Autophagic Pathways to AD Pathogenesis

Although previous sections describe macroautophagy, selective autophagy, and CMA based on their mechanisms, it is crucial to recognize that not all autophagic pathways contribute equally to AD pathogenesis. Emerging evidence suggests that specific autophagy types, particularly macroautophagy and CMA, play dominant roles in the development and progression of AD.

Macroautophagy, particularly mitophagy, is a crucial pathway for clearing aggregated proteins and dysfunctional organelles. In AD, impairment of macroautophagy has been well documented and is linked to the accumulation of autophagic vacuoles, misfolded tau and Aβ aggregates, and defective axonal transport [141,142]. Specifically, mitophagy dysfunction causes oxidative stress and energy deficits in neurons, which accelerates neurodegeneration [143,144]. Recent reviews also highlighted that research in transgenic AD models has demonstrated that enhancing macroautophagy can decrease Aβ accumulation and enhance synaptic function [145].

CMA also contributes to proteostasis by targeting specific cytosolic proteins, including tau, for lysosomal degradation. In AD, CMA dysfunction has been linked to reduced expression and impaired assembly of the lysosomal receptor LAMP-2A, resulting in inefficient clearance of tau and other neurotoxic proteins [146]. Notably, the ApoE4 isoform, a major genetic risk factor for sporadic AD, has been shown to inhibit CMA by disrupting lysosomal stability and receptor availability [147,148]. Thus, CMA dysfunction not only contributes to tauopathy but may also interact with genetic vulnerabilities in AD [146].

In contrast, the roles of microautophagy and endosomal microautophagy in AD remain less well-defined. While these pathways participate in vesicle turnover and cytoplasmic content recycling, their direct involvement in AD-related proteinopathies is still under investigation [149]. Nevertheless, alterations in endo-lysosomal trafficking observed in AD suggest that these processes may play indirect or modulatory roles in disease progression [150]. In summary, the macroautophagy–lysosomal system and CMA are the most critically involved autophagic pathways in AD pathophysiology. Understanding their differential roles is essential for developing targeted therapeutic strategies aimed at restoring autophagic flux and proteostasis in affected neurons.

7. Future Research

Despite significant advancements in understanding the pathological mechanisms of AD, the role of peripheral organs, particularly the liver, in modulating cerebral Aβ burden remains an underexplored but potentially transformative area. Future research must investigate the functional relationship between hepatic autophagy and brain Aβ homeostasis, aiming to delineate the molecular and systemic pathways that link liver metabolism to neurodegeneration. This could be achieved through the use of sophisticated in vivo models, such as liver-specific autophagy knockout or overexpression systems in genetically engineered AD mice. These models would allow for a temporal and spatial dissection of how altered hepatic autophagic flux influences Aβ production, clearance, and deposition in the brain [151–153].

Importantly, while hepatic autophagy is known to play a role in proteostasis and systemic metabolic regulation, its direct impact on circulating Aβ levels and subsequent brain pathology has not been thoroughly quantified. The hypothesis that impaired hepatic clearance contributes to cerebral Aβ accumulation represents a paradigm shift from brain-centric AD models to a more systemic disease framework. Such an approach is not only biologically plausible given the liver's central role in metabolic detoxification and protein clearance, but also therapeutically attractive, as hepatic interventions may circumvent the BBB, which remains a formidable challenge in AD drug delivery [154].

Several treatments have been studied to activate autophagy, offering promising opportunities for AD treatment. For example, rapamycin, an mTOR inhibitor, enhances autophagy, promoting the clearance of Aβ and tau aggregates in AD models, which may offer promising outcomes [155]. Similarly, resveratrol activates SIRT1, upregulating autophagy, reducing Aβ accumulation, and neuroinflammation [156]. Moreover, curcumin induces autophagy via AMPK activation and mTOR inhibition and has shown potential in reducing Aβ plaque burden and warrants deeper investigation [157]. Valproic acid and suberoylanilide hydroxamic acid (SAHA), compounds that inhibit HDAC, can improve autophagy by modifying histone acetylation, thereby enhancing protein clearance and potentially treating neurodegenerative diseases [158].

Besides pharmaceutical treatments, natural stimuli provide a non-invasive method for activating autophagy, which may yield potential benefits in Alzheimer's disease. Physical activity has been demonstrated to stimulate autophagy in peripheral tissues, including the liver, muscle, and brain. This contributes to enhanced Aβ clearance. Interestingly, exercise also enhances neural synaptic plasticity, fosters adult neurogenesis, mitigates cognitive loss associated with aging, and postpones the onset of neurodegenerative disorders [159]. Caloric restriction similarly enhances autophagy, a cytoplasmic recycling process, by decreasing protein acetylation [160]. Thus, these findings highlight the therapeutic significance of lifestyle modifications in regulating autophagy for neuroprotection.

To this end, future research should also focus on the development of pharmacological agents or gene therapy techniques aimed at selectively enhancing hepatic autophagy. Small molecule modulators of autophagy regulators (e.g., TFEB activators, mTOR inhibitors) or targeted delivery of autophagy-related genes (such as Atg7, p62/SQSTM1, and NBR1) via liver-specific vectors could be explored. Such approaches should be rigorously tested in preclinical models to evaluate their capacity not only to upregulate autophagy but also to reduce Aβ burden and improve cognitive outcomes.

Another important research avenue involves the identification of reliable systemic biomarkers of autophagy. Most current autophagy studies rely on brain tissue analysis, which is not feasible in living patients. Peripheral blood-based or liver-derived biomarkers (e.g., circulating p62, LC3-II levels, exosomal Aβ or tau) could enable early detection, risk stratification, and dynamic monitoring of disease progression or therapeutic response [161]. However, distinguishing disease-specific autophagy signals from those associated with normal aging or comorbidities remains a major challenge and necessitates longitudinal human cohort studies.

Critically, any proposed intervention that targets peripheral autophagy must also consider potential trade-offs and unintended consequences. For instance, excessive autophagic activation could lead to hepatocellular stress or dysfunction, particularly in aging or metabolically compromised individuals [162]. The efficiency of autophagy-stimulating therapies may differ according to the stage of AD and the age of the individual, as older adults often have diminished autophagy, which may impair treatment effects. Furthermore, systemic modulation of autophagy might influence immune responses, which are increasingly recognized as key players in AD pathophysiology.

Adding to these concerns, one of the notable consequences of increased autophagic activity is the initiation of autophagic cell death, referred to as autosis. This is a non-apoptotic autophagy-dependent cell death, which is a unique morphological feature that depends on the cellular Na^+, K^+-ATPase. Autosis can occur during starvation, autophagy-inducing peptide treatment, and cerebral hypoxia-ischemia, leading to cellular self-destruction. Its discovery could stimulate molecular mechanisms, understand autophagic cell death's role, and provide novel therapeutic targets [163].

In summary, targeting autophagy beyond the central nervous system, particularly in the liver, represents a promising but complex frontier in AD research. Future studies should aim to integrate multi-organ models, systems biology approaches, and translational pipelines to fully harness the therapeutic potential of peripheral autophagy modulation. This shift toward a holistic, systems-level understanding of AD could pave the way for more effective and accessible interventions in the prevention and management of this devastating neurodegenerative disease.

8. Conclusions

Autophagy is a fundamental cellular process that maintains homeostasis by degrading misfolded proteins and damaged organelles. In AD, the disruption of autophagy contributes to hallmark pathological features, including Aβ accumulation, tau hyperphosphorylation, and progressive neurodegeneration. While much of the existing literature has focused on neuronal autophagy and its dysregulation within the brain, emerging evidence highlights a critical and previously underappreciated role for hepatic autophagy in modulating systemic Aβ dynamics. This review underscores the liver's involvement in peripheral Aβ metabolism, possibly through presenilin-mediated pathways and autophagy-dependent clearance mechanisms. Aberrant liver function, particularly impaired autophagic flux, may lead to increased circulating Aβ levels and contribute to its accumulation in the brain. The recognition that the liver is not merely a bystander but an active player in AD pathophysiology challenges the traditional neurocentric view of the disease and opens up new avenues for therapeutic intervention. By integrating central and peripheral autophagy into the broader framework of AD pathology, we advance a more systemic perspective of the disease. This conceptual shift not only broadens our understanding of disease mechanisms but also suggests novel treatment strategies, especially those aimed at restoring hepatic autophagic function, to reduce brain Aβ burden. Continued research in this direction could lead to the development of multi-organ targeted therapies and innovative biomarker platforms, potentially transforming the diagnosis and treatment of AD.

Author Contributions: Conceptualization, Z.M.H., T.V., C.M.N.C.M.N. and P.K.G.; methodology, Z.M.H.; software, T.V.; validation, Z.M.H., T.V., C.M.N.C.M.N. and P.K.G.; formal analysis, Z.M.H., C.M.N.C.M.N. and P.K.G.; investigation, Z.M.H., T.V., C.M.N.C.M.N. and P.K.G.; resources, Z.M.H., T.V., C.M.N.C.M.N. and P.K.G.; data curation, Z.M.H. and T.V.; writing—original draft preparation, Z.M.H. and T.V.; writing—review and editing, Z.M.H., T.V., B.K., A.A., F.H., S.J., M.A.K., M.D.C.R., C.M.N.C.M.N. and P.K.G.; visualization, T.V.; supervision, Z.M.H. and P.K.G.; project administration, Z.M.H., C.M.N.C.M.N. and P.K.G.; funding acquisition, Z.M.H. All authors have read and agreed to the published version of the manuscript.

Funding: The APC was funded by Ajman University.

Institutional Review Board Statement: Not applicable.

Informed Consent Statement: Not applicable.

Data Availability Statement: Not applicable.

Acknowledgments: We would also like to acknowledge Ajman University for the APC support. The authors would also like to acknowledge that the assistance of ChatGPT 4.0, an AI language model developed by OpenAI, was used to help in the refinement of certain and limited sections of this manuscript, particularly for the language editing and refinement. Further, all AI-assisted content has been duly reviewed and thoroughly edited by the authors to ensure accuracy, scientific rigor, and adherence to the standards of academic writing. Authors are fully responsible for the content of this manuscript, even those parts improved by an AI tool, and are thus liable for any breach of publication ethics.

Conflicts of Interest: The authors declare no conflicts of interest.

Abbreviations

The following abbreviations are used in this manuscript:

AAA-ATPase VCP/p9	ATP-driven chaperone valosin-containing protein
AD	Alzheimer's disease
AMPK	AMP-activated protein kinase
ApoE4	Apolipoprotein E4
APP	Amyloid precursor protein
ATG	Autophagy-related genes
AVs	Autophagic vacuoles
Aβ	Amyloid-beta
BBB	Blood–brain barrier
BECN	Beclin
BNIP3L	BCL2/adenovirus E1B 19 kDa protein-interacting protein 3-like
CCPG1	Cell cycle progression gene 1
CMA	Chaperone-mediated autophagy
DFCPI	Double FYVE-containing protein 1
EF1α	Elongation factor-1α
ESCRT	Endosomal sorting complex required for transport
FAM134B	Family with Sequence Similarity 134 Member B
FIP200	FAK family kinase-interacting protein of 200 kDa
GFAP	Glial fibrillary acidic protein
HIF-1α	Hypoxia-inducible factor-1α
HSC70	Cytosolic chaperone heat shock cognate 71-kDa protein
JLK	UNC-51-like autophagy-activating kinase
LAMP-2A	Lysosome-associated membrane protein type 2A

LANDO	LC3-associated endocytosis
LAP	LC3-associated phagocytosis
LC3	Microtubule-associated protein 1A/1B light chain 3
LE/MVBs	Late endosomes/multivesicular bodies
LLOMe	L-leucyl-L-leucine methyl ester
LRP1	Low-density lipoprotein receptor-related protein 1
mTORC1	Mechanistic target of rapamycin complex 1
NDP52	Nuclear dot protein 52 kDa
NF-κB	Nuclear factor kappa B
NFAT	Nuclear factor of activated T-cells
NIX	NIP3-like protein X
NRF2	Nuclear factor erythroid 2-related factor 2
OPTN	Optineurin
PARP	Poly(ADP-ribose) polymerase
PAS	Phagophore assembly site
PE	Phosphatidyl-ethanolamine
PEX3	Peroxisomal biogenesis factor 3
PINK1	PTEN-induced kinase 1
PS-1 or 2	Presenilin-1 or 2
PtdIns3P	Phosphatidylinositol 3-phosphate
RAGE	Receptor for advanced glycation end-products
RARα	Retinoic acid receptor alpha
SNARE	Soluble N-ethylmaleimide-sensitive-factor attachment protein receptor
SQSTM1	Sequestosome 1
STX17	Syntaxin 17
TAX1BP1	Tax1 (Human T-cell Leukemia Virus Type I) binding protein 1
TBK1	Tank-binding kinase 1
TPD52	Tumor protein D52
TEX264	Testis expressed 264
TFEB	Transcription factor EB
TLR	Toll-like receptors
TNKS1	Tankyrase 1
UPS	Ubiquitin-proteasome system
VHL	von Hippel–Lindau protein
VPS34	Vacuolar protein sorting 34
WIPI	WD repeat domain, phosphoinositide interacting

References

1. DeTure, M.A.; Dickson, D.W. The Neuropathological Diagnosis of Alzheimer's Disease. *Mol. Neurodegener.* **2019**, *14*, 32. [CrossRef] [PubMed]
2. Vishnumukkala, T.; Gopalakrishna, P.K.; Jagadeesan, S.; Chiroma, S.M.; Mohd Nor, N.H.; Baharuldin, M.T.H.; Thomas, W.; Mohd Moklas, M.A. Herbal Medicine: A Promising Approach for the Treatment of Alzheimer's Disease. *Int. J. Ayurveda Pharma Res.* **2024**, *12*, 142–150. [CrossRef]
3. Mat Nuri, T.H.; Hong, Y.H.; Ming, L.C.; Mohd Joffry, S.; Othman, M.F.; Neoh, C.F. Knowledge on Alzheimer's Disease among Public Hospitals and Health Clinics Pharmacists in the State of Selangor, Malaysia. *Front. Pharmacol.* **2017**, *8*, 739. [CrossRef] [PubMed]
4. Ren, R.; Qi, J.; Lin, S.; Liu, X.; Yin, P.; Wang, Z.; Tang, R.; Wang, J.; Huang, Q.; Li, J.; et al. The China Alzheimer Report 2022. *Gen. Psychiatry* **2022**, *35*, e100751. [CrossRef]
5. Liu, Y.; Wu, Y.; Chen, Y.; Lobanov-Rostovsky, S.; Liu, Y.; Zeng, M.; Bandosz, P.; Xu, D.R.; Wang, X.; Liu, Y.; et al. Projection for Dementia Burden in China to 2050: A Macro-Simulation Study by Scenarios of Dementia Incidence Trends. *Lancet Reg. Health West. Pac.* **2024**, *50*, 101158. [CrossRef]
6. Nakahori, N.; Sekine, M.; Yamada, M.; Tatsuse, T.; Kido, H.; Suzuki, M. Future Projections of the Prevalence of Dementia in Japan: Results from the Toyama Dementia Survey. *BMC Geriatr.* **2021**, *21*, 602. [CrossRef]

7. Tsering, W.; Prokop, S. Neuritic Plaques—Gateways to Understanding Alzheimer's Disease. *Mol. Neurobiol.* **2024**, *61*, 2808–2821. [CrossRef]
8. Azargoonjahromi, A. The Duality of Amyloid-β: Its Role in Normal and Alzheimer's Disease States. *Mol. Brain* **2024**, *17*, 44. [CrossRef]
9. Guo, F.; Liu, X.; Cai, H.; Le, W. Autophagy in Neurodegenerative Diseases: Pathogenesis and Therapy. *Brain Pathol.* **2018**, *28*, 3–13. [CrossRef]
10. Yuan, M.; Wang, Y.; Huang, Z.; Jing, F.; Qiao, P.; Zou, Q.; Li, J.; Cai, Z. Impaired Autophagy in Amyloid-Beta Pathology: A Traditional Review of Recent Alzheimer's Research. *J. Biomed. Res.* **2022**, *37*, 30–46. [CrossRef]
11. Yao, R.Q.; Ren, C.; Xia, Z.F.; Yao, Y.M. Organelle-Specific Autophagy in Inflammatory Diseases: A Potential Therapeutic Target Underlying the Quality Control of Multiple Organelles. *Autophagy* **2021**, *17*, 385–401. [CrossRef] [PubMed]
12. Harnet, M.M. From Christian de Duve to Yoshinori Ohsumi: More to Autophagy Than Just Dining at Home. *Biomed. J.* **2017**, *40*, 9–22. [CrossRef] [PubMed]
13. Deretic, V.; Kroemer, G. Autophagy in Metabolism and Quality Control: Opposing, Complementary or Interlinked Functions? *Autophagy* **2022**, *18*, 283–292. [CrossRef] [PubMed]
14. Riebisch, A.K.; Mühlen, S.; Beer, Y.Y.; Schmitz, I. Autophagy—A Story of Bacteria Interfering with the Host Cell Degradation Machinery. *Pathogens* **2021**, *10*, 110. [CrossRef]
15. Maiuolo, J.; Gliozzi, M.; Musolino, V.; Carresi, C.; Scarano, F.; Nucera, S.; Scicchitano, M.; Bosco, F.; Ruga, S.; Zito, M.C.; et al. From Metabolic Syndrome to Neurological Diseases: Role of Autophagy. *Front. Cell Dev. Biol.* **2021**, *9*, 651021. [CrossRef]
16. Choi, A.M.; Ryter, S.W.; Levine, B. Autophagy in Human Health and Disease. *N. Engl. J. Med.* **2013**, *368*, 1845–1846. [CrossRef]
17. Kaushik, S.; Cuervo, A.M. Chaperone-Mediated Autophagy: A Unique Way to Enter the Lysosome World. *Trends Cell Biol.* **2012**, *22*, 407–417. [CrossRef]
18. Mijaljica, D.; Prescott, M.; Devenish, R.J. Microautophagy in Mammalian Cells: Revisiting a 40-Year-Old Conundrum. *Autophagy* **2011**, *7*, 673–682. [CrossRef]
19. Takeshige, K.; Baba, M.; Tsuboi, S.; Noda, T.; Ohsumi, Y. Autophagy in Yeast Demonstrated with Proteinase-Deficient Mutants and Conditions for Its Induction. *J. Cell Biol.* **1992**, *119*, 301–311. [CrossRef]
20. Feng, Y.; He, D.; Yao, Z.; Klionsky, D.J. The Machinery of Macroautophagy. *Cell Res.* **2014**, *24*, 24–41. [CrossRef]
21. Kamada, Y.; Funakoshi, T.; Shintani, T.; Nagano, K.; Ohsumi, M.; Ohsumi, Y. Tor-Mediated Induction of Autophagy via an Apg1 Protein Kinase Complex. *J. Cell Biol.* **2000**, *150*, 1507–1513. [CrossRef] [PubMed]
22. Cai, Y.Y.; Li, L.; Zhu, X.M.; Lu, J.P.; Liu, X.H.; Lin, F.C. The Crucial Role of the Regulatory Mechanism of the Atg1/ULK1 Complex in Fungi. *Front. Microbiol.* **2022**, *13*, 1019543. [CrossRef] [PubMed]
23. Wang, C.; Wang, H.; Zhang, D.; Luo, W.; Liu, R.; Xu, D.; Diao, L.; Liao, L.; Liu, Z. Phosphorylation of ULK1 Affects Autophagosome Fusion and Links Chaperone-Mediated Autophagy to Macroautophagy. *Nat. Commun.* **2018**, *9*, 3492. [CrossRef] [PubMed]
24. Kawamata, T.; Kamada, Y.; Suzuki, K.; Kuboshima, N.; Akimatsu, H.; Ota, S.; Ohsumi, M.; Ohsumi, Y. Characterization of a Novel Autophagy-Specific Gene, ATG29. *Biochem. Biophys. Res. Commun.* **2005**, *338*, 1884–1889. [CrossRef]
25. Kihara, A.; Noda, T.; Ishihara, N.; Ohsumi, Y. Two Distinct Vps34 Phosphatidylinositol 3-Kinase Complexes Function in Autophagy and Carboxypeptidase Y Sorting in *Saccharomyces cerevisiae*. *J. Cell Biol.* **2001**, *152*, 519–530. [CrossRef]
26. Fleming, A.; Bourdenx, M. The Different Autophagy Degradation Pathways and Neurodegeneration. *Neuron* **2022**, *110*, 935–966. [CrossRef]
27. Puri, C.; Vicinanza, M.; Ashkenazi, A.; Gratian, M.J.; Zhang, Q.; Bento, C.F.; Renna, M.; Menzies, F.M.; Rubinsztein, D.C. The RAB11A-Positive Compartment Is a Primary Platform for Autophagosome Assembly Mediated by WIPI2 Recognition of PI3P–RAB11A. *Dev. Cell* **2018**, *45*, 114–131.e8. [CrossRef]
28. Kimura, S.; Noda, T.; Yoshimori, T. Dynein-Dependent Movement of Autophagosomes Mediates Efficient Encounters with Lysosomes. *Cell Struct. Funct.* **2008**, *33*, 109–122. [CrossRef]
29. Kristensen, A.R.; Schandorff, S.; Høyer-Hansen, M.; Nielsen, M.O.; Jäättelä, M.; Dengjel, J.; Andersen, J.S. Ordered Organelle Degradation During Starvation-Induced Autophagy. *Mol. Cell. Proteomics* **2008**, *7*, 2419–2428. [CrossRef]
30. Gomes, L.C.; Di Benedetto, G.; Scorrano, L. During Autophagy Mitochondria Elongate, Are Spared from Degradation and Sustain Cell Viability. *Nat. Cell Biol.* **2011**, *13*, 589–598. [CrossRef]
31. Rambold, A.S.; Kostelecky, B.; Elia, N.; Lippincott-Schwartz, J. Tubular Network Formation Protects Mitochondria from Autophagosomal Degradation During Nutrient Starvation. *Proc. Natl. Acad. Sci. USA* **2011**, *108*, 10190–10195. [CrossRef] [PubMed]
32. Anding, A.L.; Baehrecke, E.H. Cleaning House: Selective Autophagy of Organelles. *Dev. Cell* **2017**, *41*, 10–22. [CrossRef] [PubMed]
33. Rogov, V.; Dötsch, V.; Johansen, T.; Kirkin, V. Interactions Between Autophagy Receptors and Ubiquitin-Like Proteins Form the Molecular Basis for Selective Autophagy. *Mol. Cell* **2014**, *53*, 167–178. [CrossRef] [PubMed]
34. Cen, X.; Li, Z.; Chen, X. Ubiquitination in the Regulation of Autophagy. *Acta Biochim. Biophys. Sin.* **2023**, *55*, 1348–1357. [CrossRef]
35. Marshall, R.S.; Hua, Z.; Mali, S.; McLoughlin, F.; Vierstra, R.D. ATG8-Binding UIM Proteins Define a New Class of Autophagy Adaptors and Receptors. *Cell* **2019**, *177*, 766–781. [CrossRef]

36. Ravenhill, B.J.; Boyle, K.B.; von Muhlinen, N.; Ellison, C.J.; Masson, G.R.; Otten, E.G.; Foeglein, A.; Williams, R.; Randow, F. The Cargo Receptor NDP52 Initiates Selective Autophagy by Recruiting the ULK Complex to Cytosol-Invading Bacteria. *Mol. Cell* **2019**, *74*, 320–329. [CrossRef]
37. Turco, E.; Fracchiolla, D.; Martens, S. Recruitment and Activation of the ULK1/Atg1 Kinase Complex in Selective Autophagy. *J. Mol. Biol.* **2019**, *431*, 822–836. [CrossRef]
38. Lazarou, M.; Sliter, D.A.; Kane, L.A.; Sarraf, S.A.; Wang, C.; Burman, J.L.; Sideris, D.P.; Fogel, A.I.; Youle, R.J. The Ubiquitin Kinase PINK1 Recruits Autophagy Receptors to Induce Mitophagy. *Nature* **2015**, *524*, 309–314. [CrossRef]
39. Li, J. PINK1/Parkin-Mediated Mitophagy in Neurodegenerative Diseases. *Ageing Res. Rev.* **2022**, *73*, 101817. [CrossRef]
40. Kane, L.A.; Lazarou, M.; Fogel, A.I.; Li, Y.; Yamano, K.; Sarraf, S.A.; Banerjee, S.; Youle, R.J. PINK1 Phosphorylates Ubiquitin to Activate Parkin E3 Ubiquitin Ligase Activity. *J. Cell Biol.* **2014**, *205*, 143–153. [CrossRef]
41. Narendra, D.P.; Jin, S.M.; Tanaka, A.; Suen, D.-F.; Gautier, C.A.; Shen, J.; Cookson, M.R.; Youle, R.J. PINK1 Is Selectively Stabilized on Impaired Mitochondria to Activate Parkin. *PLoS Biol.* **2010**, *8*, e1000298. [CrossRef] [PubMed]
42. Matsuda, N.; Sato, S.; Shiba, K.; Okatsu, K.; Saisho, K.; Gautier, C.A.; Sou, Y.-S.; Saiki, S.; Kawajiri, S.; Sato, F.; et al. PINK1 Stabilized by Mitochondrial Depolarization Recruits Parkin to Damaged Mitochondria and Activates Latent Parkin for Mitophagy. *J. Cell Biol.* **2010**, *189*, 211–221. [CrossRef] [PubMed]
43. Papadopoulos, C.; Kravic, B.; Meyer, H. Repair or Lysophagy: Dealing with Damaged Lysosomes. *J. Mol. Biol.* **2020**, *432*, 231–239. [CrossRef] [PubMed]
44. Fujita, N.; Morita, E.; Itoh, T.; Tanaka, A.; Nakaoka, M.; Osada, Y.; Umemoto, T.; Saitoh, T.; Nakatogawa, H.; Kobayashi, S.; et al. Recruitment of the autophagic machinery to endosomes during infection is mediated by ubiquitin. *J. Cell Biol.* **2013**, *203*, 115–128. [CrossRef]
45. Bussi, C.; Peralta Ramos, J.M.; Arroyo, D.S.; Gallea, J.I.; Ronchi, P.; Kolovou, A.; Wang, J.M.; Florey, O.; Celej, M.S.; Schwab, Y.; et al. Alpha-synuclein fibrils recruit TBK1 and OPTN to lysosomal damage sites and induce autophagy in microglial cells. *J. Cell Sci.* **2018**, *131*, jcs226241. [CrossRef]
46. Law, K.B.; Bronte-Tinkew, D.; Di Pietro, E.; Snowden, A.; Jones, R.O.; Moser, A.; Brumell, J.H.; Braverman, N.; Kim, P.K. The peroxisomal AAA ATPase complex prevents pexophagy and development of peroxisome biogenesis disorders. *Autophagy* **2017**, *13*, 868–884. [CrossRef]
47. Nuttall, J.M.; Motley, A.M.; Hettema, E.H. Deficiency of the exportomer components Pex1, Pex6, and Pex15 causes enhanced pexophagy in *Saccharomyces cerevisiae*. *Autophagy* **2014**, *10*, 835–845. [CrossRef]
48. Germain, K.; Kim, P.K. Pexophagy: A Model for Selective Autophagy. *Int. J. Mol. Sci.* **2020**, *21*, 578. [CrossRef]
49. Sarraf, S.A.; Shah, H.V.; Kanfer, G.; Pickrell, A.M.; Holtzclaw, L.A.; Ward, M.E.; Youle, R.J. Loss of TAX1BP1-directed autophagy results in protein aggregate accumulation in the brain. *Mol. Cell* **2020**, *80*, 779–795.e10. [CrossRef]
50. Cui, Z.; Napolitano, G.; de Araujo, M.E.G.; Esposito, A.; Monfregola, J.; Huber, L.A.; Ballabio, A.; Hurley, J.H. Structure of the lysosomal mTORC1-TFEB-Rag-Ragulator megacomplex. *Nature* **2023**, *614*, 572–579. [CrossRef]
51. Malampati, S.; Song, J.-X.; Tong, B.C.-K.; Nalluri, A.; Yang, C.-B.; Wang, Z.; Sreenivasmurthy, S.G.; Zhu, Z.; Liu, J.; Su, C.; et al. Targeting Aggrephagy for the Treatment of Alzheimer's Disease. *Cells* **2020**, *9*, 311. [CrossRef] [PubMed]
52. Eapen, V.V.; Swarup, S.; Hoyer, M.J.; Paulo, J.A.; Harper, J.W. Quantitative proteomics reveals the selectivity of ubiquitin-binding autophagy receptors in the turnover of damaged lysosomes by lysophagy. *ELife* **2021**, *10*, e72328. [CrossRef] [PubMed]
53. Saxton, R.A.; Sabatini, D.M. mTOR signaling in growth, metabolism, and disease. *Cell* **2017**, *169*, 361–371. [CrossRef] [PubMed]
54. Dibble, C.C.; Manning, B.D. Signal integration by mTORC1 coordinates nutrient input with biosynthetic output. *Nat. Cell Biol.* **2013**, *15*, 555–564. [CrossRef]
55. Son, S.M.; Park, S.J.; Stamatakou, E.; Vicinanza, M.; Menzies, F.M.; Rubinsztein, D.C. Leucine regulates autophagy via acetylation of the mTORC1 component raptor. *Nat. Commun.* **2020**, *11*, 3148. [CrossRef]
56. Karabiyik, C.; Vicinanza, M.; Son, S.M.; Rubinsztein, D.C. Glucose starvation induces autophagy via ULK1-mediated activation of PIKfyve in an AMPK-dependent manner. *Dev. Cell* **2021**, *56*, 1961–1975. [CrossRef]
57. Vicinanza, M.; Korolchuk, V.I.; Ashkenazi, A.; Puri, C.; Menzies, F.M.; Clarke, J.H.; Rubinsztein, D.C. PI(5)P regulates autophagosome biogenesis. *Mol. Cell* **2015**, *57*, 219–234. [CrossRef]
58. Walter, K.M.; Schönenberger, M.J.; Trötzmüller, M.; Horn, M.; Elsässer, H.-P.; Moser, A.B.; Lucas, M.S.; Schwarz, T.; Gerber, P.A.; Faust, P.L.; et al. Hif- promotes degradation of mammalian peroxisomes by selective autophagy. *Cell Metab.* **2014**, *20*, 882–897. [CrossRef]
59. Schönenberger, M.J.; Kovacs, W.J. Hypoxia signaling pathways: Modulators of oxygen-related organelles. *Front. Cell Dev. Biol.* **2015**, *3*, 42. [CrossRef]
60. Li, X.; Han, H.; Zhou, M.-T.; Yang, B.; Ta, A.P.; Li, N.; Chen, J.; Wang, W. Proteomic Analysis of the Human Tankyrase Protein Interaction Network Reveals Its Role in Pexophagy. *Cell Rep.* **2017**, *20*, 737–749. [CrossRef]

61. Yamaguchi, H.; Honda, S.; Torii, S.; Shimizu, K.; Katoh, K.; Miyake, K.; Miyake, N.; Fujikake, N.; Sakurai, H.T.; Arakawa, S.; et al. Wipi3 is essential for alternative autophagy and its loss causes neurodegeneration. *Nat. Commun.* **2020**, *11*, 5311. [CrossRef] [PubMed]
62. Galluzzi, L.; Green, D.R. Autophagy-independent functions of the autophagy machinery. *Cell* **2019**, *177*, 1682–1699. [CrossRef] [PubMed]
63. Heckmann, B.L.; Boada-Romero, E.; Cunha, L.D.; Magne, J.; Green, D.R. LC3-associated phagocytosis and inflammation. *J. Mol. Biol.* **2017**, *429*, 3561–3576. [CrossRef] [PubMed]
64. Nishida, Y.; Arakawa, S.; Fujitani, K.; Yamaguchi, H.; Mizuta, T.; Kanaseki, T.; Komatsu, M.; Otsu, K.; Tsujimoto, Y.; Shimizu, S. Discovery of Atg5/Atg7-independent alternative macroautophagy. *Nature* **2009**, *461*, 654–658. [CrossRef]
65. Honda, S.; Arakawa, S.; Yamaguchi, H.; Torii, S.; Sakurai, H.T.; Tsujioka, M.; Murohashi, M.; Shimizu, S. Association between Atg5-independent alternative autophagy and neurodegenerative diseases. *J. Mol. Biol.* **2020**, *432*, 2622–2632. [CrossRef]
66. Yin, Z.; Popelka, H.; Lei, Y.; Yang, Y.; Klionsky, D.J. The roles of ubiquitin in mediating autophagy. *Cells* **2020**, *9*, 2025. [CrossRef]
67. Seibenhener, M.L.; Babu, J.R.; Geetha, T.; Wong, H.C.; Krishna, N.R.; Wooten, M.W. Sequestosome 1/p62 is a polyubiquitin chain binding protein involved in ubiquitin proteasome degradation. *Mol. Cell. Biol.* **2004**, *24*, 8055–8068. [CrossRef]
68. Sun-Wang, J.L.; Ivanova, S.; Zorzano, A. The dialogue between the ubiquitin-proteasome system and autophagy: Implications in ageing. *Ageing Res. Rev.* **2020**, *64*, 101203. [CrossRef]
69. Kaushik, S.; Cuervo, A.M. The Coming of Age of Chaperone-Mediated Autophagy. *Nat. Rev. Mol. Cell Biol.* **2018**, *19*, 365–381. [CrossRef]
70. Bandyopadhyay, U.; Sridhar, S.; Kaushik, S.; Kiffin, R.; Cuervo, A.M. Identification of Regulators of Chaperone-Mediated Autophagy. *Mol. Cell* **2010**, *39*, 535–547. [CrossRef]
71. Endicott, S.J.; Ziemba, Z.J.; Beckmann, L.J.; Boynton, D.N.; Miller, R.A. Inhibition of Class I PI3K Enhances Chaperone-Mediated Autophagy. *J. Cell Biol.* **2020**, *219*, e202001031. [CrossRef] [PubMed]
72. Anguiano, J.; Garner, T.P.; Mahalingam, M.; Das, B.C.; Gavathiotis, E.; Cuervo, A.M. Chemical Modulation of Chaperone-Mediated Autophagy by Retinoic Acid Derivatives. *Nat. Chem. Biol.* **2013**, *9*, 374–382. [CrossRef] [PubMed]
73. Schuck, S. Microautophagy—Distinct Molecular Mechanisms Handle Cargoes of Many Sizes. *J. Cell Sci.* **2020**, *133*, jcs246322. [CrossRef] [PubMed]
74. Krause, G.J.; Cuervo, A.M. Assessment of Mammalian Endosomal Microautophagy. *Methods Cell Biol.* **2021**, *164*, 167–185. [CrossRef]
75. Kirchner, P.; Bourdenx, M.; Madrigal-Matute, J.; Tiano, S.; Diaz, A.; Bartholdy, B.A.; Will, B.; Cuervo, A.M. Proteome-Wide Analysis of Chaperone-Mediated Autophagy Targeting Motifs. *PLoS Biol.* **2019**, *17*, e3000301. [CrossRef]
76. Mejlvang, J.; Olsvik, H.; Svenning, S.; Bruun, J.A.; Abudu, Y.P.; Larsen, K.B.; Brech, A.; Hansen, T.E.; Brenne, H.; Hansen, T.; et al. Starvation Induces Rapid Degradation of Selective Autophagy Receptors by Endosomal Microautophagy. *J. Cell Biol.* **2018**, *217*, 3640–3655. [CrossRef]
77. Suzuki, K.; Terry, R.D. Fine structural localization of acid phosphatase in senile plaques in Alzheimer's presenile dementia. *Acta Neuropathol.* **1967**, *8*, 276–284. [CrossRef]
78. Ruan, Z.; Pathak, D.; Kalavai, S.V.; Yoshii-Kitahara, A.; Muraoka, S.; Bhatt, N.; Takamatsu-Yukawa, K.; Hu, J.; Wang, Y.; Hersh, S.; et al. Alzheimer's Disease Brain-Derived Extracellular Vesicles Spread Tau Pathology in Interneurons. *Brain* **2021**, *144*, 288–309. [CrossRef]
79. Nixon, R.A.; Wegiel, J.; Kumar, A.; Yu, W.H.; Peterhoff, C.; Cataldo, A.; Cuervo, A.M. Extensive involvement of autophagy in Alzheimer disease: An immuno-electron microscopy study. *J. Neuropathol. Exp. Neurol.* **2005**, *64*, 113–122. [CrossRef]
80. Zhang, Z.; Yang, X.; Song, Y.Q.; Tu, J. Autophagy in Alzheimer's Disease Pathogenesis: Therapeutic Potential and Future Perspectives. *Ageing Res. Rev.* **2021**, *72*, 101464. [CrossRef]
81. Lee, J.H.; Yang, D.S.; Goulbourne, C.N.; Im, E.; Stavrides, P.; Pensalfini, A.; Chan, H.; Bouchet-Marquis, C.; Bleiwas, C.; Berg, M.J.; et al. Faulty Autolysosome Acidification in Alzheimer's Disease Mouse Models Induces Autophagic Build-Up of Aβ in Neurons, Yielding Senile Plaques. *Nat. Neurosci.* **2022**, *25*, 688–701. [CrossRef] [PubMed]
82. Goel, P.; Chakrabarti, S.; Goel, K.; Bhutani, K.; Chopra, T.; Bali, S. Neuronal Cell Death Mechanisms in Alzheimer's Disease: An Insight. *Front. Mol. Neurosci.* **2022**, *15*, 937133. [CrossRef] [PubMed]
83. Zhao, Y.; Zhang, Y.; Zhang, J.; Zhang, X.; Yang, G. Molecular Mechanism of Autophagy: Its Role in the Therapy of Alzheimer's Disease. *Curr. Neuropharmacol.* **2020**, *18*, 720–739. [CrossRef] [PubMed]
84. Wen, X.; An, P.; Li, H.; Zhou, Z.; Sun, Y.; Wang, J.; Ma, L.; Lu, B. Tau Accumulation via Reduced Autophagy Mediates GGGGCC Repeat Expansion-Induced Neurodegeneration in Drosophila Model of ALS. *Neurosci. Bull.* **2020**, *36*, 1414–1428. [CrossRef]
85. Chatterjee, S.; Sealey, M.; Ruiz, E.; Pegasiou, C.M.; Brookes, K.; Green, S.; Crisford, A.; Duque-Vasquez, M.; Luckett, E.; Robertson, R.; et al. Age-Related Changes in Tau and Autophagy in Human Brain in the Absence of Neurodegeneration. *PLoS ONE* **2023**, *18*, e0262792. [CrossRef]

86. Deng, Z.; Dong, Y.; Zhou, X.; Lu, J.H.; Yue, Z. Pharmacological Modulation of Autophagy for Alzheimer's Disease Therapy: Opportunities and Obstacles. *Acta Pharm. Sin. B* **2022**, *12*, 1688–1706. [CrossRef]
87. Andrade-Guerrero, J.; Santiago-Balmaseda, A.; Jeronimo-Aguilar, P.; Vargas-Rodríguez, I.; Cadena-Suárez, A.R.; Sánchez-Garibay, C.; Pozo-Molina, G.; Méndez-Catalá, C.F.; Cardenas-Aguayo, M.D.; Diaz-Cintra, S.; et al. Alzheimer's Disease: An Updated Overview of Its Genetics. *Int. J. Mol. Sci.* **2023**, *24*, 3754. [CrossRef]
88. Root, J.; Merino, P.; Nuckols, A.; Johnson, M.; Kukar, T. Lysosome Dysfunction as a Cause of Neurodegenerative Diseases: Lessons from Frontotemporal Dementia and Amyotrophic Lateral Sclerosis. *Neurobiol. Dis.* **2021**, *154*, 105360. [CrossRef]
89. Koutsodendris, N.; Nelson, M.R.; Rao, A.; Huang, Y. Apolipoprotein E and Alzheimer's Disease: Findings, Hypotheses, and Potential Mechanisms. *Annu. Rev. Pathol.* **2022**, *17*, 73–99. [CrossRef]
90. Pires, M.; Rego, A.C. Apoe4 and Alzheimer's Disease Pathogenesis—Mitochondrial Deregulation and Targeted Therapeutic Strategies. *Int. J. Mol. Sci.* **2023**, *24*, 778. [CrossRef]
91. Journal of Healthcare Engineering. Retracted: Inhibiting Autophagy Pathway of PI3K/AKT/mTOR Promotes Apoptosis in SK-N-SH Cell Model of Alzheimer's Disease. *J. Healthc. Eng.* **2023**, *2023*, 9780432. [CrossRef] [PubMed]
92. Suelves, N.; Saleki, S.; Ibrahim, T.; Palomares, D.; Moonen, S.; Koper, M.J.; Vrancx, C.; Vadukul, D.M.; Papadopoulos, N.; Viceconte, N.; et al. Senescence-Related Impairment of Autophagy Induces Toxic Intraneuronal Amyloid-β Accumulation in a Mouse Model of Amyloid Pathology. *Acta Neuropathol. Commun.* **2023**, *11*, 82. [CrossRef] [PubMed]
93. Park, H.; Kang, J.H.; Lee, S. Autophagy in Neurodegenerative Diseases: A Hunter for Aggregates. *Int. J. Mol. Sci.* **2020**, *21*, 3369. [CrossRef] [PubMed]
94. Reed, E.G.; Keller-Norrell, P.R. Minding the Gap: Exploring Neuroinflammatory and Microglial Sex Differences in Alzheimer's Disease. *Int. J. Mol. Sci.* **2023**, *24*, 17377. [CrossRef]
95. Subramanian, A.; Tamilanban, T.; Alsayari, A.; Ramachawolran, G.; Wong, L.S.; Sekar, M.; Gan, S.H.; Subramaniyan, V.; Chinni, S.V.; Izzati Mat Rani, N.N.; et al. Trilateral Association of Autophagy, mTOR and Alzheimer's Disease: Potential Pathway in the Development for Alzheimer's Disease Therapy. *Front. Pharmacol.* **2022**, *13*, 1094351. [CrossRef]
96. Krishnan, S.; Shrestha, Y.; Jayatunga, D.P.W.; Rea, S.; Martins, R.; Bharadwaj, P. Activate or Inhibit? Implications of Autophagy Modulation as a Therapeutic Strategy for Alzheimer's Disease. *Int. J. Mol. Sci.* **2020**, *21*, 6739. [CrossRef]
97. Feng, Q.; Luo, Y.; Zhang, X.N.; Yang, X.F.; Hong, X.Y.; Sun, D.S.; Li, X.C.; Hu, Y.; Li, X.G.; Zhang, J.F.; et al. MAPT/Tau Accumulation Represses Autophagy Flux by Disrupting IST1-Regulated ESCRT-III Complex Formation: A Vicious Cycle in Alzheimer Neurodegeneration. *Autophagy* **2020**, *16*, 641–658. [CrossRef]
98. Ye, J.; Zhang, J.; Zhu, Y.; Wang, L.; Jiang, X.; Liu, B.; He, G. Targeting Autophagy and Beyond: Deconvoluting the Complexity of Beclin-1 from Biological Function to Cancer Therapy. *Acta Pharm. Sin. B* **2023**, *13*, 4688–4714. [CrossRef]
99. He, Z.; Yang, Y.; Xing, Z.; Zuo, Z.; Wang, R.; Gu, H.; Qi, F.; Yao, Z. Intraperitoneal Injection of IFN-γ Restores Microglial Autophagy, Promotes Amyloid-β Clearance and Improves Cognition in APP/PS1 Mice. *Cell Death Dis.* **2020**, *11*, 440. [CrossRef]
100. Deng, M.; Huang, L.; Zhong, X. β-Asarone Modulates Beclin 1, LC3 and p62 Expression to Attenuate Aβ40 and Aβ42 Levels in APP/PS1 Transgenic Mice with Alzheimer's Disease. *Mol. Med. Rep.* **2020**, *21*, 2095–2102. [CrossRef]
101. Lachance, V.; Wang, Q.; Sweet, E.; Choi, I.; Cai, C.Z.; Zhuang, X.X.; Zhang, Y.; Jiang, J.L.; Blitzer, R.D.; Bozdagi-Gunal, O.; et al. Autophagy Protein NRBF2 Has Reduced Expression in Alzheimer's Brains and Modulates Memory and Amyloid-Beta Homeostasis in Mice. *Mol. Neurodegener.* **2019**, *14*, 43. [CrossRef] [PubMed]
102. McLeod, I.X.; Saxena, R.; Carico, Z.; He, Y.W. Class I PI3K Provide Lipid Substrate in T Cell Autophagy through Linked Activity of Inositol Phosphatases. *Front. Cell Dev. Biol.* **2021**, *9*, 709398. [CrossRef] [PubMed]
103. Jalin, A.M.A.; Jin, R.; Wang, M.; Li, G. EPPS Treatment Attenuates Traumatic Brain Injury in Mice by Reducing Aβ Burden and Ameliorating Neuronal Autophagic Flux. *Exp. Neurol.* **2019**, *314*, 20–33. [CrossRef]
104. Zhao, Y.G.; Codogno, P.; Zhang, H. Machinery, Regulation and Pathophysiological Implications of Autophagosome Maturation. *Nat. Rev. Mol. Cell Biol.* **2021**, *22*, 733–750. [CrossRef] [PubMed]
105. Li, Y.Y.; Qin, Z.H.; Sheng, R. The Multiple Roles of Autophagy in Neural Function and Diseases. *Neurosci. Bull.* 2023, in press. [CrossRef]
106. Kulkarni, V.V.; Stempel, M.H.; Anand, A.; Sidibe, D.K.; Maday, S. Retrograde Axonal Autophagy and Endocytic Pathways Are Parallel and Separate in Neurons. *J. Neurosci.* **2022**, *42*, 8524–8541. [CrossRef]
107. Fahimi, A.; Noroozi, M.; Salehi, A. Enlargement of Early Endosomes and Traffic Jam in Basal Forebrain Cholinergic Neurons in Alzheimer's Disease. *Handb. Clin. Neurol.* **2021**, *179*, 207–218. [CrossRef]
108. Conway, O.; Akpinar, H.A.; Rogov, V.V.; Kirkin, V. Selective Autophagy Receptors in Neuronal Health and Disease. *J. Mol. Biol.* **2020**, *432*, 2483–2509. [CrossRef]
109. Ganesan, D.; Cai, Q. Understanding Amphisomes. *Biochem. J.* **2021**, *478*, 1959–1976. [CrossRef]
110. Jiang, S.; Bhaskar, K. Degradation and Transmission of Tau by Autophagic-Endolysosomal Networks and Potential Therapeutic Targets for Tauopathy. *Front. Mol. Neurosci.* **2020**, *13*, 586731. [CrossRef]

111. Caballero, B.; Bourdenx, M.; Luengo, E.; Diaz, A.; Sohn, P.D.; Chen, X.; Wang, C.; Juste, Y.R.; Wegmann, S.; Patel, B.; et al. Acetylated Tau Inhibits Chaperone-Mediated Autophagy and Promotes Tau Pathology Propagation in Mice. *Nat. Commun.* **2021**, *12*, 2238. [CrossRef] [PubMed]
112. Kent, S.A.; Spires-Jones, T.L.; Durrant, C.S. The Physiological Roles of Tau and Aβ: Implications for Alzheimer's Disease Pathology and Therapeutics. *Acta Neuropathol.* **2020**, *140*, 417–447. [CrossRef] [PubMed]
113. Lin, D.S.; Ho, C.S.; Huang, Y.W.; Wu, T.Y.; Lee, T.H.; Huang, Z.D.; Wang, T.J.; Yang, S.J.; Chiang, M.F. Impairment of Proteasome and Autophagy Underlying the Pathogenesis of Leukodystrophy. *Cells* **2020**, *9*, 1124. [CrossRef] [PubMed]
114. Zhao, Y.; Long, Z.; Ding, Y.; Jiang, T.; Liu, J.; Li, Y.; Liu, Y.; Peng, X.; Wang, K.; Feng, M.; et al. Dihydroartemisinin Ameliorates Learning and Memory in Alzheimer's Disease through Promoting Autophagosome–Lysosome Fusion and Autolysosomal Degradation for Aβ Clearance. *Front. Aging Neurosci.* **2020**, *12*, 47. [CrossRef]
115. Mary, A.; Eysert, F.; Checler, F.; Chami, M. Mitophagy in Alzheimer's Disease: Molecular Defects and Therapeutic Approaches. *Mol. Psychiatry* **2023**, *28*, 202–216. [CrossRef]
116. Kendall, R.L.; Holian, A. The Role of Lysosomal Ion Channels in Lysosome Dysfunction. *Inhal. Toxicol.* **2021**, *33*, 41–54. [CrossRef]
117. Mustaly-Kalimi, S.; Gallegos, W.; Marr, R.A.; Gilman-Sachs, A.; Peterson, D.A.; Sekler, I.; Stutzmann, G.E. Protein Mishandling and Impaired Lysosomal Proteolysis Generated through Calcium Dysregulation in Alzheimer's Disease. *Proc. Natl. Acad. Sci. USA* **2022**, *119*, e2211999119. [CrossRef]
118. Sose, P.M.; Doshi, G.M.; Kale, P.P. An Update on Autophagy as a Target in the Treatment of Alzheimer's Disease. *Curr. Drug Targets* **2023**, *24*, 547–567. [CrossRef]
119. Chen, S.; Zhou, Q.; Ni, Y.; Le, W. Autophagy and Alzheimer's Disease. In *Advances Experimental Medicine and Biology*; Springer: Singapore, Singapore, 2020; Volume 1207, pp. 3–19. [CrossRef]
120. Murariu, M.; Drochioiu, G. Biostructural theory of the living systems. *BioSystems* **2012**, *109*, 126–132. [CrossRef]
121. Trivedi, P.C.; Bartlett, J.J.; Pulinilkunnil, T. Lysosomal Biology and Function: Modern View of Cellular Debris Bin. *Cells* **2020**, *9*, 1131. [CrossRef]
122. Han, X.; Tang, Y.; Zhang, Y.; Zhang, J.; Hu, Z.; Xu, W.; Xu, S.; Niu, Q. Impaired V-ATPase leads to increased lysosomal pH, results in disrupted lysosomal degradation and autophagic flux blockage, contributes to fluoride-induced developmental neurotoxicity. *Ecotoxicol. Environ. Saf.* **2022**, *236*, 113500. [CrossRef] [PubMed]
123. Nixon, R.A. Autophagy, amyloidogenesis and Alzheimer disease. *J. Cell Sci.* **2007**, *120*, 4081–4091. [CrossRef] [PubMed]
124. Kuang, H.; Tan, C.Y.; Tian, H.Z.; Liu, L.H.; Yang, M.W.; Hong, F.F.; Yang, S.L. Exploring the Bi-Directional Relationship between Autophagy and Alzheimer's Disease. *CNS Neurosci. Ther.* **2020**, *26*, 155–166. [CrossRef] [PubMed]
125. Bharadwaj, P.R.; Martins, R.N. Autophagy Modulates Aβ Accumulation and Formation of Aggregates in Yeast. *Mol. Cell. Neurosci.* **2020**, *104*, 103466. [CrossRef]
126. Bassendine, M.F.; Taylor-Robinson, S.D.; Fertleman, M.; Khan, M.; Neely, D. Is Alzheimer's Disease a Liver Disease of the Brain? *J. Alzheimers Dis.* **2020**, *75*, 1–14. [CrossRef]
127. Rostagno, A.A. Pathogenesis of Alzheimer's Disease. *Int. J. Mol. Sci.* **2022**, *24*, 107. [CrossRef]
128. Cheng, Y.; Tian, D.Y.; Wang, Y.J. Peripheral Clearance of Brain-Derived Aβ in Alzheimer's Disease: Pathophysiology and Therapeutic Perspectives. *Transl. Neurodegener.* **2020**, *9*, 16. [CrossRef]
129. Limanaqi, F.; Biagioni, F.; Gambardella, S.; Familiari, P.; Frati, A.; Fornai, F. Promiscuous Roles of Autophagy and Proteasome in Neurodegenerative Proteinopathies. *Int. J. Mol. Sci.* **2020**, *21*, 3028. [CrossRef]
130. Xu, W.; Ocak, U.; Gao, L.; Tu, S.; Lenahan, C.J.; Zhang, J.; Shao, A. Selective Autophagy as a Therapeutic Target for Neurological Diseases. *Cell. Mol. Life Sci.* **2021**, *78*, 1369–1392. [CrossRef]
131. Le Guerroué, F.; Youle, R.J. Ubiquitin Signaling in Neurodegenerative Diseases: An Autophagy and Proteasome Perspective. *Cell Death Differ.* **2021**, *28*, 439–454. [CrossRef]
132. Chen, Y.; Li, Q.; Li, Q.; Xing, S.; Liu, Y.; Liu, Y.; Chen, Y.; Liu, W.; Feng, F.; Sun, H. p62/SQSTM1, a Central but Unexploited Target: Advances in Its Physiological/Pathogenic Functions and Small Molecular Modulators. *J. Med. Chem.* **2020**, *63*, 10135–10157. [CrossRef] [PubMed]
133. Niture, S.; Lin, M.; Rios-Colon, L.; Qi, Q.; Moore, J.T.; Kumar, D. Emerging Roles of Impaired Autophagy in Fatty Liver Disease and Hepatocellular Carcinoma. *Int. J. Hepatol.* **2021**, *2021*, 6675762. [CrossRef] [PubMed]
134. Wu, Y.H.; Mo, S.T.; Chen, I.T.; Hsieh, F.-Y.; Hsieh, S.-L.; Zhang, J.; Lai, M.-Z. Caspase-8 Inactivation Drives Autophagy-Dependent Inflammasome Activation in Myeloid Cells. *Sci. Adv.* **2022**, *8*, eabn9912. [CrossRef] [PubMed]
135. Hennig, P.; Fenini, G.; Di Filippo, M.; Karakaya, T.; Beer, H.-D. The Pathways Underlying the Multiple Roles of p62 in Inflammation and Cancer. *Biomedicines* **2021**, *9*, 707. [CrossRef]
136. Qian, H.; Chao, X.; Williams, J.; Fulte, S.; Li, T.; Yang, L.; Ding, W.-X. Autophagy in Liver Diseases: A Review. *Mol. Asp. Med.* **2021**, *82*, 100973. [CrossRef]

137. Gao, P.Y.; Ou, Y.N.; Wang, H.F.; Wang, Z.B.; Fu, Y.; He, X.Y.; Ma, Y.H.; Feng, J.F.; Cheng, W.; Tan, L.; et al. Associations of liver dysfunction with incident dementia, cognition, and brain structure: A prospective cohort study of 431 699 adults. *J. Neurochem.* **2024**, *168*, 26–38. [CrossRef]
138. Zhang, J.; Wang, W.; Hou, X.; Wu, J.; Wang, Y.; Fan, J.; Zhang, Z.; Yuan, Z.; Sun, C.; Lu, B.; et al. Metabolic-associated steatotic liver disease and risk of Alzheimer's disease: A real-world retrospective cohort study. *Front. Endocrinol.* **2024**, *15*, 1451908. [CrossRef]
139. Weinstein, G.; Schonmann, Y.; Yeshua, H.; Zelber-Sagi, S. The association between liver fibrosis score and incident dementia: A nationwide retrospective cohort study. *Alzheimers Dement.* **2024**, *20*, 5385–5397. [CrossRef]
140. Bajaj, J.S.; Silvey, S.G.; Rogal, S.; O'Leary, J.G.; Patton, H.; Morgan, T.R.; Kanagalingam, G.; Gentili, A.; Godschalk, M.; Patel, N. Undiagnosed Cirrhosis and Hepatic Encephalopathy in a National Cohort of Veterans With Dementia. *JAMA Netw. Open* **2024**, *7*, e2353965. [CrossRef]
141. Tammineni, P.; Ye, X.; Feng, T.; Aikal, D.; Cai, Q. Impaired retrograde transport of axonal autophagosomes contributes to autophagic stress in Alzheimer's disease neurons. *ELife* **2017**, *6*, e21776. [CrossRef]
142. Son, J.; Shim, J.; Kim, K.H.; Ha, J.Y.; Han, J.Y. Neuronal autophagy and neurodegenerative diseases. *Exp. Mol. Med.* **2012**, *44*, 89–98. [CrossRef] [PubMed]
143. Wang, S.; Long, H.; Hou, L.; Feng, B.; Ma, Z.; Wu, Y.; Zeng, Y.; Cai, J.; Zhang, D.-W.; Zhao, G. The mitophagy pathway and its implications in human diseases. *Signal Transduct. Target. Ther.* **2023**, *8*, 304. [CrossRef] [PubMed]
144. Kerr, J.S.; Adriaanse, B.A.; Greig, N.H.; Mattson, M.P.; Cader, M.Z.; Bohr, V.A.; Fang, E.F. Mitophagy and Alzheimer's Disease: Cellular and Molecular Mechanisms. *Trends Neurosci.* **2017**, *40*, 151–166. [CrossRef] [PubMed]
145. Hamano, T.; Enomoto, S.; Shirafuji, N.; Ikawa, M.; Yamamura, O.; Yen, S.-H.; Nakamoto, Y. Autophagy and Tau Protein. *Int. J. Mol. Sci.* **2021**, *22*, 7475. [CrossRef]
146. Bourdenx, M.; Martín-Segura, A.; Scrivo, A.; Rodriguez-Navarro, J.A.; Kaushik, S.; Tasset, I.; Diaz, A.; Storm, N.J.; Xin, Q.; Juste, Y.R.; et al. Chaperone-mediated autophagy prevents collapse of the neuronal metastable proteome. *Cell* **2021**, *184*, 2696–2714. [CrossRef]
147. Fote, G.M.; Geller, N.R.; Efstathiou, N.E.; Hendricks, N.; Vavvas, D.G.; Reidling, J.C.; Thompson, L.M.; Steffan, J.S. Isoform-dependent lysosomal degradation and internalization of apolipoprotein E requires autophagy proteins. *J. Cell Sci.* **2022**, *135*, jcs258687. [CrossRef]
148. Valdor, R.; Martinez-Vicente, M. The Role of Chaperone-Mediated Autophagy in Tissue Homeostasis and Disease Pathogenesis. *Biomedicines* **2024**, *12*, 257. [CrossRef]
149. Cecarini, V.; Bonfili, L.; Cuccioloni, M.; Mozzicafreddo, M.; Angeletti, M.; Keller, J.N.; Eleuteri, A.M. The fine-tuning of proteolytic pathways in Alzheimer's disease. *Cell. Mol. Life Sci.* **2016**, *73*, 3433–3451. [CrossRef]
150. Limone, A.; Veneruso, I.; D'Argenio, V.; Sarnataro, D. Endosomal trafficking and related genetic underpinnings as a hub in Alzheimer's disease. *J. Cell. Physiol.* **2022**, *237*, 3803–3815. [CrossRef]
151. Xiang, Y.; Bu, X.L.; Liu, Y.H.; Zhu, C.; Shen, L.L.; Jiao, S.S.; Zhu, X.Y.; Giunta, B.; Tan, J.; Song, W.H.; et al. Physiological Amyloid-Beta Clearance in the Periphery and Its Therapeutic Potential for Alzheimer's Disease. *Acta Neuropathol.* **2015**, *130*, 487–499. [CrossRef]
152. Maarouf, C.L.; Walker, J.E.; Sue, L.I.; Dugger, B.N.; Beach, T.G.; Serrano, G.E. Impaired Hepatic Amyloid-Beta Degradation in Alzheimer's Disease. *PLoS ONE* **2018**, *13*, e0203659. [CrossRef] [PubMed]
153. Jiang, T.; Yu, J.T.; Zhu, X.C.; Tan, M.S.; Wang, H.F.; Cao, L.; Zhang, Q.Q.; Shi, J.Q.; Gao, L.; Qin, H.; et al. Temsirolimus Promotes Autophagic Clearance of Amyloid-β and Provides Protective Effects in Cellular and Animal Models of Alzheimer's Disease. *Pharmacol. Res.* **2014**, *81*, 54–63. [CrossRef] [PubMed]
154. Huang, Z.; Lin, H.W.K.; Zhang, Q.; Zong, X. Targeting Alzheimer's Disease: The Critical Crosstalk between the Liver and Brain. *Nutrients* **2022**, *14*, 4298. [CrossRef] [PubMed]
155. Spilman, P.; Podlutskaya, N.; Hart, M.J.; Debnath, J.; Gorostiza, O.; Bredesen, D.; Galvan, V. Inhibition of mTOR by rapamycin abolishes cognitive deficits and reduces amyloid-β levels in a mouse model of Alzheimer's disease. *PLoS ONE* **2010**, *5*, e9979. [CrossRef]
156. Vingtdeux, V.; Giliberto, L.; Zhao, H.; Chandakkar, P.; Wu, Q.; Simon, J.E.; Marambaud, P. AMP-activated protein kinase signaling activation by resveratrol modulates amyloid-β peptide metabolism. *J. Biol. Chem.* **2010**, *285*, 9100–9113. [CrossRef]
157. Reddy, P.H.; Manczak, M.; Yin, X.; Grady, M.C.; Mitchell, A.; Kandimalla, R.; Kuruva, C.S. Protective effects of a natural product, curcumin, against amyloid β induced mitochondrial and synaptic toxicities in Alzheimer's disease. *J. Investig. Med.* **2016**, *64*, 1220–1234. [CrossRef]
158. Chuang, D.M.; Leng, Y.; Marinova, Z.; Kim, H.J.; Chiu, C.T. Multiple roles of HDAC inhibition in neurodegenerative conditions. *Trends Neurosci.* **2009**, *32*, 591–601. [CrossRef]
159. He, C.; Sumpter, R., Jr.; Levine, B. Exercise induces autophagy in peripheral tissues and in the brain. *Autophagy* **2012**, *8*, 1548–1551. [CrossRef]

160. Madeo, F.; Carmona-Gutierrez, D.; Hofer, S.J.; Kroemer, G. Caloric restriction mimetics against age-associated disease: Targets, mechanisms, and therapeutic potential. *Cell Metab.* **2019**, *29*, 592–610. [CrossRef]
161. Jiang, Y.; Uhm, H.; Ip, F.C.; Ouyang, L.; Lo, R.M.N.; Cheng, E.Y.L.; Cao, X.; Tan, C.M.C.; Law, B.C.H.; Ortiz-Romero, P.; et al. A Blood-Based Multi-Pathway Biomarker Assay for Early Detection and Staging of Alzheimer's Disease across Ethnic Groups. *Alzheimers Dement.* **2024**, *20*, 2000–2015. [CrossRef]
162. Ren, J.; Sowers, J.R.; Zhang, Y. Metabolic Stress, Autophagy, and Cardiovascular Aging: From Pathophysiology to Therapeutics. *Trends Endocrinol. Metab.* **2018**, *29*, 699–711. [CrossRef]
163. Liu, Y.; Levine, B. Autosis and autophagic cell death: The dark side of autophagy. *Cell Death Differ.* **2015**, *22*, 367–376. [CrossRef]

cells

MDPI

Review

Modulating Cognition-Linked Histone Acetyltransferases (HATs) as a Therapeutic Strategy for Neurodegenerative Diseases: Recent Advances and Future Trends

Huong Anh Mai †, Christina M. Thomas †, Gu Gu Nge † and Felice Elefant *

Department of Biology, Drexel University, Philadelphia, PA 19104, USA; am4829@drexel.edu (H.A.M.); cmt397@drexel.edu (C.M.T.); gn345@drexel.edu (G.G.N.)

* Correspondence: fe22@drexel.edu; Tel.: +1-215-895-0220; Fax: +1-215-895-1273

† These authors contributed equally to this work.

Abstract: Recent investigations into the neuroepigenome of the brain are providing unparalleled understanding into the impact of post-translational modifications (PTMs) of histones in regulating dynamic gene expression patterns required for adult brain cognitive function and plasticity. Histone acetylation is one of the most well-characterized PTMs shown to be required for neuronal function and cognition. Histone acetylation initiates neural circuitry plasticity via chromatin control, enabling neurons to respond to external environmental stimuli and adapt their transcriptional responses accordingly. While interplay between histone acetylation and deacetylation is critical for these functions, dysregulation during the aging process can lead to significant alterations in the neuroepigenetic landscape. These alterations contribute to impaired cognitive functions, neuronal cell death, and brain atrophy, all hallmarks of age-related neurodegenerative disease. Significantly, while age-related generation of DNA mutations remains irreversible, most neuroepigenetic PTMs are reversible. Thus, manipulation of the neural epigenome is proving to be an effective therapeutic strategy for neuroprotection in multiple types of age-related neurodegenerative disorders (NDs) that include Alzheimer's disease (AD), Parkinson's disease (PD), Amyotrophic lateral sclerosis (ALS) and Huntington's disease (HD). Here, we highlight recent progress in research focusing on specific HAT-based neuroepigenetic mechanisms that underlie cognition and pathogenesis that is hallmarked in age-related NDs. We further discuss how these findings have potential to be translated into HAT-mediated cognitive-enhancing therapeutics to treat these debilitating disorders.

Keywords: neuroepigenetics; histone acetyltransferases (HATs); histone deacetylases (HDACs); neurodegenerative diseases (NDs); histone code; learning and memory; cell death; synaptic plasticity

1. Introduction

Within the nucleus of eukaryotic cells, the chromatin environment is organized into nucleosomes, which comprise DNA wrapped around a histone octamer core consisting of two copies each of histone proteins H2A, H2B, H3, H4 [1,2]. Such compact chromatin organizations present a major physical hurdle that transcriptional machinery must overcome to initiate gene transcription in vivo. Chromatin packaging is tightly controlled by tissue and site-specific covalent epigenetic post-translational modifications (PTMs) of the core histone proteins that can trigger chromatin packaging changes to activate or repress genes. Thus, epigenetic PTMs elicit heritable phenotypic variation without altering the

https://doi.org/10.3390/cells14120873

nucleotide sequence that comprise genes [3]. Epigenetic PTMs include acetylation, methylation, phosphorylation, ubiquitination, sumoylation, and ADP-ribosylation [4]. Histones are often concurrently modified on several residues that allow for an active interplay between histone PTMs and DNA methylation, thus creating diverse combinatorial possibilities for transcriptional control [5]. These modifications not only can modulate chromatin structure, but can also serve as distinct docking sites for other protein complexes that activate or repress transcription [6,7]. Regulation of these PTMs is governed by enzymes that function as "writers" to add specific PTMs to histones, recognized by "readers" that comprise regulatory proteins that control chromatin packaging, and removed by enzymes that serve as "erasers". Importantly, the recruitment of chromatin-interacting proteins often depends not on a single PTM, but rather on a combination of local marks [7]. These chromatin regulatory proteins work in concert to determine the transcriptional programs employed during the cell differentiation process, timing-specific developmental stages, or stably maintained gene control across generations [8].

Recent explorations into the brain neuroepigenome are providing unprecedented insights into the impact of PTMs of histones in controlling gene expression, not only in early brain development but also in adult brain cognitive function [9]. Together, with studies implicating altered chromatin structure and histone PTM patterns in multiple age-related neurodegenerative diseases, these findings have generated intense interest into investigating how the aging process can detrimentally affect the brain neuroepigenome. While age-related generation of DNA mutations is irreversible, significantly, the majority of neuroepigenetic modification marks are reversible. One of the most extensively studied reversible PTM critical for mediating cognition is histone acetylation, which serves to initiate a neural circuitry plasticity process via chromatin control, enabling neurons to adapt their transcriptional responses to external environmental cues [10]. Importantly, many compelling studies support the concept that cognitive ability in the ND brain is not entirely lost and can be recovered by therapeutic reversal of histone acetylation-mediated gene transcription. In one such groundbreaking study, researchers demonstrated that cognitive capacities in the neurodegenerating brain in an AD-associated mouse model are constrained by an epigenetic blockade of gene transcription. Such transcriptional repression is caused by an increase in the histone deacetylation enzyme HDAC2 that triggers inappropriate hypoacetylation of chromatin at synaptic plasticity gene loci, thereby reducing expression [11]. Importantly, reducing HDAC2 buildup reinstates histone acetylation to restore plasticity gene expression, promoting renewal of structural plasticity in the brain and memory restoration. Similarly, medicinal activation of HAT p300/CBP to increase histone acetylation in the ND brain induces neuronal survival and stimulation of neurite growth by upregulation of neuroprotective gene transcription [12,13]. Together, these findings highlight the reversible nature of histone acetylation-mediated neuroepigenetic modification marks and their role in modulating memory-related gene expression. Thus, manipulating the neuroepigenome is proving to be a promising strategy for neuroprotection of multiple types of age-related neurodegenerative disorders (NDs) that include Alzheimer's disease (AD), Parkinson's disease (PD), Amyotrophic lateral sclerosis (ALS), and Huntington's disease (HD) [14–17]. Here, we will discuss recent research findings that link neuroepigenetic histone acetylation to cognition and pathogenesis hallmarked in age-associated NDs. We then provide insights into how these findings have potential to be translated into effective cognitive-enhancing therapeutic strategies to treat these debilitating neurodegenerative diseases. This review synthesizes peer-reviewed studies from 2000–2025.

2. HAT Families in Brain Function and Cognition

One of the most intensely researched PTMs to date that is critical for brain function and cognition is histone acetylation, which regulates neural gene expression by controlling chromatin packaging in neurons [10,11,17–22]. This epigenetic modification is governed by family members of histone acetyltransferase (HAT) enzymes that catalyze the transfer of an acyl moiety from acyl coenzyme A to lysine residues on the n-terminal tails of the histones [23]. Due to the rapid and reversible nature of histone acetylation, this modification serves as a highly dynamic epigenetic mechanism that links environmental stimuli to the molecular adaptations underlying experience-dependent behavioral changes. Thus, cognition relies on coordinated neural gene expression programs to achieve synthesis of new synaptic proteins, which is dependent on histone acetylation control of chromatin structure in neurons [24]. Acetylation alters the chromatin state in favor of a more relaxed, open euchromatin conformation that is conducive to transcriptional activation [8,25]. To achieve this, acetylation is believed to neutralize the positive charge on lysine residues, thus reducing the interaction affinity between histones and DNA, allowing for a more relaxed chromatin state [1,26]. Recent evidence points to acetylated lysine residues serving as molecular docking sites for proteins, including bromodomain-containing ATP-dependent chromatin remodeling complexes and components of the transcriptional machinery, that further facilitate relaxation of chromatin structure [25,27]. The relaxed hyperacetylated chromatin conformation is characterized by increased sensitivity to DNase digestion and enhanced flexibility of the nucleosomal DNA, due to reduced conformational hindrance from nucleosomal compaction [28]. This chromatin unwinding allows access of the transcriptional machinery to the regulatory elements of the genes, facilitating their expression. Further, the TBP-associated factor (TAF_{II}) also possesses intrinsic HAT activity, enabling the unwinding of regions around specific genes to allow TFIID access to previously repressed genomic regions [29]. Other groups have demonstrated increased access to gene promoter regions via histone acetylation, which facilitates the ability of RNA PolII to initiate transcription at specific genetic loci [29]. Based on sequence similarity, substrate preferences, and functions, HATs are divided into distinct families. The three most notable HAT families that are required for brain function and cognition are GCN5/PCAF, MYST, and CBP/p300 [30]. Their functional roles in the learning and memory process are discussed in the following sections (Figure 1).

CBP/p300 Family. CBP (also known as CREB binding protein or KAT3A) and p300 (also known as EP300 or KAT3b) are members of the CBP/p300 (also known as KAT3) HAT superfamily [31]. These HATs share highly similar sequence and structural organizations with overlapping yet distinct roles in brain function. They contain protein interaction domains, including the nuclear receptor interaction domain (RID), the CREB and MYB interaction KIX domain, the cysteine/histidine regions (CH1 and CH3), and the interferon response-binding domain (IBiD) [4]. These HATs also contain a histone acetyltransferase domain, a bromodomain that interacts with acetylated lysine residues, and a plant homeodomain (PHD)-type zinc finger motif that is a part of the integral enzymatic core of the acetyltransferase domain. [32]. CBP and p300 were each initially identified in protein interaction screens for CREB (cAMP response element) and E1A, respectively. Several key studies revealed that both CBP and p300 are expressed ubiquitously in the cortex and hippocampus [12]. Although both CBP and p300 have overlapping expression patterns, their distinct roles in memory formation arise from the different protein partners they interact with and the preferential histone lysine residues they target. The first evidence that CBP played a role in cognition was revealed by the discovery that Rubinstein–Taybi Syndrome (RTS), a severe neurodevelopmental disorder accompanied by mental retardation, was caused by mutations in the gene encoding CBP [33]. Subsequent studies revealed that mice with a haploid knockout for CBP (cbp+/−) showed reduced histone acetylation and impairment of hippocampus long-term potentiation (L-LTP) and long-term memory (LTM) [34]. Notably, CBP's HAT function was necessary for these functions, solidifying a critical role for CBP HAT function in cognition [35]. Additionally, HATs such as the E1A-binding protein p300 (p300) are shown to be critical for memory processes [36,37]. To clarify specific roles of CBP and p300 in brain function, studies have utilized p300 and CBP KIX mutant mice and demonstrated that p300 and CBP play different roles in motor-skill learning [38]. In contrast, studies using HAT mutant p300 and CBP mutant mice revealed that both HATs are specifically necessary for long-term recognition and contextual fear memory [37]. More recently, the CBP KIX domain has also been shown to regulate long-term memory and circadian activity in mouse models [39]. Thus, CBP and p300 appear to play functionally overlapping roles in some but not all types of memory processes.

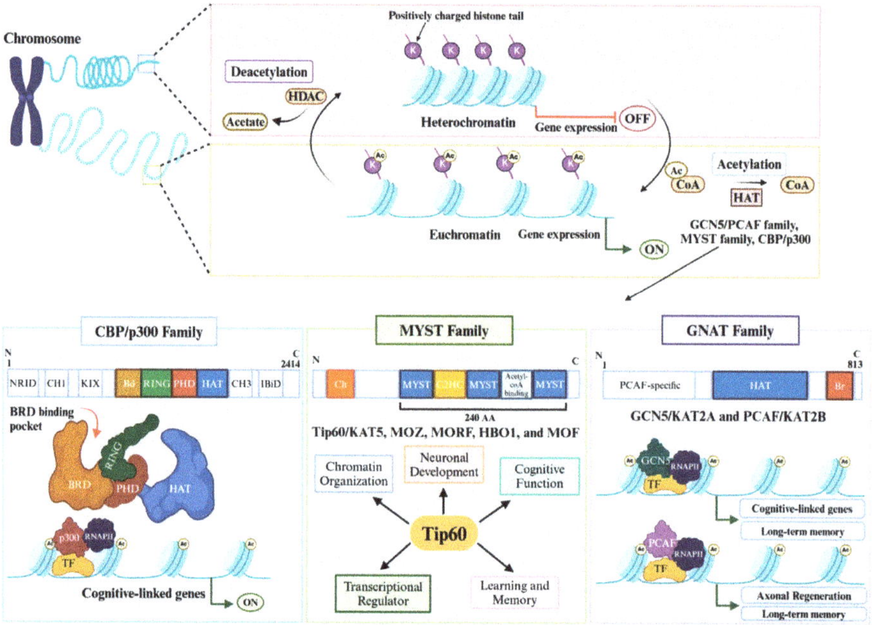

Figure 1. HAT families in brain function and cognition. The three most notable HAT families that are required for brain function and cognition are the CBP/p300, MYST, and GNAT. Their functional roles in the specific cognition-linked neuronal cellular processes they control are illustrated.

MYST Family. The MYST family compromises five HATs: Tip60 (KAT5), MOZ (KAT6A, MORF (KAT6B), HBO1 (KAT7), and MOF (KAT8) [40]. MYST family members each contain a MYST region of about 240 amino acids, a canonical acetyl-CoA binding site, and a C2HC-type zinc finger motif [41]. They also possess a chromodomain involved in protein–protein interactions and, more recently discovered, in neural RNA interactions [40]. Tip60/KAT5 (Tat-interactive protein-60 kDa) is one of the best characterized MYST family HATs concerning its role in cognition, providing substantial insights into histone acetylation-mediated gene control required for neural function in learning and memory [22]. Tip60 is a member of the Tip60 protein complex shown to possess not only HAT activity, but also ATPase, DNA helicase, and DNA binding activities [27]. In an expression analysis of the 18 mammalian HATs in the learning and memory hippocampal CA1 region in adult mice, Tip60 was identified as the second highest expressed HAT [42]. Similarly, Tip60 is robustly expressed in the *Drosophila* mushroom body, a brain region functionally analogous to the mammalian hippocampus and essential for learning and memory, supporting species conservation of Tip60 brain localization [22]. A plethora of research using the *Drosophila* model system has defined a critical role for Tip60 in multiple neuronal brain functions. For example, reducing Tip60 HAT levels within the fly mushroom body causes significantly shorter mushroom body lobes in the adult fly brain, indicative of defects in axonal outgrowth [22]. Further, loss of Tip60 HAT activity in motor neurons causes axonal transport stalling. Additionally, reduction of Tip60 in the small lateral neuron ventral (sLNv) that regulate sleep/wake cycles in the fly causes repressed axonal outgrowth [43,44]. Functional consequences of these Tip60-mediated defects include impairment in short-term memory, sleep/wake cycles, and locomotor activity [16,22,43]. Additionally, Tip60 HAT activity has been shown to regulate proper synaptic bouton expansion at the *Drosophila* neuromuscular junction through modulation of the synaptic microtubule cytoskeleton [45]. Although the identity of the genes responsible for the morphological and molecular changes underlying learning and memory remain incompletely characterized, advances in high-throughput sequencing technology have begun to uncover epigenetically regulated genes involved in synaptic plasticity and neuronal development. For instance, microarray analysis of Tip60 HAT-mutant flies revealed that Tip60 regulates the expression of genes crucial for neuronal development and synaptic function [46]. Further studies have since revealed that several synaptic plasticity genes are bona fide direct targets of Tip60, and the loss of Tip60 activity in mutant flies results in reduced Tip60 chromatin enrichment and synaptic gene repression [47]. Moreover, chromatin immunoprecipitation sequencing (ChIP-Seq) analysis in the *Drosophila* brain reveals that Tip60 target genes are enriched for functions in cognitive processes and, accordingly, key genes representing these pathways are dysregulated in the Tip60 HAT mutant fly brain [22]. Similarly, Tip60 (KAT5) reduction in the forebrain of adult mice causes substantial transcriptional impairments with concomitant neuronal cell loss in the hippocampal CA1 region that is accompanied by mild memory impairment [48]. Additional studies revealed a role for Tip60 within the hippocampus during strong memory training [49]. More recently, studies characterizing dysregulated gene sets (DEGs) in cKO Tip60 mice identified significant overlap with DEGs in AD patients, highlighting the importance of these genes in neuronal function and AD [50]. Together, these studies support a critical role for Tip60 in regulating transcriptional networks required for neuronal function and cognition in both *Drosophila* and mammalian systems.

While fewer studies have investigated potential roles for MOZ(KAT8), MOF(KAT6B), and HBO1(KAT7) in brain function, each of these HATs are expressed in adult mouse hippocampus, and their knockdown causes defects in neuronal progenitor cell function and repression of neuronal-patterning gene expression [15,51]. MORF (KAT6B) is also highly expressed in both the embryonic and adult brain and thought to function in adult neurogenesis [51]. More recently, sociability, learning, and memory were shown to be impaired in MORF (KAT6B) haploinsufficiency mice, supporting a role for this HAT in cognitive function as well [52]. Additionally, gene expression analysis in MORF (KAT6B) haploinsufficiency mice revealed reduced expression of genes required for brain development, and neuron differentiation and expansion [53]. Additional studies investigating MOF (KAT8) identified embryonic lethality in mice with a conditional MOF deletion that was caused by neural inflammation with metabolic impairment of neuronal cells [54]. Lastly, research focused on HBO1 (KAT7) identified a novel cellular pathway where KO of neurite outgrowth inhibitor-B receptor (NgBR) inhibits KAT7, causing inhibition of downstream transcription and regulatory factors that trigger neuronal damage [55].

GNAT Family. The Gcn5-related N-acetyltransferases (GNAT) HAT family contains two primary representative members, GCN5/KAT2A and PCAF/KAT2B, which each play critical roles in the developing brain [56]. The GNAT family HATs contain a C-terminal bromodomain that can bind specifically to acetylated lysine residues and have up to four conserved motifs (A-D) that are within the catalytic HAT domain [57]. Gcn5 has been shown to be the most highly expressed of the 18 HATs present in the learning and memory hippocampal CA1 region in adult mice, and was identified as one of the HAT genes that is upregulated in response to fear-conditional training, implicating a critical role for Gcn5 in cognition [42,58]. Indeed, conditional deletion of Gcn5 causes substantial loss of brain mass, resulting in dysregulation of neural progenitor differentiation [58]. Furthermore, elegant studies demonstrated that knockout of *Gcn5/Kat2a* in mice show impaired hippocampal synaptic plasticity and long-term memory consolidation [42]. The study demonstrated that Gcn5/Kat2a controls a hippocampal gene expression network that is highly interconnected, supporting Gcn5/Kat2a as a critical HAT in the control of memory formation [42].

While there are limited studies regarding PCAF function in the brain, research shows that PCAF-dependent epigenetic changes promote axonal regeneration in the central nervous system. The research showed that extracellular signal-regulated kinase (ERK) axonal retrograde signaling triggers PCAF-dependent acetylation at promoters of axonal regeneration genes, and that overexpression of PCAF itself is sufficient to promote axonal regeneration [59]. Moreover, mice carrying a homozygous knockout of the PCAF gene are viable and show short-term memory impairment at 2 months of age, which is the adolescent stage for mice [36]. Interestingly, memory deficits exhibited in mice carrying the PCAF gene KO become exacerbated during aging with mice exhibiting long-term contextual impairments at 6 and 12 months [36]. Together, these studies support the concept that Gcn5 and PCAF each play distinct roles in particular memory formation processes.

3. HAT: HDAC Interplay in Activity-Dependent Cognition-Linked Gene Control

The generation of chromatin histone acetylation by HATs is reversible and is conversely controlled by the antagonistic activity of histone deacetylase (HDAC) enzymes. HDAC enzymes function to catalyze removal of acetyl groups from specific and conserved histone lysine residues, causing a compacted and transcriptionally repressive chromatin environment [60]. Under healthy conditions, maintaining histone acetylation homeostasis within the central nervous system via the interplay between HAT and HDAC action is required for neuroepigenetic gene control involved in neuronal survival and function. This concept is supported by numerous studies showing that altering HAT/HDAC homeostasis in neurons causes detrimental effects. For example, treatment of dopaminergic neural cell culture lines using rat, mouse, and human cell lines treated with the HDAC inhibitor trichostatin A (TSA) triggers apoptosis of neuronal cells [61]. Furthermore, increasing either HDAC6 or HDAC2 levels in neuronal nuclei results in deleterious consequences for transcriptional regulation required for synaptic function. Similarly, increasing CBP HAT levels in primary cerebellar granule neurons (CGN) causes chromatin condensation and cell death [62,63]. Additionally, appropriate levels of Tip60 HAT activity in the *Drosophila* learning and memory mushroom body center in the brain were shown to be crucial for immediate recall memory [44]. Conversely, targeted overexpression of the HDAC2 (Rpd3) and HDAC4 to reduce acetylation levels within the adult *Drosophila* mushroom body resulted in severely impaired long-term courtship memory [64,65]. These findings highlight the importance of appropriate HAT/HDAC levels and balance in promoting neural health and function.

The human brain contains a multitude of neuronal networks that form synaptic connections that change in response to experiences; however, many of these changes are not long-lasting. The basis for long-term memory formation is dependent upon activity-dependent gene control that is mediated by neuroepigenetic histone acetylation-mediated mechanisms. Activity-dependent gene regulation is a fundamental mechanism by which neurons adapt and fine-tune their transcriptional responses to behavioral, environmental, and neural cues to promote sustained neural plasticity and higher-order brain function [66]. Support for this concept was first demonstrated by compelling studies showing that mice subjected to contextual fear conditioning exhibited significantly enriched histone acetylation marks at distinct sites on H3 and H4 with a concomitant increase in multiple learning-associated genes [58]. Indeed, the specific HATs and HDACs that generate distinct chromatin histone acetylation patterns and the mechanisms by which they respond to external stimuli still remains to be determined. The first HAT identified to be critical in activity-dependent gene control is CBP, which was shown to use Ca2+-mediated signaling pathways in response to external neuronal stimuli. Research using primary mouse neuronal cells in culture dosed with potassium chloride (KCl) [67] showed that Ca^{2+} influx promotes transcription at activity-dependent neuronal enhancers via CBP recruitment to these gene loci. Neuronal stimuli also control chromatin histone acetylation by regulating nucleocytoplasmic transport of specific class II HDACs and Tip60 HAT into and out of the nucleus. In DIV rat hippocampal neurons, stimulation with KCl caused the translocation of Tip60 into the nucleus and transcriptional activation of bona fide Tip60 target genes [68]. Additional studies showed that class II HDACs accumulated in the nucleus when neuronal cell cultures were treated with a nuclear calcium-signaling inhibitor, hindering neuronal activation [69]. This regulation of HDAC intracellular localization is critical in mediating histone acetylation and transcriptional activity of neuronal genes that sense and respond to environmental stimuli [20].

Neuronal activity-mediated sensory experiences also have differential effects on synaptic plasticity at excitatory or inhibitory synapses, triggering long-term potentiation (LTP) or long-term depression (LTD), whereby synaptic transmission is up- or downregulated, respectively [10]. Particular forms of LTP and LTD require long-lasting changes in gene expression that is mediated by histone acetylation. In support of this concept, an elegant study using sensory motor neurons of *Aplysia* showed that the coordinated interplay between CBP HAT and HDAC5 regulates histone acetylation levels to promote the switch between LTP and LTD at the same synapses [70]. Intriguingly, the activity-dependent remodeling events that shape the formation of neural circuits during development mirror those underlying changes in synaptic connectivity necessary for learning and memory formation [71]. Experience-dependent neuroadaptation requires an increase in synaptic bouton densities and dendrite extensions, similar to those essential during development [72]. Therefore, it has been hypothesized that the signaling cascades activated by experience-driven changes in cognitive function modulate epigenetically mediated transcriptional control of genes that promote synaptic growth and neuronal homeostasis.

Post-mitotic neurons within the brain have the remarkable ability to shape and refine their synaptic connections in response to ever-changing environmental external stimuli. Multiple studies have shown that environmental enrichment (EE) that includes diversity in housing, play toys, and exercise apparati significantly enhances memory in mice. Indeed, mice raised under EE conditions show changes in the expression of genes in the brain involved in formation of new synapses, strengthening of existing synapses, neurotransmission, and cytoskeletal changes involved in promoting neurogenesis [73]. Elegant work demonstrated the epigenetic mechanisms underlying these beneficial morphological changes by showing that EE conditions trigger hippocampal induction of histone acetylation specifically at marks H3 (K9, K14) and H4 (K5, K8, K12) [74]. Moreover, work with *Drosophila* demonstrated that EE also enhances histone acetylation marks in the fly mushroom body EE-exposed *Drosophila* brains show significant increase in H3K14ac, H4K5ac, and H4K12ac under the EE condition that is mediated by Tip60 HAT, suggesting that the EE-induced histone acetylation induction response is tightly conserved in both flies and mice [75]. Further, work by [76] demonstrated that rats that were housed in new open-field environments that promoted their exploratory nature caused PTM-associated chromatin remodeling mediated by histone H3 Ser10 phosphorylation and H3 Lys14 acetylation with concomitant activation of the intermediate early genes (IEG) c-Fos and zif/268 protein expression within the hippocampal CA1 location in the brain required for learning and memory. Together, these studies support that HAT/HDAC homeostatic balance plays a critical role in mediating EE transcriptional control that contributes to cognition.

4. Neuroepigenetic Alterations Drive Multiple Age-Related Neurodegenerative Diseases

Appropriate histone acetylation homeostasis is maintained by the antagonistic activity of HAT and HDAC [17,77]. Disruption of this finely tuned epigenetic balance in the aged brain is characterized by reduced histone acetylation and triggers an epigenetic blockade of transcription. This impairment is associated with detrimental consequences on neural function and has been proposed to be a key early step in age-related cognitive decline and multiple age-associated neurodegenerative diseases. Accordingly, cases of reduced histone acetylation levels are found in the brains of animal models for multiple types of neurodegenerative diseases [17,74,78–81]. Pharmacological therapies focused on elevating histone acetylation by HDAC inhibition in ND animal models have shown alleviation of cognitive impairments and are thus a topic of intense research [77,79,82]. This section reviews recent findings that reveal an association between impairment of HAT and HDAC balance and function to alterations in the neuroepigenome that contribute to multiple neurodegenerative diseases (Figure 2).

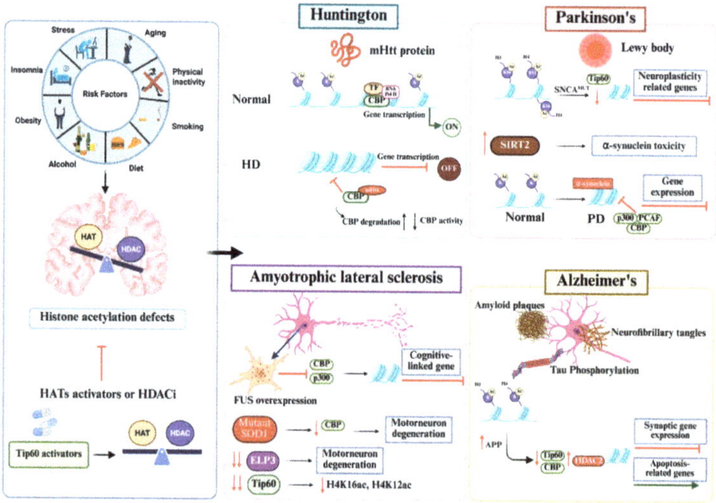

Figure 2. HAT-mediated neuroepigenetic alterations drive multiple age-related neurodegenerative diseases. Reduced HAT and histone acetylation levels are found in the brains of animal models for multiple types of neurodegenerative diseases. Pharmacological treatments aimed at increasing histone acetylation by inhibiting HDAC action or activating specific HATs in these models have shown reversal of cognitive deficits and are thus a topic of intense research. This schematic provides an overview of data linking impaired control of specific HATs and HDACs to alterations in the neuroepigenome that contribute to multiple neurodegenerative diseases.

4.1. Huntington's Disease

Huntington's disease (HD) is a fatal neurodegenerative disorder characterized by motor and cognitive symptoms caused by significantly extended and unstable CAG repeats (polyglutamine or polyQ expansion) that occur within the coding region of the huntingtin (Htt) gene [83]. This expansion results in production of a mutant HTT protein (mHtt) that contains a toxic polyglutamine (polyQ) tract that causes degeneration of the caudate nucleus and the putamen within the basal ganglia of the brain [83]. Such deficits typically appear in adulthood with severity dependent on the length of the polyQ tract. Transcriptional dysregulation plays a central role in HD pathogenesis with alterations being reported in the brains of both HD patients and multiple HD mouse models [84]. Although the precise mechanism underlying HD-related neurotoxicity remains poorly understood, increasing evidence suggests that the expanded glutamine tract within mHtt confers it with new toxic properties that include aberrant interactions with proteins that include HATs required for chromatin acetylation and gene expression [85]. For example, increased HD with expanded CAG repeats was shown to induce redistribution of CBP into inclusions within the nucleus or cytoplasm in multiple cellular models, thereby reducing HAT activity of CBP, causing global deacetylation and cellular death [86]. Additional studies have shown that a polyglutamine peptide that corresponds to exon 1 of Htt (Httex1p) inappropriately interacts with CBP and PCAF HAT domains in vitro, thereby titrating out HAT activity [87,88]. Such reduced histone acetylation in HD patients, mouse, and *Drosophila* HD models causes chromatin hypoacetylation with concomitant transcriptional dysregulation of multiple genes, including those in involved in long-term memory [89,90].

Studies have shown that polyQ-expanded Htt induces ubiquitylation of CBP and its degradation [91,92]. As such, HDAC inhibitor (HDACi) treatments show restoration of decreased histone acetylation levels and alleviate disease pathology in mouse and *Drosophila* HD models. In support of this concept, induction of HTTex1P in cells induces reduction in H3 and H4 acetylation that is reversed by treating cells with pan-HDAC inhibitors that include trichostatin (TSA) and suberoylanilide hydroxamic acid (SAHA) [90]. Further, neurodegeneration of photoreceptor neurons in *Drosophila* induced by Httex1p can be mitigated through treatment with HDAC inhibitors such as sodium butyrate (NaB) and SAHA. Additionally, suppression of class I HDAC Rpd3 and class III HDAC sirtuin 2 (SIRT2), either alone or in combination by genetic and/or pharmacological manipulation, confers neuroprotective effects [93]. Similarly, treatments of SAHA reduce HDAC4 in the cortex and brain stem in an HD mouse model [94]. In a study exploring the effects of HDAC1 inhibitor 4b, researchers found that within fly, cell, and mouse models of HD, HDAC3 was selectively targeted, and the inhibitor was able to suppress HD-related symptoms [95]. More recent studies demonstrate that a selective inhibitor of HDAC3 (selective inhibitor RGFP966) suppressed striatal CAG repeat expansions and prevented cognitive defects in Huntington's disease mice [96]. Taken together, these research findings highlight the concept that HD is at least in part caused by reduced histone acetylation levels.

4.2. Parkinson's Disease

Parkinson's disease (PD) is a progressive neurodegenerative disorder, characterized by degeneration and death of dopaminergic (DA) neurons in the substantia nigra pars compacta (SNc) within the ventral midbrain [97]. Molecular studies have characterized that the pathological hallmark of PD is by misfolded protein aggregation (Lewy body), which is mainly composed of alpha-synuclein (a-SYN) in PD patients [98]. PD progresses through multiple interconnected mechanisms that include mitochondrial dysfunction, energy depletion, oxidative stress, excitotoxicity, protein misfolding and aggregation, disruption of protein clearance mechanisms, cell-intrinsic disturbances, and the prion-like spread of misfolded proteins [99]. Approximately 10–15% of PD cases are familial and known to be caused by genetically inherited mutations in PD-associated genes such as SNCA, Parkin, and LRRK2, and likely genes that have yet to be identified. Unfortunately, most PD cases are classified as sporadic (sPD) with no family history, making it challenging to identify specific genes and environmental triggers. PD pathology is characterized by bradykinesia, muscular rigidity, unstable posture and gate, and resting tremor [100]. Additionally, non-motor features including dementia and depression have been described. A plethora of research has demonstrated a link between epigenetic dysregulation and Parkinson's disease [101]. The initial link between PD and dysregulation of histone acetylation was established from findings that the PD-linked α-syn protein interacts with histones that caused inactivation of HATs that include CBP, p300, and PCAF, triggering enhanced apoptotic cell death in human neuroblastoma cells [102,103]. The A53T mutation in SNCA also significantly decreased acetylation of histone H3 in cultured SH-SY5Y cells and reduced histone H3 acetylation is found in the striatum and substantia nigra of PD mice injected with a-synuclein folded fibrils (PFFs). PFF also reduces CBP/p300-associated factor TADA2A, which is required for acetylation of histone H3/H4 [104]. Additionally, in primary cortex neurons of transgenic mice that express human wild-type α-synuclein, upregulation of α-synuclein is linked with reduced H3K9ac and increased H3K14ac and H3K18ac, where both H3K14 and H3K18 are important histone modifications that are known to facilitate regulation of active transcription [105]. Moreover, brain-specific Tip60 HAT levels are reduced in the *Drosophila* PD model, causing H4K16 and H4K12 acetylation reduction at cognition-linked gene loci with concomitant transcriptional repression [89].

Both HDAC6 and SIRT2 are histone deacetylases (HDACs) that are primarily localized in the cytoplasm and have both been implicated in PD pathogenesis. HDAC6 and SIRT2 both play a role in deacetylation of α -tubulin and additional proteins that mediates microtubule dynamics and trafficking. Multiple findings demonstrate that age-dependent accumulation of SIRT2 in the brain and spinal cord in mice correlates with decreased α-tubulin acetylation in cortical neurons that reduces acetylation of α synuclein, thereby increasing its aggregation and toxicity in neurons [106,107]. Additional research has demonstrated that genetic or pharmacological inhibition of SIRT2 rescued α-synuclein toxicity in dopaminergic neurons. Accordingly, studies showed that reducing SIRT2 activity in human neuroglioma cells inhibits α-synuclein toxicity by promoting enlargement of α-synuclein inclusions [108]. Interestingly, reduced histone acetylation in the PD brain has been shown to be brain-region specific. For example, assessment of histone acetylation in CNS of post-mortem patients with PD and age-matched controls using high resolution studies such as ChIP and RNA-Seq reveals a significant increase in acetylated histone residues in PD patients compared to control individuals. In contrast, the histone acetylation of AcH2A-K3, AcH3-K9, and AcH3-K5 did not increase in the cerebellar cortex in PD patients [109]. Further, studies have identified H3K27-acetylated regions in over 20 genes associated with familial or sporadic forms of PD and are also involved in dysregulating PD associated genes that include SNCA, PARK7, and MAPT [110]. Together, these studies support dysregulation of histone acetylation, alongside inherited genetic mutations and environmental factors, as contributing factors for PD.

Significant efforts have focused on developing HDAC inhibitors for potential therapeutic use in PD. In both cultured cells and transgenic fly lines, the administration of two different HDAC inhibitors, NaB or SAHA, which showed efficacy for HD, also showed significant protection against α-synuclein-dependent neurotoxicity in PD [102]. For instance, valproic acid (VPA), a class I/IIa HDAC inhibitor, displays neuroprotective effects by increasing acetylation in tyrosine hydroxylase (TH)-positive dopaminergic neurons, while decreasing activated microglia in the striatum, improving motor function and cell survival in PD genetic model mice [111]. NaB was also found to have beneficial effects in brain function and reduced dopaminergic neuron death [112]. However, despite its neuroprotective potential, VPA is associated with various adverse effects, including reduced platelet count or an increased risk of hyperammonemia [113,114]. Notably, it has also caused impairments in patients that mimic PD [115,116]. Patients undergoing long-term VPA therapy reported to have symptoms such as bradykinesia, rigidity, postural instability, and resting tremor [117]. These findings highlight the need for alternative therapeutic strategies, potentially through HAT-targeted therapy to enhance treatment efficacy while minimizing adverse effects.

4.3. Amyotrophic Lateral Sclerosis (ALS)

ALS is defined as a fatal ND disease caused by the deterioration of neurons within the brain cortex, brainstem, and spinal motor neurons resulting in generalized weakness, muscle atrophy and paralysis [118,119]. Prognosis is poor and patients typically succumb to the disorder within three years of diagnosis [118]. ALS can be genetically inherited and familial with autosomal dominant transmission, although 90% of cases are sporadic with no family history. Several genes have been implicated in the pathogenesis of ALS, including superoxide dismutase 1 (*SOD1*), Fused in Sarcoma (*FUS*), chromosome 9 open reading frame 72 (*C9orf72*), and TAR DNA binding protein 43 (*TDP-43*) [120,121]. Histone hypoacetylation has been repeatedly shown to be associated with ALS. For example, the FUS gene that encodes a nuclear Fused in Sarcoma protein, which is associated with motor dysfunction in ALS and has been shown to inhibit CBP/p300 when overexpressed in HeLa cells, causing cell cycle disruption [122,123]. Moreover, due to the conserved cellular pathways between yeast and humans, researchers have found that yeast models overexpressing human FUS exhibit reduction in histone H3 acetylation at lysine residues 14 and 56 (H3K14ac and H3K56ac) [124]. Further, the ELP3 HAT has been linked to ALS-associated motor neurodegeneration in a microsatellite-based genetic association study [125,126]. Using zebrafish and mouse ALS models, other researchers found that expression of human ELP3 was neuroprotective and helped increase longevity and delayed onset of the disease [127]. Additionally, studies using a well-characterized ALS *Drosophila* model reveal brain-specific reduction in Tip60 HAT that resulted in reduced H4K16 and H4K12 acetylation at cognition linked gene loci with concomitant transcriptional repression [89].

Therapeutic restoration of histone acetylation has been explored in ALS by utilizing HDACi [128,129]. ALS has been attributed to gain-of-function mutations in the gene-encoding Cu/Zn superoxide dismutase 1 (*SOD1*) in which *SOD1* point mutation in an ALS mouse model causes motor neuron degeneration [130]. Treatment with HDAC inhibitor, trichostatin A (TSA), reduced motor neuron degeneration, prolonged lifespan, and reduced gliosis in the spinal cord of a SOD1-G93A mouse model [131]. Pharmacological inhibition and/or genetic silencing of HDAC6 rescues axonal transport defects in an ALS-associated mutant FUS mouse model [132]. Perplexingly, gene expression levels for class I, II, and IV HDACs in the ALS brain and spinal cord revealed that mRNA and protein levels of HDAC2 and HDAC11 were activated and repressed, respectively [133]. Accordingly, loss of HDAC function is associated with ALS in some models whereas HDAC inhibition is neuroprotective in others. Thus, future studies are exploring these seemingly opposite acetylation landscapes to elucidate how they contribute to ALS.

4.4. Alzheimer's Disease

Alzheimer's disease (AD) pathology is hallmarked by a progressive decline in cognition and memory, neuronal cell death, and the accumulation of neurofibrillary tangles and amyloid beta plaques [134]. AD is the leading cause of dementia with 95% of all AD cases arising as late-onset sporadic cases (LOAD), with only 5% being linked to disease-causing genes that include apolipoprotein E (*ApoE4*), presenilin (*PSEN*), and amyloid precursor protein (*APP*) [135]. Sporadic AD is complex, resulting from an interplay of genetics, environmental, and aging conditions. Several common hallmarks have been identified in AD, including the miscleavage of APP into amyloid beta plaques and neurofibrillary tangles (NFTs) in the brain caused by intraneuronal aggregates tau, which is a protein associated with microtubules and neurodegeneration [136]. During early stages of AD progression, neuronal signaling is disrupted and causes mild cognitive impairment (MCI). It is now widely accepted that alterations in histone acetylation with concomitant transcriptional dysregulation are critical contributing factors to MCI-associated early neuronal impairments [122,134,137–139]. Evidence supporting an association between histone acetylation and cognitive impairment in AD originally stemmed from observations that histone acetylation was reduced in multiple AD mouse models [78,140]. For example, decreased H4 acetylation but not H3 acetylation was found in a well-characterized tg2576 mouse model for amyloid pathology [141]. Similarly, research using well-characterized *Drosophila* models for AD have shown reduced Tip60 HAT levels in the AD brain with concomitant disruption of histone acetylation marks and cognition-linked gene repression that is recapitulated in the brains of AD patients [47]. Intriguingly, recent studies investigating histone acetylation in specific subregions of the AD brain revealed that histone acetylation dysregulation may differentially affect pathology stages and brain subregions. In support of this concept, researchers researobserved that histone acetylation was differentially altered in AD patients for H3K27ac in post-mortem entorhinal cortex brain region, with most sites depleted in acetylation, while some were enriched [142]. In contrast, researchers found dramatic loss of H4K16ac in the proximity of genes associated with aging, and AD in the lateral temporal lobe of post-mortem AD patients' brains [143]. It is important to note that H4K16 acetylation has been directly implicated in cognition [47,81]. Accordingly, researchers showed APP-mediated hypoacetylation of H4 at critical learning and memory genes in the prefrontal cortex of mouse brain [144].

A substantial body of evidence shows that inhibition of HDACs can be protective and beneficial in AD. Indeed, APP overexpression in cultured cortical neurons leads to H3 and H4 hypoacetylation, accompanied by a decreased in CBP levels [63]. Additionally, in the hippocampus of APP/PS1 transgenic mice, researchers found a significant reduction in CBP mRNA and protein levels, which resulted in reduced H3 acetylation in hippocampal neurons [145]. Loss of function mutations in genes encoding PS1 and PS2 reduces CBP expression and its target genes such as *c-fos* and *BDNF*, along with increased levels of p25/Cdk5 activator result in negative consequences on spatial and contextual memory and synaptic plasticity [146]. Thus, treatment with the broad HDACi NaB increased H3 and H4 acetylation levels, and caused restoration of learning abilities, including access to long-term memories that was ablated by prior hyperactivation of p25/Cdk5 [74]. Additionally, both general and class I-selective HDAC inhibitors have been demonstrated to ameliorate cognitive defects in an AD mouse model containing a hereditary genetic AD mutation [147,148]. Recent work shows that pan-HDAC inhibitor VPA, as well as WT161, a novel HDAC6 selective inhibitor primarily used in tumor treatment, effectively reduced Aβ deposition in both AD cell and mouse models [134]. These treatments also improved cognitive function in AD mice, further supporting the therapeutic potential of HDACis in AD. Substantial progress has been made in enhancing treatments for NDs, emphasizing the enhancing of selective inhibitor drugs. Among these, HAT-targeted therapies represent a promising avenue, offering potential for more effective clinical translation, and is discussed in the next section.

5. HAT Activation as a Promising Therapeutic Strategy for Age-Related Neurodegenerative Disease

It is widely accepted that histone acetylation levels can become inappropriately reduced during neuronal development via reduction in specific HAT activity, with harmful consequences on neuronal activity that culminate in multiple types of neurodegenerative disorders [149] (Table 1). As histone acetylation homeostasis is tilted by inappropriate HAT reduction and HDAC gain, the HAT: HDAC homeostasis ratio tilts in favor of HDACs regarding their availability and enzymatic functionality. While current epigenetic therapies that utilize HDACi treatments to restore histone acetylation have shown promise in reversing cognitive deficits, their therapeutic use is limited due to their non-specific modes of inhibiting multiple HDACs, some of which are critical for cognition. Additionally, HDACi treatments can elicit non-specific chromatin acetylation patterns that might be more harmful than beneficial. For example, superpolyamide hydroxamic acid (SAHA) and Nicotinamide (vitamin B3) have been advanced to Phase I and Phase II clinical trials as representative HDAC inhibitors for AD treatment [150,151]. Despite the initial promise, their poor blood–brain barrier (BBB) permeability, lack of specificity by targeting several HDACs, and low HDAC inhibitory potency, make these pan-HDAC inhibitors face limited and negative outcomes because they produce global hyperacetylation and widespread transcriptional dysregulation in clinical trials [152–154]. Similarly, SAHA is a pan-inhibitor that targets class I and class II HDACs (HDACs 1-10), and it has been shown to have positive results in AD animal models. However, despite its demonstrated efficacy in animal models, determining an optimal dose–response curve in clinical trials remains a significant challenge, thereby limiting its therapeutic potential in humans [152]. Similarly, Nicotinamide, another pan inhibitor of all class III sirtuin HDACs (SIRT1-7), failed to demonstrate significant cognitive outcomes in Phase II clinical trials involving patients with mild to moderate AD [155]. Additional evidence suggests that Nicotinamide may exacerbate several neuropathological features in mouse models of PD, which further raises concerns about the use of this inhibitor's safety profile in other neurodegenerative contexts [156].

Table 1. HAT/HDAC modulation and dysregulation within neurodegenerative diseases.

HAT/HDAC	ND	Modulation	Reference
CBP (HAT)	Huntington's	Redistributed in cytoplasm vs nuclear	[86]
		⇧ ubiquitylation and degradation	[87,91,92]
		⇧ Inhibition caused by inappropriate binding of first exon of HTT	[88]
	Parkinson's	Inactivation caused by alpha-syn protein	[102,103]
		Unaltered levels	
	Amyotrophic Lateral Sclerosis	⇩ Inhibition caused by overexpression of FUS gene	[63,123,130]
	Alzheimer's	⇩ levels by loss of function mutations in PS1 and PS2	[146]
PCAF (HAT)	Huntington's disease	⇧ Inhibition caused by inappropriate binding of first exon of HTT	[87,88]

Table 1. *Cont.*

HAT/HDAC	ND	Modulation	Reference
	Parkinson's	Inactivation caused by alpha-syn protein	[102,103]
P300 (HAT)	Parkinson's	Inactivation caused by alpha-syn protein	[102,103]
	Amyotrophic Lateral Sclerosis	Inhibition caused by overexpression of FUS gene	[122,123]
SIRT2 (HDAC)	Parkinson's	⇧ levels and accumulation	[108]
	Huntington's	Suppression leads to neuroprotection	[93]
ELP3 (HAT)	Amyotrophic Lateral Sclerosis	⇩ levels	[127]
Tip60 (HAT)	Huntington's	⇧ levels	[157]
	Parkingson's	⇩ levels	[157]
	Amyotrophic Lateral Sclerosis	⇩ levels	[157]
	Alzheimer's	⇩ protein levels	[47]
RPD3 (HDAC)	Huntington's	Suppression leads to neuroprotection	[93]

In contrast, targeting HATs over HDACs presents benefits since HATs, unlike HDACs, play non-redundant critical roles in "re-writing" the specific chromatin histone acetylation patterns required for cognition gene expression and higher-order brain function. For example, reduced Tip60 protein levels are observed in the brains of human post-mortem AD hippocampus and in a well-characterized AD *Drosophila* model, causing decreased Tip60-mediated site-specific cognition-associated histone acetylation marks with concomitant repression of critical neuroplasticity genes [47]. Importantly, genetically increasing Tip60 in the *Drosophila* AD brain prevents site-specific histone hypoacetylation and restores activation of repressed neuroplasticity genes [47,157]. Increasing Tip60 levels also restore multiple neural processes impaired in early stages of AD, including axonal outgrowth/transport, learning/memory, circadian rhythm, synaptic plasticity, locomotor function, apoptotic driven neurodegeneration, and neuroplasticity gene control [16,22,43,46,47,75,89,158,159]. Together, these studies support an overall neuroprotective role for Tip60 in multiple neural cognitive circuits impaired in AD. Thus, activation of specific HAT using therapeutic modulators can potentially have more direct and causative positive effects for treatment of NDs.

Although HDACi treatment induces general hyperacetylation in multiple NDs, there are additional limitations to this treatment in that it is often not sufficient to produce absolute beneficial effects alone and requires restoration of the specific HATs that "write" the site-specific acetylation marks. For instance, increased p300, but not HDAC inhibition, promotes axonal regeneration in mature retinal ganglion cells after optic nerve injury. These beneficial effects are mediated by p300-induced hyperacetylation of histone H3 and p53, causing increased expression of selected pro-axonal outgrowth genes [160]. Further, additional research demonstrated that general induction of histone acetylation by HDACi treatment alone failed to rescue synaptic plasticity and memory loss in the hippocampus of a mouse model lacking the learning and memory-linked HAT CBP, and that a wild-type CBP allele in conjunction with HDACi treatment was required [161]. Such incomplete rescue by HDACi treatment alone is likely due to the necessity of CBP for site-specific histone acetylation and the recruitment of basal transcriptional machinery required for the activation of key synaptic plasticity target genes. Consistent with these findings, other researchers showed that enhancing Histone H4 acetylation by using the HDAC inhibitor, MS-275, in a *Drosophila* AD-associated APP model was insufficient to fully restore the expression of certain Tip60 target genes [16]. This is likely since Tip60 is still required for site-specific acetylation and interactions with other transcription factors. In summary, these support the promising concept that enhancing the activity of specific HATs like CBP or Tip60 might serve as alternative therapeutic strategy for alleviating cognitive impairments that characterize multiple neurodegenerative disorders.

Importantly, certain HATs also play critical roles in non-chromatin-mediated cellular processes critical for cognitive function that can also be exploited in ND therapeutics by specific HAT activation (Table 2). For example, Elp3 HAT acetylates microtubules critical for synaptic bouton increase during neurogenesis [125]. More recent evidence demonstrates that disruptions in ELP3-mediated microtubule acetylation are a common feature in multiple neurological disorders, presenting ELP3 action as a potential target for acetylation-based therapeutic interventions [162]. Similarly, *Drosophila* Tip60 functions in regulating synaptic plasticity in the neuromuscular junction, in part via microtubule acetylation [158]. More recently, it was reported that Tip60 carries out a novel RNA-binding function in addition to its HAT function that modulates alternative splicing of pre-mRNA targets implicated in AD [163]. Indeed, defects in alternative RNA splicing have emerged as a widespread mechanism in multiple neurodegenerative diseases including AD, bringing another layer of complexity to solving the neurodegenerative disease puzzle. Alterations in RNA splicing culminate in aberrant protein isoform variations that interfere with neuronal cellular processes. Intriguingly, Tip60 HAT action and RNA splicing function may be co-dependent such that modulating one function may also have an impact on the other function. Although the precise mechanism by which Tip60 targets RNA is yet to be elucidated, Tip60 RNA immunoprecipitation (RIP) and chromatin immunoprecipitation (ChIP) studies reveal an extensive overlap between Tip60's gene targets at the chromatin and RNA level. These results support a model by which Tip60-RNA targeting occurs very close to its target chromatin loci [163]. Tip60's specificity for pre-mRNA may arise from its proximity to the new transcripts. After Tip60 is recruited it to specific chromatin loci for HAT mediated gene activation, it can then switch to the newly synthesized pre-mRNA strand to control its splicing. Thus, increasing Tip60's HAT action in NDs such as AD using Tip60-specific HAT activators might also restore the Tip60 RNA splicing function due to enhancement of transcriptional activation that might also increase Tip60's RNA-targeting and splicing function [163].

Importantly, activation of specific HATs using therapeutic modulators has been shown to have direct and causative positive effects for treatment of NDs. For example, CBP/P300 activators CSP-TTK21 and CTPB, and recently discovered Tip60 HAT activators can cross the blood–brain barrier (BBB) to promote neuroprotection in multiple ND animal models [12,13,164]. The HAT activator CSP-TTK21 was demonstrated to promote adult neurogenesis and spatial memory formation with significant increases in histone acetylation within the hippocampus and frontal cortex in adult mice [12]. Additionally, CTPB was shown to have neurotrophic effects when administered to SH-SY5Y cells to study PD, enhancing cell survival and growth [13]. Notably, a recent study by Bhatnagar et al. (2025) [164] identified novel small molecule Tip60 HAT activator compounds that exhibited efficacy in rescuing functional neuronal deficits in vivo in both Tip60 knockdown *Drosophila* and well-characterized AD-associated APP *Drosophila* models. These compounds also restored repressed plasticity gene expression and significantly extended lifespan in AD flies. These exciting findings represent a significant advance over previously explored pan-HDACi treatments and highlight the potential for the use of HAT activators as a therapeutic strategy for multiple NDs.

Table 2. Therapeutic Treatments: HDAC inhibitors and HAT Activators.

Therapeutic	Modulation	ND Applicable	Clinical Application	Reference
TSA	Small molecule HDAC Inhibitor	Huntington	Pre-clinical/experimental	[90]
		ALS	Pre-clinical/ experimental	[131]
Vorinostat (SAHA)	Small molecule pan-HDAC inhibitor	Huntington	Pre-clinical/experimental	[90]
		Parkinson's	Pre-clinical/ experimental	[102]
		Alzheimer's	Pre-clinical/experimental	[147]
Sodium Butyrate	Short fatty acid HDAC inhibitor	Alzheimer's	Pre-clinical/experimental	[147]
		Parkinson's	Pre-clinical/experimental	[112]
		Huntington	Pre-clinical/experimental	[102]
Sodium Valproate (VPA)	Small molecule HDAC inhibitor Negatively impacts Acetyl-CoA action	Parkinson's	Pre-clinical/experimental	[111]
		ALS	Pre-clinical/experimental	[63]
		Alzheimer's	Pre-clinical/experimental	[134,147]
4B	Pimelic diphenylamide HDAC inhibitor	Huntington's	Pre-clinical/experimental	[95]
RGFP966	Selective HDAC 3 inhibitor	Huntington's	Pre-clinical/ experimental	[96]
CSP-TTK21	P300/CBP Hat activator	Alzheimer's	Pre-clinical/experimental	[12,13]
CTPB	HAT Activator	Alzheimer's	Pre-clinical/experimental	[13]
P-compounds (P10, P13)	Specific Tip60 HAT Activator	Alzheimer's	Pre-clinical/experimental	[163]
WT161	Selective HDAC6 inhibitor	Alzheimer's	Preclinical/ experimental	[134]

6. Conclusions

In conclusion, histone acetylation is currently highlighted as a critical mechanism that controls gene regulatory programs required for cognition. Although tightly regulated histone acetylation and deacetylation is critical for maintaining proper neuronal function, age-related dysregulation leads to alterations in the neuroepigenetic environment that impairs cognitive abilities. Dysregulation of histone acetylation chromatin writers can lead to cognitive deficits, neuronal apoptosis, and neurodegeneration, all hallmarks of ND pathology. Future research will be critical in unraveling the precise mechanisms underlying how specific HATs positively impact cognition and slow or prevent neurodegeneration in different age-associated NDs. Indeed, the differences in mechanisms underlying HAT/HDAC contribution to AD, PD, HD, and ALS, along with the varying types of histone acetylation alterations in different subregions of the brain and at various stages of disease progression observed in the multiple types of NDs, highlight the inadequacy of a one-size-fits-all therapeutic strategy. Thus, further investigation into disease stage and brain region-specific acetylation landscapes between multiple NDs is necessary for advancing the development of HAT-specific therapeutic strategies with improved precision and efficacy. This knowledge will be informative for developing specific HAT activation-targeted therapeutic strategies in the early intervention of NDs. Furthermore, there are limitations that must be noted regarding any drug discovery, since the development of compounds routinely involves multiple rounds of testing and optimization before finding the most clinically effective compounds. Throughout the drug discovery process, it is also favorable to select for optimal pharmacokinetic properties, when feasible, to design drugs that display favorable absorption, distribution, metabolism, and excretion (ADME) characteristics, together with preferential blood–brain barrier (BBB) transport. Thus, future directions for HAT activator drug discovery involve lead compound optimization that would then necessitate testing for in vivo pharmacokinetics and efficacy in mammalian models of ND. Additionally, since NDs are complex, HAT activators may be more suitable as a combinatorial therapy in conjunction with genetic and/or other neuroepigenetic modulator therapeutic strategies. In summary, exploration into the generation of small molecular compounds that selectively target and enhance the activity of neuroprotective HATs like CBP, p300, and Tip60 should facilitate the development of HAT-based therapeutics with reduced adverse effects for patients afflicted with these debilitating neurodegenerative disorders.

Funding: This research was funded by NIH NINDS grant RF1NS095799 to F.E.

Institutional Review Board Statement: Not applicable.

Informed Consent Statement: Not applicable.

Data Availability Statement: Not applicable.

Conflicts of Interest: The authors declare no conflict of interest.

Abbreviations

Abbreviations	Meaning
PTM	Post-translational modification
ND	Neurodegenerative disease
AD	Alzheimer's disease
PD	Parkinson's disease
ALS	Amyotrophic lateral sclerosis
HD	Huntington's disease
HAT	Histone acetyltransferase
TAF_{II}	TBP-associated Factor
TFIID	Transcription Factor IID
TBP	TATA-binding protein
RNA PolII	RNA Polymerase II
GCN5	General Control non-depressible 5 (HAT2)
PCAF	p300/CREB-binding protein-associated factor
MYST	MOF,Ybf2/Sas3,Sas2,Tip60
MOF	Monocytic leukemia zinc finger protein
Ybf2	Yeast Binding Factor two (HAT)
Tip60	Tat-interacting protein 60
CBP	CREB Binding Protein
CREB	cAMP Response Element-Binding protein
RID	Receptor interaction domain
KIX	Kinase-inducible domain
PHD	Plant homeodomain
RTS	Rubinstein–Taybi Syndrome
LTM	Long-term memory
MOZ	Monocytic leukemia zinc finger protein (KAT6A)
MORF	MOZ-related factor
HBO1	Histone acetyl transferase
sLNv	Small lateral neurons ventral
KAT	Lysine acetyl transferase
ChIP-seq	Chromatin immunoprecipitation sequencing
GNAT	Gcn5-related N-acetyltransferase
ERK	Extra-cellular signal related kinase

KO	Knockout
HDAC	Histone deacetylase
TSA	Trichostatin A
KCL	Potassium Chloride
NMDA	N-methyl D-aspartate receptor (glutamate receptor)
LTP	Long-term potentiation
LTD	Long-term depression
EE	Environmental enrichment
HDACi	Histone Deacetylase inhibitor
SIRT2	Sirtuin 2(NAD+-dependent deacetylase)
SNCA	Gene that encodes alpha-synuclein
APP	Amyloid precursor protein
HTT	Huntington protein
mHTT	Mutant Huntington protein
polyQ	Polyglutamine
SAHA	Suberoylanilide hydroxamic acid
Sirtuin 2	Sirtuin 2
a-syn	Alpha synuclein
PFFs	a-syn folded fibrils
CNS	Central nervous system
FUS	Fused in Sarcoma
PARK7	Gene name for Parkinsonism associate deglycase
MAPT	Microtubule-associated Protein tau
ELP3	Elongator Protein 3
HeLa cells	Cell line derived of cancerous tissues
SOD1	Superoxide dismutase 1
VPA	Valproic acid
mRNA	Messenger RNA
LOAD	Late-onset Alzheimer's disease
ApoE4	Apolioprotein E4
PSEN	Presenilin
AB	Amyloid Beta
NTF	Neurofibrillary Tangle
MCI	Mild cognitive impairment
PS1/2	Presenilin 1/2
BDNF	Brain-derived neurotrophic factor
ZBGs	Zinc-binding groups
NaB	Sodium burytate
Httex1p	Polyglutamine-containing domain of Htt-htt exon 1 protein
TH	Tyrosine hydroxylase

References

1. Boros, I.M. Histone modification in Drosophila. *Brief. Funct. Genom.* **2012**, *11*, 319–331. [CrossRef] [PubMed]
2. Peterson, C.L.; Laniel, M.A. Histones and histone modifications. *Curr. Biol.* **2004**, *14*, R546–R551. [CrossRef] [PubMed]
3. Levenson, J.M.; Sweatt, J.D. Epigenetic mechanisms in memory formation. *Nat. Rev. Neurosci.* **2005**, *6*, 108–118. [CrossRef]
4. Schneider, A.; Chatterjee, S.; Bousiges, O.; Selvi, B.R.; Swaminathan, A.; Cassel, R.; Blanc, F.; Kundu, T.K.; Boutillier, A.L. Acetyltransferases (HATs) as targets for neurological therapeutics. *Neurotherapeutics* **2013**, *10*, 568–588. [CrossRef]
5. Wood, M.A.; Hawk, J.D.; Abel, T. Combinatorial chromatin modifications and memory storage: A code for memory? *Learn. Mem.* **2006**, *13*, 241–244. [CrossRef]
6. Kuo, M.H.; Allis, C.D. Roles of histone acetyltransferases and deacetylases in gene regulation. *Bioessays* **1998**, *20*, 615–626. [CrossRef]
7. Scharf, A.N.; Imhof, A. Every methyl counts--epigenetic calculus. *FEBS Lett.* **2011**, *585*, 2001–2007. [CrossRef]
8. Day, J.J.; Sweatt, J.D. Epigenetic treatments for cognitive impairments. *Neuropsychopharmacology* **2012**, *37*, 247–260. [CrossRef]

9. Lardenoije, R.; Iatrou, A.; Kenis, G.; Kompotis, K.; Steinbusch, H.W.; Mastroeni, D.; Coleman, P.; Lemere, C.A.; Hof, P.R.; van den Hove, D.L.; et al. The epigenetics of aging and neurodegeneration. *Prog. Neurobiol.* **2015**, *131*, 21–64. [CrossRef]
10. Borrelli, E.; Nestler, E.J.; Allis, C.D.; Sassone-Corsi, P. Decoding the epigenetic language of neuronal plasticity. *Neuron* **2008**, *60*, 961–974. [CrossRef]
11. Graff, J.; Rei, D.; Guan, J.S.; Wang, W.Y.; Seo, J.; Hennig, K.M.; Nieland, T.J.; Fass, D.M.; Kao, P.F.; Kahn, M.; et al. An epigenetic blockade of cognitive functions in the neurodegenerating brain. *Nature* **2012**, *483*, 222–226. [CrossRef] [PubMed]
12. Chatterjee, S.; Mizar, P.; Cassel, R.; Neidl, R.; Selvi, B.R.; Mohankrishna, D.V.; Vedamurthy, B.M.; Schneider, A.; Bousiges, O.; Mathis, C.; et al. A novel activator of CBP/p300 acetyltransferases promotes neurogenesis and extends memory duration in adult mice. *J. Neurosci.* **2013**, *33*, 10698–10712. [CrossRef] [PubMed]
13. Hegarty, S.V.; O'Leary, E.; Solger, F.; Stanicka, J.; Sullivan, A.M.; O'Keeffe, G.W. A Small Molecule Activator of p300/CBP Histone Acetyltransferase Promotes Survival and Neurite Growth in a Cellular Model of Parkinson's Disease. *Neurotox. Res.* **2016**, *30*, 510–520. [CrossRef]
14. Jakovcevski, M.; Akbarian, S. Epigenetic mechanisms in neurological disease. *Nat. Med.* **2012**, *18*, 1194–1204. [CrossRef]
15. Kabir, F.; Atkinson, R.; Cook, A.L.; Phipps, A.J.; King, A.E. The role of altered protein acetylation in neurodegenerative disease. *Front. Aging Neurosci.* **2022**, *14*, 1025473. [CrossRef]
16. Johnson, A.A.; Sarthi, J.; Pirooznia, S.K.; Reube, W.; Elefant, F. Increasing Tip60 HAT levels rescues axonal transport defects and associated behavioral phenotypes in a Drosophila Alzheimer's disease model. *J. Neurosci.* **2013**, *33*, 7535–7547. [CrossRef]
17. Pirooznia, S.K.; Elefant, F. Targeting specific HATs for neurodegenerative disease treatment: Translating basic biology to therapeutic possibilities. *Front. Cell Neurosci.* **2013**, *7*, 30. [CrossRef]
18. Meaney, M.J.; Ferguson-Smith, A.C. Epigenetic regulation of the neural transcriptome: The meaning of the marks. *Nat. Neurosci.* **2010**, *13*, 1313–1318. [CrossRef]
19. Nelson, E.D.; Monteggia, L.M. Epigenetics in the mature mammalian brain: Effects on behavior and synaptic transmission. *Neurobiol. Learn. Mem.* **2011**, *96*, 53–60. [CrossRef]
20. Riccio, A. Dynamic epigenetic regulation in neurons: Enzymes, stimuli and signaling pathways. *Nat. Neurosci.* **2010**, *13*, 1330–1337. [CrossRef]
21. Sweatt, J.D. Neuroscience. Epigenetics and cognitive aging. *Science* **2010**, *328*, 701–702. [CrossRef] [PubMed]
22. Xu, S.; Wilf, R.; Menon, T.; Panikker, P.; Sarthi, J.; Elefant, F. Epigenetic control of learning and memory in Drosophila by Tip60 HAT action. *Genetics* **2014**, *198*, 1571–1586. [CrossRef] [PubMed]
23. Grunstein, M. Histone acetylation in chromatin structure and transcription. *Nature* **1997**, *389*, 349–352. [CrossRef] [PubMed]
24. Chen, L.F.; Lin, Y.T.; Gallegos, D.A.; Hazlett, M.F.; Gomez-Schiavon, M.; Yang, M.G.; Kalmeta, B.; Zhou, A.S.; Holtzman, L.; Gersbach, C.A.; et al. Enhancer Histone Acetylation Modulates Transcriptional Bursting Dynamics of Neuronal Activity-Inducible Genes. *Cell Rep.* **2019**, *26*, 1174–1188. [CrossRef]
25. Peixoto, L.; Abel, T. The role of histone acetylation in memory formation and cognitive impairments. *Neuropsychopharmacology* **2013**, *38*, 62–76. [CrossRef]
26. Krajewski, W.A.; Becker, P.B. Reconstitution of hyperacetylated, DNase I-sensitive chromatin characterized by high conformational flexibility of nucleosomal DNA. *Proc. Natl. Acad. Sci. USA* **1998**, *95*, 1540–1545. [CrossRef]
27. Ikura, T.; Ogryzko, V.V.; Grigoriev, M.; Groisman, R.; Wang, J.; Horikoshi, M.; Scully, R.; Qin, J.; Nakatani, Y. Involvement of the TIP60 histone acetylase complex in DNA repair and apoptosis. *Cell* **2000**, *102*, 463–473. [CrossRef]
28. Mizzen, C.A.; Yang, X.J.; Kokubo, T.; Brownell, J.E.; Bannister, A.J.; Owen-Hughes, T.; Workman, J.; Wang, L.; Berger, S.L.; Kouzarides, T.; et al. The TAF(II)250 subunit of TFIID has histone acetyltransferase activity. *Cell* **1996**, *87*, 1261–1270. [CrossRef]
29. Liu, X.; Yang, S.; Zhao, M.; Luo, M.; Yu, C.W.; Chen, C.Y.; Tai, R.; Wu, K. Transcriptional repression by histone deacetylases in plants. *Mol. Plant* **2014**, *7*, 764–772. [CrossRef]
30. Roth, S.Y.; Denu, J.M.; Allis, C.D. Histone acetyltransferases. *Annu. Rev. Biochem.* **2001**, *70*, 81–120. [CrossRef]
31. Bedford, D.C.; Brindle, P.K. Is histone acetylation the most important physiological function for CBP and p300? *Aging* **2012**, *4*, 247–255. [CrossRef] [PubMed]
32. Dyson, H.J.; Wright, P.E. Role of Intrinsic Protein Disorder in the Function and Interactions of the Transcriptional Coactivators CREB-binding Protein (CBP) and p300. *J. Biol. Chem.* **2016**, *291*, 6714–6722. [CrossRef] [PubMed]
33. Petrij, F.; Giles, R.H.; Dauwerse, H.G.; Saris, J.J.; Hennekam, R.C.; Masuno, M.; Tommerup, N.; van Ommen, G.J.; Goodman, R.H.; Peters, D.J.; et al. Rubinstein-Taybi syndrome caused by mutations in the transcriptional co-activator CBP. *Nature* **1995**, *376*, 348–351. [CrossRef] [PubMed]
34. Alarcon, J.M.; Malleret, G.; Touzani, K.; Vronskaya, S.; Ishii, S.; Kandel, E.R.; Barco, A. Chromatin acetylation, memory, and LTP are impaired in CBP+/- mice: A model for the cognitive deficit in Rubinstein-Taybi syndrome and its amelioration. *Neuron* **2004**, *42*, 947–959. [CrossRef]
35. Korzus, E.; Rosenfeld, M.G.; Mayford, M. CBP histone acetyltransferase activity is a critical component of memory consolidation. *Neuron* **2004**, *42*, 961–972. [CrossRef]

36. Maurice, T.; Duclot, F.; Meunier, J.; Naert, G.; Givalois, L.; Meffre, J.; Celerier, A.; Jacquet, C.; Copois, V.; Mechti, N.; et al. Altered memory capacities and response to stress in p300/CBP-associated factor (PCAF) histone acetylase knockout mice. *Neuropsychopharmacology* **2008**, *33*, 1584–1602. [CrossRef]
37. Oliveira, A.M.; Wood, M.A.; McDonough, C.B.; Abel, T. Transgenic mice expressing an inhibitory truncated form of p300 exhibit long-term memory deficits. *Learn. Mem.* **2007**, *14*, 564–572. [CrossRef]
38. Oliveira, A.M.; Abel, T.; Brindle, P.K.; Wood, M.A. Differential role for CBP and p300 CREB-binding domain in motor skill learning. *Behav. Neurosci.* **2006**, *120*, 724–729. [CrossRef]
39. Chatterjee, S.; Angelakos, C.C.; Bahl, E.; Hawk, J.D.; Gaine, M.E.; Poplawski, S.G.; Schneider-Anthony, A.; Yadav, M.; Porcari, G.S.; Cassel, J.C.; et al. The CBP KIX domain regulates long-term memory and circadian activity. *BMC Biol.* **2020**, *18*, 155. [CrossRef]
40. Sapountzi, V.; Cote, J. MYST-family histone acetyltransferases: Beyond chromatin. *Cell Mol. Life Sci.* **2011**, *68*, 1147–1156. [CrossRef]
41. Shukla, S.; Levine, C.; Sripathi, R.P.; Elson, G.; Lutz, C.S.; Leibovich, S.J. The Kat in the HAT: The Histone Acetyl Transferase Kat6b (MYST4) Is Downregulated in Murine Macrophages in Response to LPS. *Mediat. Inflamm.* **2018**, *2018*, 7852742. [CrossRef] [PubMed]
42. Stilling, R.M.; Ronicke, R.; Benito, E.; Urbanke, H.; Capece, V.; Burkhardt, S.; Bahari-Javan, S.; Barth, J.; Sananbenesi, F.; Schutz, A.L.; et al. K-Lysine acetyltransferase 2a regulates a hippocampal gene expression network linked to memory formation. *EMBO J.* **2014**, *33*, 1912–1927. [CrossRef] [PubMed]
43. Pirooznia, S.K.; Sarthi, J.; Johnson, A.A.; Toth, M.S.; Chiu, K.; Koduri, S.; Elefant, F. Tip60 HAT activity mediates APP induced lethality and apoptotic cell death in the CNS of a Drosophila Alzheimer's disease model. *PLoS ONE* **2012**, *7*, e41776. [CrossRef] [PubMed]
44. Xu, S.; Elefant, F. Tip off the HAT- Epigenetic control of learning and memory by Drosophila Tip60. *Fly* **2015**, *9*, 22–28. [CrossRef]
45. Hofmann, A.; Brunner, M.; Korge, G. The winged-helix transcription factor JUMU is a haplo-suppressor/triplo-enhancer of PEV in various tissues but exhibits reverse PEV effects in the brain of Drosophila melanogaster. *Chromosome Res.* **2009**, *17*, 347–358. [CrossRef]
46. Lorbeck, M.; Pirooznia, K.; Sarthi, J.; Zhu, X.; Elefant, F. Microarray analysis uncovers a role for Tip60 in nervous system function and general metabolism. *PLoS ONE* **2011**, *6*, e18412. [CrossRef]
47. Panikker, P.; Xu, S.J.; Zhang, H.; Sarthi, J.; Beaver, M.; Sheth, A.; Akhter, S.; Elefant, F. Restoring Tip60 HAT/HDAC2 Balance in the Neurodegenerative Brain Relieves Epigenetic Transcriptional Repression and Reinstates Cognition. *J. Neurosci.* **2018**, *38*, 4569–4583. [CrossRef]
48. Urban, I.; Kerimoglu, C.; Sakib, M.S.; Wang, H.; Benito, E.; Thaller, C.; Zhou, X.; Yan, J.; Fischer, A.; Eichele, G. TIP60/KAT5 is required for neuronal viability in hippocampal CA1. *Sci. Rep.* **2019**, *9*, 16173. [CrossRef]
49. Uchida, S.; Teubner, B.J.W.; Hevi, C.; Hara, K.; Kobayashi, A.; Dave, R.M.; Shintaku, T.; Jaikhan, P.; Yamagata, H.; Suzuki, T.; et al. CRTC1 Nuclear Translocation Following Learning Modulates Memory Strength via Exchange of Chromatin Remodeling Complexes on the Fgf1 Gene. *Cell Rep.* **2017**, *18*, 352–366. [CrossRef]
50. Wiley, J.C.; Young, J.E.; Jayadev, S.; Darvas, M.; Swerdlow, R.H.; Wolfe, M.S.; Heath, L.M.; Gockley, J.; Cary, G.A.; Poehlman, W.L.; et al. Alternative Theory of AD pathogenesis: Membrane Delimitation of the histone Acetyltransferase Tip60/Kat5. *Alzheimers Dement.* **2022**, *S4*, 068190. [CrossRef]
51. Kueh, A.J.; Dixon, M.P.; Voss, A.K.; Thomas, T. HBO1 is required for H3K14 acetylation and normal transcriptional activity during embryonic development. *Mol. Cell Biol.* **2011**, *31*, 845–860. [CrossRef]
52. Bergamasco, M.I.; Abeysekera, W.; Garnham, A.L.; Hu, Y.; Li-Wai-Suen, C.S.; Sheikh, B.N.; Smyth, G.K.; Thomas, T.; Voss, A.K. KAT6B is required for histone 3 lysine 9 acetylation and SOX gene expression in the developing brain. *Life Sci. Alliance* **2025**, *8*, e202402969. [CrossRef] [PubMed]
53. Bergamasco, M.I.; Ozturk, E.; Casillas-Espinosa, P.M.; Garnham, A.L.; Abeysekera, W.; Wimmer, V.C.; Rajasekhar, P.; Vanyai, H.K.; Whitehead, L.; Blewitt, M.E.; et al. KAT6B overexpression in mice causes aggression, anxiety, and epilepsy. *iScience* **2025**, *28*, 111953. [CrossRef] [PubMed]
54. Sheikh, B.N.; Guhathakurta, S.; Tsang, T.H.; Schwabenland, M.; Renschler, G.; Herquel, B.; Bhardwaj, V.; Holz, H.; Stehle, T.; Bondareva, O.; et al. Neural metabolic imbalance induced by MOF dysfunction triggers pericyte activation and breakdown of vasculature. *Nat. Cell Biol.* **2020**, *22*, 828–841. [CrossRef]
55. Hu, Y.; Ma, Y.; Liu, L.; Hong, Y.; Wang, G.; Tang, B.; Guo, J.; Yang, P.; Cao, Y.; Ren, H. Loss of NgBR causes neuronal damage through decreasing KAT7-mediated RFX1 acetylation and FGF1 expression. *Cell Mol. Life Sci.* **2025**, *82*, 140. [CrossRef]
56. Grant, P.A.; Duggan, L.; Cote, J.; Roberts, S.M.; Brownell, J.E.; Candau, R.; Ohba, R.; Owen-Hughes, T.; Allis, C.D.; Winston, F.; et al. Yeast Gcn5 functions in two multisubunit complexes to acetylate nucleosomal histones: Characterization of an Ada complex and the SAGA (Spt/Ada) complex. *Genes. Dev.* **1997**, *11*, 1640–1650. [CrossRef]
57. Sterner, D.E.; Berger, S.L. Acetylation of histones and transcription-related factors. *Microbiol. Mol. Biol. Rev.* **2000**, *64*, 435–459. [CrossRef]

58. Peleg, S.; Sananbenesi, F.; Zovoilis, A.; Burkhardt, S.; Bahari-Javan, S.; Agis-Balboa, R.C.; Cota, P.; Wittnam, J.L.; Gogol-Doering, A.; Opitz, L.; et al. Altered histone acetylation is associated with age-dependent memory impairment in mice. *Science* **2010**, *328*, 753–756. [CrossRef]
59. Puttagunta, R.; Tedeschi, A.; Soria, M.G.; Hervera, A.; Lindner, R.; Rathore, K.I.; Gaub, P.; Joshi, Y.; Nguyen, T.; Schmandke, A.; et al. PCAF-dependent epigenetic changes promote axonal regeneration in the central nervous system. *Nat. Commun.* **2014**, *5*, 3527. [CrossRef]
60. Das, C.; Lucia, M.S.; Hansen, K.C.; Tyler, J.K. CBP/p300-mediated acetylation of histone H3 on lysine 56. *Nature* **2009**, *459*, 113–117. [CrossRef]
61. Wang, Y.; Wang, X.; Liu, L.; Wang, X. HDAC inhibitor trichostatin A-inhibited survival of dopaminergic neuronal cells. *Neurosci. Lett.* **2009**, *467*, 212–216. [CrossRef]
62. Contreras, P.S.; Gonzalez-Zuniga, M.; Gonzalez-Hodar, L.; Yanez, M.J.; Dulcey, A.; Marugan, J.; Seto, E.; Alvarez, A.R.; Zanlungo, S. Neuronal gene repression in Niemann-Pick type C models is mediated by the c-Abl/HDAC2 signaling pathway. *Biochim. Biophys. Acta* **2016**, *1859*, 269–279. [CrossRef] [PubMed]
63. Rouaux, C.; Jokic, N.; Mbebi, C.; Boutillier, S.; Loeffler, J.P.; Boutillier, A.L. Critical loss of CBP/p300 histone acetylase activity by caspase-6 during neurodegeneration. *EMBO J.* **2003**, *22*, 6537–6549. [CrossRef] [PubMed]
64. Fitzsimons, H.L.; Scott, M.J. Genetic modulation of Rpd3 expression impairs long-term courtship memory in Drosophila. *PLoS ONE* **2011**, *6*, e29171. [CrossRef]
65. Fitzsimons, H.L.; Schwartz, S.; Given, F.M.; Scott, M.J. The histone deacetylase HDAC4 regulates long-term memory in Drosophila. *PLoS ONE* **2013**, *8*, e83903. [CrossRef]
66. Monday, H.R.; Younts, T.J.; Castillo, P.E. Long-Term Plasticity of Neurotransmitter Release: Emerging Mechanisms and Contributions to Brain Function and Disease. *Annu. Rev. Neurosci.* **2018**, *41*, 299–322. [CrossRef]
67. Kim, T.K.; Hemberg, M.; Gray, J.M.; Costa, A.M.; Bear, D.M.; Wu, J.; Harmin, D.A.; Laptewicz, M.; Barbara-Haley, K.; Kuersten, S.; et al. Widespread transcription at neuronal activity-regulated enhancers. *Nature* **2010**, *465*, 182–187. [CrossRef]
68. Karnay, A.; Karisetty, B.C.; Beaver, M.; Elefant, F. Hippocampal stimulation promotes intracellular Tip60 dynamics with concomitant genome reorganization and synaptic gene activation. *Mol. Cell. Neurosci.* **2019**, *101*, 103412. [CrossRef]
69. Schlumm, F.; Mauceri, D.; Freitag, H.E.; Bading, H. Nuclear calcium signaling regulates nuclear export of a subset of class IIa histone deacetylases following synaptic activity. *J. Biol. Chem.* **2013**, *288*, 8074–8084. [CrossRef]
70. Guan, Z.; Giustetto, M.; Lomvardas, S.; Kim, J.H.; Miniaci, M.C.; Schwartz, J.H.; Thanos, D.; Kandel, E.R. Integration of long-term-memory-related synaptic plasticity involves bidirectional regulation of gene expression and chromatin structure. *Cell* **2002**, *111*, 483–493. [CrossRef]
71. Tessier, C.R.; Broadie, K. Activity-dependent modulation of neural circuit synaptic connectivity. *Front. Mol. Neurosci.* **2009**, *2*, 8. [CrossRef]
72. Mayford, M.; Siegelbaum, S.A.; Kandel, E.R. Synapses and memory storage. *Cold Spring Harb. Perspect. Biol.* **2012**, *4*, a005751. [CrossRef] [PubMed]
73. Han, Y.; Yuan, M.; Guo, Y.S.; Shen, X.Y.; Gao, Z.K.; Bi, X. The role of enriched environment in neural development and repair. *Front. Cell Neurosci.* **2022**, *16*, 890666. [CrossRef] [PubMed]
74. Fischer, A.; Sananbenesi, F.; Wang, X.; Dobbin, M.; Tsai, L.H. Recovery of learning and memory is associated with chromatin remodelling. *Nature* **2007**, *447*, 178–182. [CrossRef] [PubMed]
75. Xu, S.; Panikker, P.; Iqbal, S.; Elefant, F. Tip60 HAT Action Mediates Environmental Enrichment Induced Cognitive Restoration. *PLoS ONE* **2016**, *11*, e0159623. [CrossRef]
76. Sarantis, K.; Antoniou, K.; Matsokis, N.; Angelatou, F. Exposure to novel environment is characterized by an interaction of D1/NMDA receptors underlined by phosphorylation of the NMDA and AMPA receptor subunits and activation of ERK1/2 signaling, leading to epigenetic changes and gene expression in rat hippocampus. *Neurochem. Int.* **2012**, *60*, 55–67. [CrossRef]
77. Fischer, A.; Sananbenesi, F.; Mungenast, A.; Tsai, L.H. Targeting the correct HDAC(s) to treat cognitive disorders. *Trends Pharmacol. Sci.* **2010**, *31*, 605–617. [CrossRef]
78. Francis, Y.I.; Fa, M.; Ashraf, H.; Zhang, H.; Staniszewski, A.; Latchman, D.S.; Arancio, O. Dysregulation of histone acetylation in the APP/PS1 mouse model of Alzheimer's disease. *J. Alzheimers Dis.* **2009**, *18*, 131–139. [CrossRef]
79. Graff, J.; Kim, D.; Dobbin, M.M.; Tsai, L.H. Epigenetic regulation of gene expression in physiological and pathological brain processes. *Physiol. Rev.* **2011**, *91*, 603–649. [CrossRef]
80. Graff, J.; Mansuy, I.M. Epigenetic dysregulation in cognitive disorders. *Eur. J. Neurosci.* **2009**, *30*, 1–8. [CrossRef]
81. Stilling, R.M.; Fischer, A. The role of histone acetylation in age-associated memory impairment and Alzheimer's disease. *Neurobiol. Learn. Mem.* **2011**, *96*, 19–26. [CrossRef]
82. Langley, B.; Gensert, J.M.; Beal, M.F.; Ratan, R.R. Remodeling chromatin and stress resistance in the central nervous system: Histone deacetylase inhibitors as novel and broadly effective neuroprotective agents. *Curr. Drug Targets CNS Neurol. Disord.* **2005**, *4*, 41–50. [CrossRef] [PubMed]

83. Tong, H.; Yang, T.; Xu, S.; Li, X.; Liu, L.; Zhou, G.; Yang, S.; Yin, S.; Li, X.J.; Li, S. Huntington's Disease: Complex Pathogenesis and Therapeutic Strategies. *Int. J. Mol. Sci.* **2024**, *25*, 3845. [CrossRef] [PubMed]
84. Kumar, A.; Vaish, M.; Ratan, R.R. Transcriptional dysregulation in Huntington's disease: A failure of adaptive transcriptional homeostasis. *Drug Discov. Today* **2014**, *19*, 956–962. [CrossRef] [PubMed]
85. Francelle, L.; Lotz, C.; Outeiro, T.; Brouillet, E.; Merienne, K. Contribution of Neuroepigenetics to Huntington's Disease. *Front. Hum. Neurosci.* **2017**, *11*, 17. [CrossRef]
86. McCampbell, A.; Fischbeck, K.H. Polyglutamine and CBP: Fatal attraction? *Nat. Med.* **2001**, *7*, 528–530. [CrossRef]
87. Gao, R.; Chakraborty, A.; Geater, C.; Pradhan, S.; Gordon, K.L.; Snowden, J.; Yuan, S.; Dickey, A.S.; Choudhary, S.; Ashizawa, T.; et al. Mutant huntingtin impairs PNKP and ATXN3, disrupting DNA repair and transcription. *Elife* **2019**, *8*, 42988. [CrossRef]
88. Cornett, J.; Friedman, M.; Li, S.H.; Li, X.J. Transcriptional Dysregulation: A therapeutic Target for Polyglutamine Diseases. *Curr. Med. Chem. Cenetral Nerv. Syst. Agents* **2005**, *5*, 119–127. [CrossRef]
89. Beaver, M.; Bhatnagar, A.; Panikker, P.; Zhang, H.; Snook, R.; Parmar, V.; Vijayakumar, G.; Betini, N.; Akhter, S.; Elefant, F. Disruption of Tip60 HAT mediated neural histone acetylation homeostasis is an early common event in neurodegenerative diseases. *Sci. Rep.* **2020**, *10*, 18265. [CrossRef]
90. Steffan, J.S.; Bodai, L.; Pallos, J.; Poelman, M.; McCampbell, A.; Apostol, B.L.; Kazantsev, A.; Schmidt, E.; Zhu, Y.Z.; Greenwald, M.; et al. Histone deacetylase inhibitors arrest polyglutamine-dependent neurodegeneration in Drosophila. *Nature* **2001**, *413*, 739–743. [CrossRef]
91. Cong, S.Y.; Pepers, B.A.; Evert, B.O.; Rubinsztein, D.C.; Roos, R.A.; van Ommen, G.J.; Dorsman, J.C. Mutant huntingtin represses CBP, but not p300, by binding and protein degradation. *Mol. Cell. Neurosci.* **2005**, *30*, 12–23. [CrossRef]
92. Jiang, H.; Nucifora, F.C., Jr.; Ross, C.A.; DeFranco, D.B. Cell death triggered by polyglutamine-expanded huntingtin in a neuronal cell line is associated with degradation of CREB-binding protein. *Hum. Mol. Genet.* **2003**, *12*, 1–12. [CrossRef] [PubMed]
93. Pallos, J.; Bodai, L.; Lukacsovich, T.; Purcell, J.M.; Steffan, J.S.; Thompson, L.M.; Marsh, J.L. Inhibition of specific HDACs and sirtuins suppresses pathogenesis in a Drosophila model of Huntington's disease. *Hum. Mol. Genet.* **2008**, *17*, 3767–3775. [CrossRef] [PubMed]
94. Mielcarek, M.; Benn, C.L.; Franklin, S.A.; Smith, D.L.; Woodman, B.; Marks, P.A.; Bates, G.P. SAHA decreases HDAC 2 and 4 levels in vivo and improves molecular phenotypes in the R6/2 mouse model of Huntington's disease. *PLoS ONE* **2011**, *6*, e27746. [CrossRef] [PubMed]
95. Jia, H.; Pallos, J.; Jacques, V.; Lau, A.; Tang, B.; Cooper, A.; Syed, A.; Purcell, J.; Chen, Y.; Sharma, S.; et al. Histone deacetylase (HDAC) inhibitors targeting HDAC3 and HDAC1 ameliorate polyglutamine-elicited phenotypes in model systems of Huntington's disease. *Neurobiol. Dis.* **2012**, *46*, 351–361. [CrossRef]
96. Suelves, N.; Kirkham-McCarthy, L.; Lahue, R.S.; Gines, S. A selective inhibitor of histone deacetylase 3 prevents cognitive deficits and suppresses striatal CAG repeat expansions in Huntington's disease mice. *Sci. Rep.* **2017**, *7*, 6082. [CrossRef]
97. Rathore, A.S.; Birla, H.; Singh, S.S.; Zahra, W.; Dilnashin, H.; Singh, R.; Keshri, P.K.; Singh, S.P. Epigenetic Modulation in Parkinson's Disease and Potential Treatment Therapies. *Neurochem. Res.* **2021**, *46*, 1618–1626. [CrossRef]
98. Srinivasan, E.; Chandrasekhar, G.; Chandrasekar, P.; Anbarasu, K.; Vickram, A.S.; Karunakaran, R.; Rajasekaran, R.; Srikumar, P.S. Alpha-Synuclein Aggregation in Parkinson's Disease. *Front. Med.* **2021**, *8*, 736978. [CrossRef]
99. Maiti, P.; Manna, J.; Dunbar, G.L. Current understanding of the molecular mechanisms in Parkinson's disease: Targets for potential treatments. *Transl. Neurodegener.* **2017**, *6*, 28. [CrossRef]
100. Kouli, A.; Torsney, K.M.; Kuan, W.L. Parkinson's Disease: Etiology, Neuropathology, and Pathogenesis. In *Parkinson's Disease: Pathogenesis and Clinical Aspects*; Stoker, T.B., Greenland, J.C., Eds.; Codon Publications: Brisbane, CA, USA, 2018. [CrossRef]
101. Song, H.; Chen, J.; Huang, J.; Sun, P.; Liu, Y.; Xu, L.; Wei, C.; Mu, X.; Lu, X.; Wang, W.; et al. Epigenetic modification in Parkinson's disease. *Front. Cell Dev. Biol.* **2023**, *11*, 1123621. [CrossRef]
102. Kontopoulos, E.; Parvin, J.D.; Feany, M.B. Alpha-synuclein acts in the nucleus to inhibit histone acetylation and promote neurotoxicity. *Hum. Mol. Genet.* **2006**, *15*, 3012–3023. [CrossRef]
103. Jin, H.; Kanthasamy, A.; Ghosh, A.; Yang, Y.; Anantharam, V.; Kanthasamy, A.G. alpha-Synuclein negatively regulates protein kinase Cdelta expression to suppress apoptosis in dopaminergic neurons by reducing p300 histone acetyltransferase activity. *J. Neurosci.* **2011**, *31*, 2035–2051. [CrossRef] [PubMed]
104. Lee, J.Y.; Kim, H.; Jo, A.; Khang, R.; Park, C.H.; Park, S.J.; Kwag, E.; Shin, J.H. alpha-Synuclein A53T Binds to Transcriptional Adapter 2-Alpha and Blocks Histone H3 Acetylation. *Int. J. Mol. Sci.* **2021**, *22*, 5392. [CrossRef] [PubMed]
105. Li, L.; Wang, Y. Cross-talk between the H3K36me3 and H4K16ac histone epigenetic marks in DNA double-strand break repair. *J. Biol. Chem.* **2017**, *292*, 11951–11959. [CrossRef]
106. Pfister, J.A.; Ma, C.; Morrison, B.E.; D'Mello, S.R. Opposing effects of sirtuins on neuronal survival: SIRT1-mediated neuroprotection is independent of its deacetylase activity. *PLoS ONE* **2008**, *3*, e4090. [CrossRef]
107. Suzuki, K.; Koike, T. Resveratrol abolishes resistance to axonal degeneration in slow Wallerian degeneration (WldS) mice: Activation of SIRT2, an NAD-dependent tubulin deacetylase. *Biochem. Biophys. Res. Commun.* **2007**, *359*, 665–671. [CrossRef]

108. Outeiro, T.F.; Kontopoulos, E.; Altmann, S.M.; Kufareva, I.; Strathearn, K.E.; Amore, A.M.; Volk, C.B.; Maxwell, M.M.; Rochet, J.C.; McLean, P.J.; et al. Sirtuin 2 inhibitors rescue alpha-synuclein-mediated toxicity in models of Parkinson's disease. *Science* **2007**, *317*, 516–519. [CrossRef]
109. Park, G.; Tan, J.; Garcia, G.; Kang, Y.; Salvesen, G.; Zhang, Z. Regulation of Histone Acetylation by Autophagy in Parkinson Disease. *J. Biol. Chem.* **2016**, *291*, 3531–3540. [CrossRef]
110. Toker, L.; Tran, G.T.; Sundaresan, J.; Tysnes, O.B.; Alves, G.; Haugarvoll, K.; Nido, G.S.; Dolle, C.; Tzoulis, C. Genome-wide histone acetylation analysis reveals altered transcriptional regulation in the Parkinson's disease brain. *Mol. Neurodegener.* **2021**, *16*, 31. [CrossRef]
111. Kim, T.; Song, S.; Park, Y.; Kang, S.; Seo, H. HDAC Inhibition by Valproic Acid Induces Neuroprotection and Improvement of PD-like Behaviors in LRRK2 R1441G Transgenic Mice. *Exp. Neurobiol.* **2019**, *28*, 504–515. [CrossRef]
112. Guo, T.T.; Zhang, Z.; Sun, Y.; Zhu, R.Y.; Wang, F.X.; Ma, L.J.; Jiang, L.; Liu, H.D. Neuroprotective Effects of Sodium Butyrate by Restoring Gut Microbiota and Inhibiting TLR4 Signaling in Mice with MPTP-Induced Parkinson's Disease. *Nutrients* **2023**, *15*, 930. [CrossRef]
113. Baddour, E.; Tewksbury, A.; Stauner, N. Valproic acid-induced hyperammonemia: Incidence, clinical significance, and treatment management. *Ment. Health Clin.* **2018**, *8*, 73–77. [CrossRef] [PubMed]
114. Buoli, M.; Serati, M.; Botturi, A.; Altamura, A.C. The Risk of Thrombocytopenia During Valproic Acid Therapy: A Critical Summary of Available Clinical Data. *Drugs R D* **2018**, *18*, 1–5. [CrossRef] [PubMed]
115. Mahmoud, F.; Tampi, R.R. Valproic acid-induced parkinsonism in the elderly: A comprehensive review of the literature. *Am. J. Geriatr. Pharmacother.* **2011**, *9*, 405–412. [CrossRef]
116. Silver, M.; Factor, S.A. Valproic acid-induced parkinsonism: Levodopa responsiveness with dyskinesia. *Park. Relat. Disord.* **2013**, *19*, 758–760. [CrossRef]
117. Muralidharan, A.; Rahman, J.; Banerjee, D.; Hakim Mohammed, A.R.; Malik, B.H. Parkinsonism: A Rare Adverse Effect of Valproic Acid. *Cureus* **2020**, *12*, e8782. [CrossRef]
118. Ranieri, F.; Mariotto, S.; Dubbioso, R.; Di Lazzaro, V. Brain Stimulation as a Therapeutic Tool in Amyotrophic Lateral Sclerosis: Current Status and Interaction With Mechanisms of Altered Cortical Excitability. *Front. Neurol.* **2021**, *11*, 605335. [CrossRef]
119. Xu, R.S.; Yuan, M. Considerations on the concept, definition, and diagnosis of amyotrophic lateral sclerosis. *Neural Regen. Res.* **2021**, *16*, 1723–1729. [CrossRef]
120. Al-Chalabi, A.; Hardiman, O. The epidemiology of ALS: A conspiracy of genes, environment and time. *Nat. Rev. Neurol.* **2013**, *9*, 617–628. [CrossRef]
121. Kaur, S.J.; McKeown, S.R.; Rashid, S. Mutant SOD1 mediated pathogenesis of Amyotrophic Lateral Sclerosis. *Gene* **2016**, *577*, 109–118. [CrossRef]
122. Wang, Z.; Zang, C.; Rosenfeld, J.A.; Schones, D.E.; Barski, A.; Cuddapah, S.; Cui, K.; Roh, T.Y.; Peng, W.; Zhang, M.Q.; et al. Combinatorial patterns of histone acetylations and methylations in the human genome. *Nat. Genet.* **2008**, *40*, 897–903. [CrossRef]
123. Bennett, S.A.; Tanaz, R.; Cobos, S.N.; Torrente, M.P. Epigenetics in amyotrophic lateral sclerosis: A role for histone post-translational modifications in neurodegenerative disease. *Transl. Res.* **2019**, *204*, 19–30. [CrossRef] [PubMed]
124. Chen, K.; Bennett, S.A.; Rana, N.; Yousuf, H.; Said, M.; Taaseen, S.; Mendo, N.; Meltser, S.M.; Torrente, M.P. Neurodegenerative Disease Proteinopathies Are Connected to Distinct Histone Post-translational Modification Landscapes. *ACS Chem. Neurosci.* **2018**, *9*, 838–848. [CrossRef] [PubMed]
125. Singh, N.; Lorbeck, M.T.; Zervos, A.; Zimmerman, J.; Elefant, F. The histone acetyltransferase Elp3 plays in active role in the control of synaptic bouton expansion and sleep in Drosophila. *J. Neurochem.* **2010**, *115*, 493–504. [CrossRef] [PubMed]
126. Simpson, C.L.; Lemmens, R.; Miskiewicz, K.; Broom, W.J.; Hansen, V.K.; van Vught, P.W.; Landers, J.E.; Sapp, P.; Van Den Bosch, L.; Knight, J.; et al. Variants of the elongator protein 3 (ELP3) gene are associated with motor neuron degeneration. *Hum. Mol. Genet.* **2009**, *18*, 472–481. [CrossRef]
127. Bento-Abreu, A.; Jager, G.; Swinnen, B.; Rue, L.; Hendrickx, S.; Jones, A.; Staats, K.A.; Taes, I.; Eykens, C.; Nonneman, A.; et al. Elongator subunit 3 (ELP3) modifies ALS through tRNA modification. *Hum. Mol. Genet.* **2018**, *27*, 1276–1289. [CrossRef]
128. Figueroa-Romero, C.; Hur, J.; Bender, D.E.; Delaney, C.E.; Cataldo, M.D.; Smith, A.L.; Yung, R.; Ruden, D.M.; Callaghan, B.C.; Feldman, E.L. Identification of epigenetically altered genes in sporadic amyotrophic lateral sclerosis. *PLoS ONE* **2012**, *7*, e52672. [CrossRef]
129. Oates, N.; Pamphlett, R. An epigenetic analysis of SOD1 and VEGF in ALS. *Amyotroph. Lateral Scler.* **2007**, *8*, 83–86. [CrossRef]
130. Joyce, P.I.; McGoldrick, P.; Saccon, R.A.; Weber, W.; Fratta, P.; West, S.J.; Zhu, N.; Carter, S.; Phatak, V.; Stewart, M.; et al. A novel SOD1-ALS mutation separates central and peripheral effects of mutant SOD1 toxicity. *Hum. Mol. Genet.* **2015**, *24*, 1883–1897. [CrossRef]
131. Yoo, Y.E.; Ko, C.P. Treatment with trichostatin A initiated after disease onset delays disease progression and increases survival in a mouse model of amyotrophic lateral sclerosis. *Exp. Neurol.* **2011**, *231*, 147–159. [CrossRef]

132. Guo, W.; Naujock, M.; Fumagalli, L.; Vandoorne, T.; Baatsen, P.; Boon, R.; Ordovas, L.; Patel, A.; Welters, M.; Vanwelden, T.; et al. HDAC6 inhibition reverses axonal transport defects in motor neurons derived from FUS-ALS patients. *Nat. Commun.* **2017**, *8*, 861. [CrossRef]
133. Janssen, C.; Schmalbach, S.; Boeselt, S.; Sarlette, A.; Dengler, R.; Petri, S. Differential histone deacetylase mRNA expression patterns in amyotrophic lateral sclerosis. *J. Neuropathol. Exp. Neurol.* **2010**, *69*, 573–581. [CrossRef] [PubMed]
134. Zhang, M.; Wang, W.; Ye, Q.; Fu, Y.; Li, X.; Yang, K.; Gao, F.; Zhou, A.; Wei, Y.; Tian, S.; et al. Histone deacetylase inhibitors VPA and WT161 ameliorate the pathological features and cognitive impairments of the APP/PS1 Alzheimer's disease mouse model by regulating the expression of APP secretases. *Alzheimers Res. Ther.* **2024**, *16*, 15. [CrossRef] [PubMed]
135. Liu, X.; Jiao, B.; Shen, L. The Epigenetics of Alzheimer's Disease: Factors and Therapeutic Implications. *Front. Genet.* **2018**, *9*, 579. [CrossRef]
136. Rajmohan, R.; Reddy, P.H. Amyloid-Beta and Phosphorylated Tau Accumulations Cause Abnormalities at Synapses of Alzheimer's disease Neurons. *J. Alzheimers Dis.* **2017**, *57*, 975–999. [CrossRef]
137. Mastroeni, D.; Grover, A.; Delvaux, E.; Whiteside, C.; Coleman, P.D.; Rogers, J. Epigenetic mechanisms in Alzheimer's disease. *Neurobiol. Aging* **2011**, *32*, 1161–1180. [CrossRef]
138. Karisetty, B.C.; Bhatnagar, A.; Armour, E.M.; Beaver, M.; Zhang, H.; Elefant, F. Amyloid-beta Peptide Impact on Synaptic Function and Neuroepigenetic Gene Control Reveal New Therapeutic Strategies for Alzheimer's Disease. *Front. Mol. Neurosci.* **2020**, *13*, 577622. [CrossRef]
139. Zhang, H.; Karisetty, B.C.; Bhatnagar, A.; Armour, E.M.; Beaver, M.; Roach, T.V.; Mortazavi, S.; Mandloi, S.; Elefant, F. Tip60 protects against amyloid-beta-induced transcriptomic alterations via different modes of action in early versus late stages of neurodegeneration. *Mol. Cell. Neurosci.* **2020**, *109*, 103570. [CrossRef]
140. Currais, A.; Huang, L.; Goldberg, J.; Petrascheck, M.; Ates, G.; Pinto-Duarte, A.; Shokhirev, M.N.; Schubert, D.; Maher, P. Elevating acetyl-CoA levels reduces aspects of brain aging. *Elife* **2019**, *8*, 47866. [CrossRef]
141. Ricobaraza, A.; Cuadrado-Tejedor, M.; Perez-Mediavilla, A.; Frechilla, D.; Del Rio, J.; Garcia-Osta, A. Phenylbutyrate ameliorates cognitive deficit and reduces tau pathology in an Alzheimer's disease mouse model. *Neuropsychopharmacology* **2009**, *34*, 1721–1732. [CrossRef]
142. Marzi, S.J.; Leung, S.K.; Ribarska, T.; Hannon, E.; Smith, A.R.; Pishva, E.; Poschmann, J.; Moore, K.; Troakes, C.; Al-Sarraj, S.; et al. A histone acetylome-wide association study of Alzheimer's disease identifies disease-associated H3K27ac differences in the entorhinal cortex. *Nat Neurosci.* **2018**, *21*, 1618–1627. [CrossRef] [PubMed]
143. Nativio, R.; Donahue, G.; Berson, A.; Lan, Y.; Amlie-Wolf, A.; Tuzer, F.; Toledo, J.B.; Gosai, S.J.; Gregory, B.D.; Torres, C.; et al. Dysregulation of the epigenetic landscape of normal aging in Alzheimer's disease. *Nat. Neurosci.* **2018**, *21*, 497–505. [CrossRef] [PubMed]
144. Hendrickx, A.; Peirrot, N.; Tasiaux, B.; Schakman, O.; Kienlen-Campard, P.; De Smet, C.; Octabe, J.-N. Epigenetic regulationsof immediate early genes experssion involved in memory formation by the amyloid precursor protein of Alzheimer Disease. *PLoS ONE* **2014**, *9*, e99467. [CrossRef] [PubMed]
145. Ettcheto, M.; Abad, S.; Petrov, D.; Pedros, I.; Busquets, O.; Sanchez-Lopez, E.; Casadesus, G.; Beas-Zarate, C.; Carro, E.; Auladell, C.; et al. Early Preclinical Changes in Hippocampal CREB-Binding Protein Expression in a Mouse Model of Familial Alzheimer's Disease. *Mol. Neurobiol.* **2018**, *55*, 4885–4895. [CrossRef]
146. Saura, C.A.; Choi, S.Y.; Beglopoulos, V.; Malkani, S.; Zhang, D.; Shankaranarayana Rao, B.S.; Chattarji, S.; Kelleher, R.J., 3rd; Kandel, E.R.; Duff, K.; et al. Loss of presenilin function causes impairments of memory and synaptic plasticity followed by age-dependent neurodegeneration. *Neuron* **2004**, *42*, 23–36. [CrossRef]
147. Kilgore, M.; Miller, C.A.; Fass, D.M.; Hennig, K.M.; Haggarty, S.J.; Sweatt, J.D.; Rumbaugh, G. Inhibitors of class 1 histone deacetylases reverse contextual memory deficits in a mouse model of Alzheimer's disease. *Neuropsychopharmacology* **2010**, *35*, 870–880. [CrossRef]
148. Ricobaraza, A.; Cuadrado-Tejedor, M.; Garcia-Osta, A. Long-term phenylbutyrate administration prevents memory deficits in Tg2576 mice by decreasing Abeta. *Front Biosci (Elite Ed)* **2011**, *3*, 1375–1384. [CrossRef]
149. Selvi, B.R.; Cassel, J.C.; Kundu, T.K.; Boutillier, A.L. Tuning acetylation levels with HAT activators: Therapeutic strategy in neurodegenerative diseases. *Biochim. Biophys. Acta* **2010**, *1799*, 840–853. [CrossRef]
150. NCT03061474. Nicotinamide as an Early Alzheimer's Disease Treatment. Available online: https://clinicaltrials.gov/study/NCT03061474 (accessed on 5 March 2025).
151. NTC03056495. Clinical Trial to Determine Tolerable Dosis of Vorinostat in Patients with Mild Alzheimer's Disease. Available online: https://clinicaltrials.gov/study/NCT03056495 (accessed on 5 March 2025).
152. Hanson, J.E.; La, H.; Plise, E.; Chen, Y.H.; Ding, X.; Hanania, T.; Sabath, E.V.; Alexandrov, V.; Brunner, D.; Leahy, E.; et al. SAHA enhances synaptic function and plasticity in vitro but has limited brain availability in vivo and does not impact cognition. *PLoS ONE* **2013**, *8*, e69964. [CrossRef]

153. Didonna, A.; Opal, P. The promise and perils of HDAC inhibitors in neurodegeneration. *Ann. Clin. Transl. Neurol.* **2015**, *2*, 79–101. [CrossRef]
154. Jeong, H.; Shin, S.; Lee, J.S.; Lee, S.H.; Baik, J.H.; Lim, S.; Kim, Y.K. Pan-HDAC Inhibitors Promote Tau Aggregation by Increasing the Level of Acetylated Tau. *Int. J. Mol. Sci.* **2019**, *20*, 4283. [CrossRef]
155. Phelan, M.; Mulnard, R.; Gillen, D.; Schreiber, S. Phase II Clinical Trial of Nicotinamide for the Treatment of Mild to Moderate Alzheimer's Disease. Available online: https://clinmedjournals.org/articles/jgmg/journal-of-geriatric-medicine-and-gerontology-jgmg-3-021.php?jid=jgmg (accessed on 5 March 2025).
156. Harrison, I.F.; Powell, N.M.; Dexter, D.T. The histone deacetylase inhibitor nicotinamide exacerbates neurodegeneration in the lactacystin rat model of Parkinson's disease. *J. Neurochem.* **2019**, *148*, 136–156. [CrossRef] [PubMed]
157. Beaver, M.; Karisetty, B.C.; Zhang, H.; Bhatnagar, A.; Armour, E.; Parmar, V.; Brown, R.; Xiang, M.; Elefant, F. Chromatin and transcriptomic profiling uncover dysregulation of the Tip60 HAT/HDAC2 epigenomic landscape in the neurodegenerative brain. *Epigenetics* **2022**, *17*, 786–807. [CrossRef] [PubMed]
158. Sarthi, J.; Elefant, F. dTip60 HAT activity controls synaptic bouton expansion at the Drosophila neuromuscular junction. *PLoS ONE* **2011**, *6*, e26202. [CrossRef] [PubMed]
159. Zhu, X.; Singh, N.; Donnelly, C.; Boimel, P.; Elefant, F. The cloning and characterization of the histone acetyltransferase human homolog Dmel\TIP60 in Drosophila melanogaster: Dmel\TIP60 is essential for multicellular development. *Genetics* **2007**, *175*, 1229–1240. [CrossRef]
160. Gaub, P.; Joshi, Y.; Wuttke, A.; Naumann, U.; Schnichels, S.; Heiduschka, P.; Di Giovanni, S. The histone acetyltransferase p300 promotes intrinsic axonal regeneration. *Brain* **2011**, *134*, 2134–2148. [CrossRef]
161. Vecsey, C.G.; Hawk, J.D.; Lattal, K.M.; Stein, J.M.; Fabian, S.A.; Attner, M.A.; Cabrera, S.M.; McDonough, C.B.; Brindle, P.K.; Abel, T.; et al. Histone deacetylase inhibitors enhance memory and synaptic plasticity via CREB:CBP-dependent transcriptional activation. *J. Neurosci.* **2007**, *27*, 6128–6140. [CrossRef]
162. Nguyen, L.; Humbert, S.; Saudou, F.; Chariot, A. Elongator—An emerging role in neurological disorders. *Trends Mol. Med.* **2010**, *16*, 1–6. [CrossRef]
163. Bhatnagar, A.; Krick, K.; Karisetty, B.C.; Armour, E.M.; Heller, E.A.; Elefant, F. Tip60's Novel RNA-Binding Function Modulates Alternative Splicing of Pre-mRNA Targets Implicated in Alzheimer's Disease. *J. Neurosci.* **2023**, *43*, 2398–2423. [CrossRef]
164. Bhatnagar, A.; Thomas, C.M.; Nge, G.G.; Zaya, A.; Dasari, R.; Chongtham, N.; Manandhar, B.; Kortagere, S.; Elefant, F. Tip60 HAT activators as therapeutic modulators for Alzheimer's disease. *Nat. Commun.* **2025**, *16*, 3347. [CrossRef]

cells

MDPI

Review

Neurobiological Mechanisms Underlying Psychological Dysfunction After Brain Injuries

Prashin Unadkat [1,2], Tania Rebeiz [2], Erum Ajmal [1,3], Vincent De Souza [4], Angela Xia [2], Julia Jinu [1,5], Keren Powell [1] and Chunyan Li [1,2,6,*]

1 Translational Brain Research Laboratory, The Feinstein Institutes for Medical Research, Manhasset, NY 11030, USA
2 Department of Neurosurgery, North Shore University Hospital at Northwell Health, Manhasset, NY 11030, USA
3 SUNY Downstate College of Medicine, Brooklyn, NY 11225, USA
4 Department of Neurosurgery, Staten Island University Hospital at Northwell Health, Staten Island, NY 10305, USA
5 Biology Department, Adelphi University, Garden City, NY 11530, USA
6 Department of Neurosurgery, Donald and Barbara Zucker School of Medicine at Hofstra/Northwell, Hempstead, NY 11549, USA
* Correspondence: cli11@northwell.edu; Tel.: +1-516-562-1078

Abstract: Despite the presentation of similar psychological symptoms, psychological dysfunction secondary to brain injury exhibits markedly lower treatment efficacy compared to injury-independent psychological dysfunction. This gap remains evident, despite extensive research efforts. This review integrates clinical and preclinical evidence to provide a comprehensive overview of the neurobiological mechanisms underlying neuropsychological disorders, focusing on the role of key brain regions in emotional regulation across various forms of brain injuries. It examines therapeutic interventions and mechanistic targets, with the primary goal of identifying pathways for targeted treatments. The review highlights promising therapeutic avenues for addressing injury-associated psychological dysfunction, emphasizing Nrf2, neuropeptides, and nonpharmacological therapies as multi-mechanistic interventions capable of modulating upstream mediators to address the complex interplay of factors underlying psychological dysfunction in brain injury. Additionally, it identifies sexually dimorphic factors as potential areas for further exploration and advocates for detailed investigations into sex-specific patterns to uncover additional contributors to these disorders. Furthermore, it underscores significant gaps, particularly the inadequate consideration of interactions among causal factors, environmental influences, and individual susceptibilities. By addressing these gaps, this review provides new insights and calls for a paradigm shift toward a more context-specific and integrative approach to developing targeted therapies for psychological dysfunction following brain injuries.

Keywords: psychological dysfunction; anxiety; depression; brain injury; cerebral ischemia; dementia

1. Introduction

Psychological dysfunction, manifesting as a complex spectrum of affective disorders, imposes a substantial global health burden [1–5]. Affecting an estimated 5–15% of the global adult population, the development of effective therapeutic strategies for psychological dysfunction is a critical research priority. This need is particularly urgent for individuals with acquired brain injury, who exhibit notably higher prevalence rates of psychological dysfunction, estimated at 30–50% [6–12]. Notably, injury-dependent psychological dysfunction

https://doi.org/10.3390/cells14020074

demonstrates lower responsiveness to pharmacotherapy, with reported efficacy rates of 30–40%, compared to 60–70% efficacy in injury-independent contexts [1,5,13–16] (Figure 1). Despite substantial research efforts and the significant societal impact of this condition, the mechanisms contributing to the lower treatment efficacy in injury-related psychological dysfunction, compared to injury-independent cases, remain insufficiently understood, highlighting the pressing need for innovative, targeted therapeutic strategies.

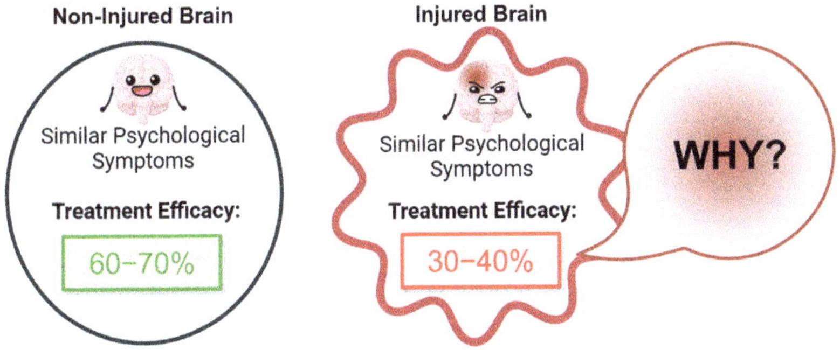

Figure 1. Therapeutic efficacy of psychological dysfunction in non-injured versus injured populations. While symptomatology is comparable, the therapeutic efficacy for psychological dysfunction is notably diminished in brain-injured populations compared to non-injured individuals. (Created with Biorender.com).

Two of the most common forms of affective disorder, anxiety and depression [1–5], are exacerbated both by brain injury and the psycho-emotional disturbances related to the acute phase of the inciting event [13,14]. The onset of these conditions reduces patient engagement in rehabilitation, impairing recovery from the brain injury and subsequently escalating psychological dysfunction, creating a self-perpetuating cycle. Therapeutic interventions are further challenged by the heterogeneity of psychological dysfunction presentations in clinical settings [17–24]. Meta-analyses on injury-dependent psychological dysfunction reveal a broad range of symptom severity, from mild to severe, regardless of the initial injury severity, influenced by various contributing factors. Research into injury-dependent psychological dysfunction has often mirrored studies on injury-independent psychological dysfunction, potentially overlooking etiologically specific factors that may contribute to reduced treatment efficacy. This gap highlights the need for more effective strategies to manage established psychological dysfunction following brain injury. It is crucial to re-evaluate the current neurobiological mechanisms targeted for therapeutic interventions addressing psychological dysfunction in various brain injury models, to develop more effective treatment strategies that enhance clinical outcomes.

This review aims to identify shared features and distinctions between injury-independent and injury-dependent psychological dysfunction, with the goal of pinpointing promising research directions specific to brain injury contexts. It assesses psychological dysfunction arising from ischemic stroke [8,10], intracerebral hemorrhage (ICH) [8], subarachnoid hemorrhage (SAH) [7], traumatic brain injury (TBI) [6,9], and vascular cognitive impairment and dementia (VCID) due to chronic cerebral hypoperfusion (CCH) or as a long-term consequence of acute brain injury [25], as well as distress observed in non-injury populations. Currently, there is no evidence suggesting fundamental differences in the mechanisms underlying psychological dysfunction across specific brain injuries [26]. However, distinct etiologies and variable damage progression patterns across these conditions indicate the potential for unique pathways influencing the development of psychological dysfunction in each injury type (Figure 2). We evaluated psychological dysfunction across diverse contexts to address three key aims: (1) to assess whether different types of brain injury influence psychological health in distinct ways, (2) to identify potential variations in targeted neurobiological mechanisms and therapeutic strategies between injury-independent and injury-dependent contexts, and (3) to evaluate the applicability of findings from non-injury ("healthy") populations to those with brain injuries. By highlighting these aspects, this review seeks to propose innovative avenues for investigating psychological dysfunction secondary to brain injury, ultimately aiming to advance therapeutic interventions.

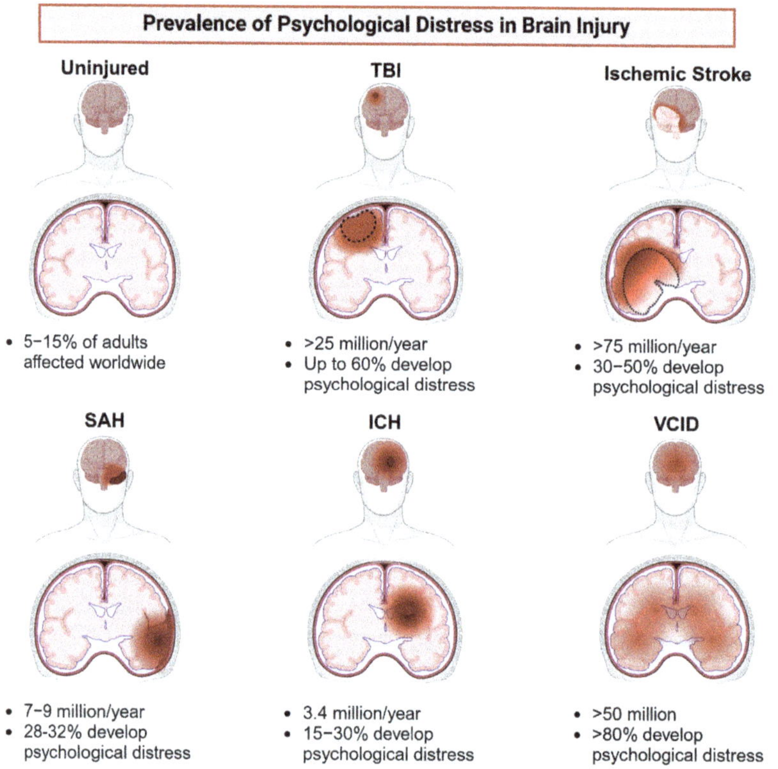

Figure 2. Different types of brain injury are associated with varying prevalences of psychological dysfunction. Psychological dysfunction is more prevalent in populations with brain injury, with each injury type exhibiting distinct rates and prevalences of psychological symptoms. Notably, vascular cognitive impairment and dementia (VCID), despite showing minimal overt damage (highlighted in red), exhibit the highest prevalence of psychological dysfunction. (Created with Biorender.com) (ICH: intracerebral hemorrhage; SAH: subarachnoid hemorrhage; TBI: traumatic brain injury).

2. Neurobiological Mechanisms and Therapeutic Strategies for Psychological Dysfunction in Injury-Independent and Injury-Dependent Contexts

To investigate the observed disparities in treatment efficacy between injury-independent and injury-dependent psychological dysfunction, we analyzed current trends in the neurobiological mechanisms explored in various brain injuries, the therapeutic interventions employed and their respective efficacies, and the brain sub-regions studied. Additionally, we compared the neurobiological mechanisms across different types of brain injuries to assess whether etiological variations should be considered in the treatment of psychological dysfunction. Overall, the mechanisms examined do not appear to be context-specific, with common themes including neuroinflammation, immunoregulation, apoptosis, oxidative stress, neurotransmission, synaptic plasticity, stress and sex hormones, amyloid and insulin metabolism, neuropeptides, and methylation. Similarly, the brain sub-regions investigated are consistent across both injured and non-injured contexts, including the hippocampus, cortex, amygdala, hypothalamus, white matter, striatum, and ventral tegmental area (VTA) (Figure 3). In injury-dependent psychological dysfunction, the primary symptoms explored include anxiety, depression, and fear responses (Table 1). This review, therefore, investigates generalized anxiety disorder (GAD), major depressive disorder (MDD), and panic disorder as examples of injury-independent psychological dysfunction, with a particular focus on comparing the neurobiological mechanisms and therapeutic interventions used in both injury-dependent and injury-independent contexts.

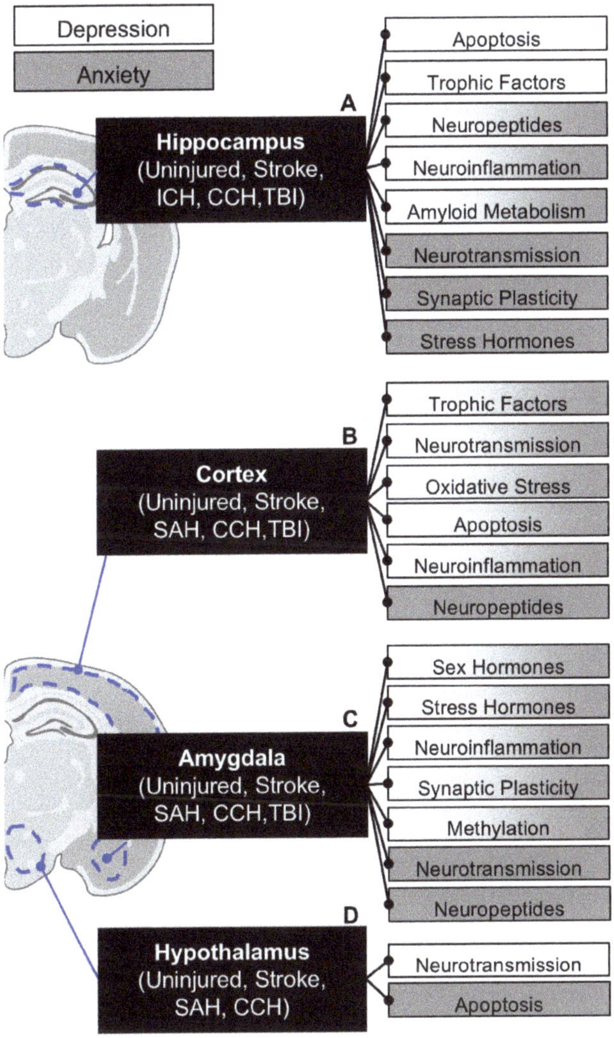

Figure 3. *Cont.*

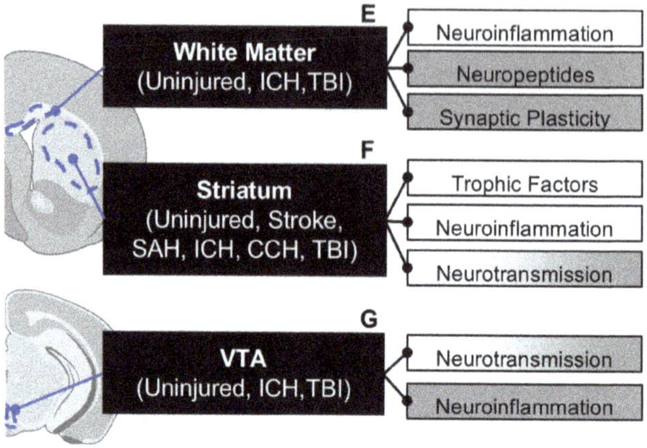

Figure 3. Psychological dysfunction is associated with alterations in specific brain regions responsible for emotional regulation, with each region contributing to distinct pathogenic cascades that collectively intensify psychological dysfunction. An analysis of studies on psychological dysfunction suggests that the brain sub-regions examined are not dependent on the location, severity, or presence of the primary injury. The primary regions investigated in relation to psychological dysfunction following brain injury include (**A**) the hippocampus, (**B**) the cortex, (**C**) the amygdala, (**D**) the hypothalamus, (**E**) white matter, (**F**) the striatum, and (**G**) the ventral tegmental area. Notably, the hippocampus, cortex, and amygdala are the main regions studied, while other areas involved in psychological regulation remain comparatively underexplored. (White boxes = depression, gray boxes = anxiety, gradient boxes = anxiety and depression) (Created with Biorender.com) (CCH: chronic cerebral hypoperfusion; ICH: intracerebral hemorrhage; SAH: subarachnoid hemorrhage; TBI: traumatic brain injury).

2.1. Psychological Dysfunction Manifesting in the Absence of Apparent Brain Injury

Previous reviews have explored the various underlying factors that influence the clinical presentation of generalized anxiety disorder, major depressive disorder, and panic disorder [27–30]. In this review, we provide a succinct overview of these factors, including neurobiological mechanisms, the roles of specific brain regions, and therapeutic interventions. This framework is used to assess whether neurobiological mechanisms and responses to therapeutic interventions differ between injured and non-injured contexts, aiming to more effectively identify potential targets for therapeutic intervention.

2.1.1. Neurobiological Mechanisms

The primary neurobiological mechanisms underlying psychological dysfunction in the absence of brain injury include neuroinflammation, which is partially influenced by stress-induced alterations in immune regulation, along with oxidative stress, DNA methylation, dysregulation of neurotransmitters and neuropeptides, disruptions in synaptic plasticity, and imbalances in sex and stress hormones. A primary theory of inflammation-mediated psychological dysfunction dictates that emotional stress can trigger peripheral immune systems, causing the release of pro-inflammatory cytokines [31]. The hypothalamic–pituitary–adrenal (HPA) axis and the sympathetic–adrenal–medullary system are activated by stress, which mediates the release of corticotropin-releasing hormone (CRH) levels by the hypothalamus, and the subsequent release of inflammatory cytokines, such as TNF-α, IL-1β, and IL-6, and chemokines, such as CXCL12. This affects the presence of serotonin, GABA, and glutamate, thus creating neurotransmitter dysregulation and impairing neurogenesis. The prolonged microglial and astrocytic activation also has an effect on oxidative stress, engendering an increase in reactive oxygen and nitrogen species [31]. The increase in reactive oxygen species (ROS) is accompanied by high levels of lipid peroxidation within the brain, as well as cyclooxygenase-2 (COX-2) activity [28,29]. This results in overexpression of the glutathione redox system and further activation of the HPA axis, forming a self-reinforcing feedback loop that promotes the production of neurotoxic substances and, thus, anxiety and depression. Hypersensitization of the HPA axis is a defining characteristic of anxiety, with stress producing an increase in plasma cortisol but not adrenocorticotropic hormone (ACTH) [32]. Depression, however, occurs with higher cerebrospinal concentrations of corticotropin-releasing factor (CRF) and a blunted ACTH response to CRF injections, suggesting the presence of sensitized glucocorticoid receptors secondary to chronic CRF elevations. Dysregulation of the stress response may be sexually mediated, with females exhibiting heightened sensitivity to CRF within the locus coeruleus [33–35], while estrogen fluctuations can increase depression development in females [36].

In addition to the interactions between inflammation, oxidative stress, and hormones, neurotransmitter and neuropeptide dysregulation, synaptic plasticity disruption, and DNA methylation can all interact to modulate psychological dysfunction. A fundamental aspect of psychological dysfunction is neurotransmitter disruption, which occurs with such a high propensity that a majority of pharmacological interventions focus on targeting it [30–32]. Disruptions in the GABA system are linked to both anxiety and panic disorder, as are disturbances in serotonin circuitry, as are disruptions in the reactions to cortisol. Depression in particular is classically associated with deficiencies in serotonin, noradrenaline, and dopamine, alongside disruptions in glutamatergic and GABAergic signaling [37]. Monoaminergic neurotransmitter disruption contributes to changes in synaptic plasticity, characterized by the alteration in neurotrophin levels, particularly those of brain-derived neurotrophic factor (BDNF). This contributes to reductions in ERK and the activity of AkT, which contributes to impairments in neuronal plasticity and survival. Concurrently, disruptions in neuropeptide expression and activity are observed in psychological dysfunction [32]. This can include hypersensitivity to cholecystokinin (CCK), the most abundant neuropeptide within the brain, decreased expression of neuropeptide Y (NPY), chronic galanin overexpression, decreased oxytocin, and increased vasopressin. These are primarily related to chronic dysregulation of the stress and pain response system, leading to emotional disturbance. Finally, DNA methylation may also contribute to observed symptomology in injury-independent psychological dysfunction [27]. Assessments have indicated that hypermethylation of BDNF and NR3C1 is associated with an increased risk of depression [38], while the methylation profiles of DNMT1/3A, EZH2, and IL-6 correlate to the severity of anxiety [39].

2.1.2. The Impact of Specific Brain Regions on Neurobiological Mechanisms

The neurobiological changes discussed earlier can occur throughout the brain; however, certain regions exhibit a stronger association with specific mechanisms. The hippocampus, which is integral to mood regulation and defensive responses, plays a crucial role in the onset of anxiety and depression [40]. In the context of psychological dysfunction, the hippocampus has been shown to undergo significant volume reduction, along with dysregulation of neurogenesis, synaptic plasticity, neurotransmission, neuroinflammation, and stress hormone activity in various studies [40–44]. Evidence suggests that the volume of the left CA1 region of the hippocampus can independently predict the duration of illness, while the CA2 through CA4 subregions are the only ones to show reductions in first-onset major depressive disorder, highlighting the hippocampus's critical role in psychological dysfunction [44]. The cortex, particularly the prefrontal cortex, demonstrates dysfunction associated with psychological dysfunction [45–48]. The impact of this dysfunction is intricately modulated by the prefrontal cortex's connectivity with other brain regions. Specifically, in primates, including humans, the specialized lateral, medial, and orbital sectors of the prefrontal cortex are linked to the amygdala and brainstem—autonomic structures that influence emotional and physiological arousal through the limbic cortex [46].

The amygdala, recognized as a central hub in the brain's stress circuitry, is another critical brain region involved in psychological dysfunction [49]. It has been proposed that the amygdala may not function as a distinct entity but rather as a network of extensions connecting to the cortices, striatum, hypothalamus, and brainstem. Within the amygdala, several neurotransmitters, including serotonin, GABA, and dopamine, have been implicated in the onset of psychological dysfunction, supporting the idea of the amygdala as a key independent regulator in emotional processing.

The hypothalamus, a key component of the hypothalamic–pituitary–adrenal (HPA) axis that regulates the body's physiological stress response, plays a crucial role in psychological dysfunction. In depression, increased hypothalamic volume has been correlated with severity [50], while dysregulation of cortisol, corticotropin-releasing hormone (CRH), and oxytocin has been linked to generalized anxiety disorder [51]. The ventral tegmental area (VTA) is another significant region, where the activity of tyrosine kinase [52] and dopamine [53] is involved in regulating depression and anxiety, respectively. Similar to the hypothalamus, individuals with anxiety and depression exhibit larger VTA volumes [54]. These regions represent only a subset of the mechanisms underlying injury-independent psychological dysfunction but highlight the interconnected pathways that contribute to the presentation of GAD, major depressive disorder, and panic disorder.

2.1.3. Therapeutic Interventions

Given the high prevalence of psychological disorders in the general population and the significant societal burden they impose, along with the complex mechanisms and pathways underlying their symptomatology, numerous therapeutic interventions have been evaluated [55–59]. These interventions can be broadly categorized into pharmacological treatments, such as benzodiazepines, selective serotonin reuptake inhibitors (SSRIs), serotonin–norepinephrine reuptake inhibitors (SNRIs), non-benzodiazepine anxiolytics, monoamine oxidase inhibitors (MAOIs), and tricyclic antidepressants (TCAs), as well as psychological therapies like cognitive behavioral therapy (CBT) and acceptance and commitment therapy (ACT). Both benzodiazepine and non-benzodiazepine anxiolytics act on the binding of the neurotransmitter gamma-aminobutyric acid (GABA), enhancing its inhibitory effects, thus calming neuronal activity and reducing anxiety [60]. SSRIs and SNRIs, on the other hand, function by preventing the reuptake of serotonin and norepinephrine, thereby increasing the relative bioavailability [61], while MAOIs block the monoamine oxidase enzyme, preventing the breakdown of neurotransmitters in the brain [62]. On the whole, the pharmacological interventions can be classified as functioning via modulation of neurotransmission within the brain. Additional non-pharmacological approaches, including physical exercise and neuromodulation, target other neurobiological mechanisms, including neurotransmission, neuroplasticity, inflammation, oxidative stress, hormonal dysregulation, peptides, and the endocrine system [56,58,63]. Despite the variety of available treatment options, a subset of individuals remains unresponsive to these therapies. Specifically, approximately 30% of individuals with major depressive disorder exhibit treatment resistance [16], and about 40% of individuals with anxiety disorders are similarly resistant to treatment [64]. However, as we will demonstrate, treatment resistance is notably higher in injury-dependent psychological dysfunction.

Table 1. Neurobiological mechanisms underlying anxiety and depression across various types of brain injuries, as investigated through animal models. (↑ = increase; ↓ = decrease; → = leads to).

Mechanism	Trends	Sex-Specificity	Brain Region Specificity	Disease/Model	Ref.
Apoptosis	P38-MAPK, ↓ERK, Akt/mTOR → ↓neurons → anxiety, compulsive-like behaviors, and decreased activity levels.	N/A	ERK, SAH: Hypothalamus Parietal–entorhinal cortex	SAH Ischemic stroke	[65–68]
	↑Caspase 3/8/9 cleavage → ↓neurons → depressive behaviors	N/A	Hypothalamus Prefrontal cortex Hippocampus	Global ischemia	[69]
	P2Y14r → ↑p54, p46, BAX, caspase 3 cleavage → depressive behaviors	N/A	Cortex	SAH	[70]
	Apoptosis-inducing factor, Cytochrome C → depressive behaviors	N/A	Ipsilateral hippocampus	mTBI	[71]
Neuroinflammation	↑TNF-α, IL-1B, IL-6, and NF-κB → Anxiety and depression	ICH: Estrogen modulates astrocytic-mediated inflammation and the presentation of depression	N/A	TBI ICH	[17,72,73]
	Astrocytosis, microgliosis → Anxiety and depression		ICH (depression): White matter Hippocampus Striatum ICH (anxiety and depression): Ventral tegmental area	ICH TBI Stroke	[17,24,73–76]
	↓pTRkB, ↓pERK, ↓pCREB → Anxiety	N/A	Hippocampus	CCH + CRS	[77]
	↓PKA/CREB → Anxiety	N/A	Amygdala	TBI	[78]
	HDAC3 → NFkB → p65 → Cox1 → Anxiety	N/A	Cortex	Photothrombotic stroke	[79]
	↓Pink1 → inflammation → Anxiety and depression	N/A	Ipsilateral hemisphere	ICH	[80]
	↓Nrf2, TrkB-BDNF, and Trbk-PI3K; ↑Nrf2-Keap1-p62 → Anxiety and depression	N/A	N/A	ICH	[81–83]
Immunoregulatory Dysfunction	↑pPERK → ↑STING → ↑IFNβ → Th1 → Anxiety and depression	N/A	N/A	TBI	[84]
	↑Neutrophil elastase → ↓sepina3n → Anxiety	N/A	Cortex	TBI	[85]
	FOXO1 → VCAN, BAX, ↓ neutrophils → IL-6, ferroptosis → Depression	N/A	N/A	TBI	[86]

Table 1. *Cont.*

Mechanism	Trends	Sex-Specificity	Brain Region Specificity	Disease/Model	Ref.
Oxidative Stress	↓SOD, GSH, CAT → Anxiety ↓SOD → Depression	ICH: Estrogen modulates SOD and the presentation of depression	N/A	ICH CCH	[73,74,87,88]
	↓Glutathione metabolism → Anxiety	N/A	Cortex	TBI	[89]
	HDAC1, HDAC3, NOX4 → prostaglandin → Anxiety	N/A	Cortex	TBI	[90]
	↑Lipid peroxidation → Anxiety and depression	N/A	N/A	ICH CCH TBI	[79]
	Keap1-Nrf2-p62 modulates oxidative stress to regulate depression	N/A	N/A	Stroke	[72,82,87]
Neurotransmission	↓Dopaminergic neurons → depression	N/A	N/A	ICH	[52]
	↓α-tubulin → ↓microtubule stability → ↓dopaminergic neurons → depression	N/A	Ventral tegmental area	ICH	[76]
	↑Dopamine → Anxiety	N/A	Striatum	CCH	[91]
	↓$GABA_{B2}$ subunits → Anxiety ↓$GABA_{B1}$, $GABA_{B2}$ → Depression ↓GABAr → Anxiety	N/A	CCH: ($GABA_{B2}$ subunits: Amygdala $GABA_{B1}$, $GABA_{B2}$: Hippocampus) TBI: Basolateral amygdala	CCH TBI	[21,22,92]
	Dysfunctional glutamate uptake and ↑glutamate → depression ↑glutamate → Anxiety, depression	N/A	SAH/ICH: Striatum TBI: central amygdala	SAH ICH TBI	[74,81,93–97]
	NMDAR → Anxiety and depression	N/A	Hippocampus	TBI: anxiety Stroke: depression	[98–100]
	↑Kynurenine → ↓serotonin → Depression	N/A	ICH: Perihematomal region CCH, Stroke: hippocampus	ICH CCH Stroke	[101–103]
	↓Nrf2/BDNF → Anxiety and depression	N/A	ICH: Perihematomal region	ICH	[81,83]
	↓BDNF → ↓TrkB → ↓ NCAM → GIRK, Kir3 → Anxiety and depression	N/A	CCH: Amygdala	CCH	[21,104]
	↑Acetylcholine → Anxiety	N/A	Basolateral amygdala	TBI	[92]
	↑HCN1 and ↓KCNQ3 → Anxiety and depression	N/A	Amygdala	CCH	[104]

Table 1. *Cont.*

Mechanism	Trends	Sex-Specificity	Brain Region Specificity	Disease/Model	Ref.
	↓CNR1, ↓COMT, ↓VEGF2r → Anxiety	N/A	Cortex	TBI	[105]
	↓CB1 receptor → Depression	N/A	Ventromedial hypothalamus	Stroke	[106]
Synaptic Plasticity	↑Lipid peroxidation → ↓BST, LTP, PPR → Anxiety	N/A	Hippocampus	CCH	[107]
	↓ mRNA β1-catenin, ↓TGF-β1, ↓NR2B → Anxiety	N/A	Hippocampus	CCH	[108]
	↑GSK-3β → Anxiety	N/A	Hippocampus	CCH	[108]
	↓BDNF → ↓PSD95 → Depression	N/A	Hippocampus	CCH	[109]
	↓PSD95, ↓SYN, ↓GAP43, SYP → Anxiety	N/A	Hippocampus	CCH + CRS	[77]
	NR2A, NR2B, NMDA proteins, GAD67, GAD65, GluA2 → Anxiety	N/A	Dorsal hippocampus Ventral hippocampus Basolateral amygdala	TBI	[110]
	↓CNPase, ↓synaptophysin → Anxiety and depression	N/A	White matter	CCH	[111]
Stress Hormones	Glucocorticoid receptor translocation → Depression	N/A	Hippocampus	CCH	[112]
	↑CRF-1, ↑CRF-1R → Anxiety	N/A	N/A	mTBI	[113]
	↑CRH1, ↑CRHR1 → ↑BDNF, ↑TrkB, and ↑pCREB-ir → Anxiety	N/A	Amygdala Hippocampus	Transient global ischemia	[114–116]
	↑CRHR1 → p62 → synaptic loss → Depression	N/A	Hippocampus	tMCAO	[82]
	CRFR2 → Anxiety	Females: ↓CRFR2 in the DH Males: ↑CRFR2 in VH, ↓ CRFR2 in amygdala	Ventral hippocampus Dorsal hippocampus Amygdala	TBI	[117]
	↑NGF → Anxiety	N/A	N/A	Transient global ischemia	[118]
Sex Hormones	Estrogen, progesterone, testosterone → α-1 $GABA_A$, ↓Dopamine → Anxiety	Diestrus and proestrus phases in females affect severity	N/A	mTBI	[19,119,120]
	Estrogen, testosterone → SERT→ serotonin → Anxiety	Proestrus females experience increased serotonin	N/A	TBI	[119,121]

Table 1. *Cont.*

Mechanism	Trends	Sex-Specificity	Brain Region Specificity	Disease/Model	Ref.
Amyloid/Insulin Metabolism	↑BACE → ↑AAP → ↑Aβ → Anxiety	N/A	Hippocampus	CCH	[122,123]
	GlcNAc dysfunction, ↓Insulin signaling pathway → Anxiety	N/A	N/A	CCH	[124]
Trophic Factors	↓BDNF → ↓Neurogenesis → Depression	N/A	Hippocampus Striatum	Stroke	[125]
	↓VEGF → dopamine → Anxiety	N/A	Pre-frontal cortex	TBI	[126]
Neuropeptides	CGRP → Nrf2, eNOS → Anxiety and depression	The presentation of decreased anxiety and depression in females is mediated by CGRP	Amygdala Thalamus Hippocampus White matter Pericontusional cortex	TBI	[127]
	↑ACE/Ang II/AT1R → ↓CBF, ↑Oxidative stress, ↑Neuroinflammation → Anxiety	N/A	Hippocampus	CCH	[128]
Methylation	↓m6A demethylase, ↓FTO → Depression	N/A	N/A	Stroke	[129]
	↑DNMT, ↑HDAC2 → GADD45 damage → Anxiety	N/A	Amygdala	TBI	[130,131]

2.2. Traumatic Brain Injury

2.2.1. Clinical Presentation and Clinical Therapeutic Efficacy

TBI affects over 25 million individuals annually [132], with its causes encompassing a wide range of events, including military combat, vehicular accidents, recreational or sports-related injuries, and falls, although mild TBI (mTBI) is often underreported [133]. TBI is defined as a disruption in brain structure or function resulting from an external force, typically acute in nature. The causes of TBI lead to varying degrees of injury, classified as mild, moderate, or severe, with a corresponding spectrum of symptoms. Individuals across this spectrum are at increased risk for developing major depressive disorder, post-traumatic stress disorder (PTSD), generalized anxiety disorder, and elevated suicidality, which significantly contributes to morbidity. In the absence of injury, approximately 5–15% of the general population is diagnosed with a DSM-IV Axis I disorder; however, within the first year following TBI, this rate increases to as high as 60% of individuals [1–5]. Treatment for individuals exhibiting depressive symptoms includes pharmacological interventions such as SSRIs, MAOIs, and TCAs, in addition to non-pharmacological approaches like CBT [134–136]. However, meta-analyses of randomized clinical trials suggest that current interventions show no significant benefit over placebo in cases of TBI-induced depression [134,136], with only certain dopaminergic agents demonstrating improvements in general behavioral outcomes [137].

2.2.2. Neurobiological Mechanisms and Pre-Clinical Therapeutic Targets

A review of the literature reveals that psychological dysfunction following TBI is mediated through dysregulation of multiple biological pathways, including apoptosis, neuroinflammation, immunoregulatory dysfunction, oxidative stress, neurotransmission, synaptic plasticity, hormonal imbalances (stress and sex hormones), trophic factors, neuropeptides, and DNA methylation (Tables 1 and 2). TBI is primarily characterized by an acute primary injury that results in a localized area of damage, followed by a secondary cascade of injury [138]. Mechanistic studies have investigated contributions across different brain regions. Although the initial TBI can lead to overt tissue damage, the role of apoptosis in psychological dysfunction remains underexplored. However, evidence suggests that apoptosis-inducing factors such as apoptosis-inducing factor and cytochrome C in the hippocampus play a role in modulating depressive behaviors following mild TBI [71]. Additionally, in the acute stage of TBI, elevation of forkhead box protein O1 (FOXO1) enhances the interaction between Versican (VCAN) and BCL-2-associated X protein (BAX), thereby inhibiting BAX translocation to the mitochondria and exerting anti-apoptotic effects [86]. This promotes neuronal degradation, later neuroinflammation, and depressive symptomology.

Table 2. Clinical studies investigating the neurobiological mechanisms underlying anxiety and depression in brains impacted by injury. (↑ = increase; ↓ = decrease; → = leads to).

Mechanism	Trends	Sex-Specificity	Global or Focal	Disease/Model	Ref.
Neuroinflammation	↓IL-6, ↑α-amylase → Depression	N/A	α-amylase: Saliva IL-6: Serum	Stroke	[139]
	↑IL-1β, ↑TNF-α → Depression	N/A	Serum	Stroke	[140]
	Elevated early CSF and sera IL-6 levels associated with fatigue	N/A	CSF/serum	SAH	[141]
Oxidative Stress	Ferritin → Depression	N/A	Serum	Stroke	[142]

Table 2. *Cont.*

Mechanism	Trends	Sex-Specificity	Global or Focal	Disease/Model	Ref.
Stress Hormones	↑early urinary free cortisol → depression, fatigue, and ↓ quality of life ↑serum cortisol, serum CRH → depression	N/A	Urine Serum	SAH Stroke	[141,142]
Synaptic Plasticity	↑Serum ApoE, ↓monocyte ApoE → Depression	N/A	Serum	Stroke	[143]
Neurotransmission	↓HTR3D, ↓NEUROG3 → Depression	N/A	Global	Stroke	[144]
	↑5-HTTLPr → Depression	N/A	Global	Stroke	[145]
	p11/tPA/BDNF → Depression	N/A	Global	Stroke	[146]
	↓Serotonin → Depression	N/A	Serum	Stroke	[147]
	↑Glutamate → Depression	N/A	Plasma	Stroke	[148]
	↓Glutamate → Depression	N/A	Plasma	Stroke	[149]
	↑Diffusivity → Anxiety and Depression	Pediatric females are more likely to develop ↑diffusivity	Amygdala	TBI	[150]
Neuropeptides	↑CGRP concentration → poorer psychological outcomes	N/A	CSF	SAH	[151]
	↓NPY → Depression	N/A	Plasma, serum	Stroke	[152,153]
	↓CCK-8, ↑Substance P → Depression	N/A	Serum	Stroke	[147]
Sex Hormones	↓Testosterone → ↓Dopamine → Anxiety	↓Testosterone is more prevalent in males	Global	TBI	[120,154]

In contrast to the relatively sparse assessment of apoptosis' effect on psychological dysfunction in TBI, neuroinflammation and immunoregulatory dysfunction, which are associated with the primary injury, have been more extensively studied. At the surface level, increased astrocytosis, microgliosis, and the release of inflammatory cytokines, including TNF-α, IL-1B, IL-6, and NF-κB, potentiate depressive and anxious symptomology [17,72]. This may be connected to the decrease in protein kinase A (PKA), leading to the decreased activation of cyclic AMP response element-binding protein (CREB), the modulation of which affects astrocyte function [72]. The observed anti-apoptotic effect of FOXO1 mediates the activation of neutrophils, resulting in an increase in IL-6 production [86]. It also promotes ferroptosis in neutrophils via the transferrin receptor (TFRC) mechanism. This disruption in iron homeostasis within oligodendrocytes leads to a reduction in myelin basic protein and contributes to depressive symptoms. Within neutrophils, the increased activation of neutrophil elastase also decreases serpina3n mRNA levels, impacting neuronal function and contributing to anxiety-like behavior [85]. Additionally, increased phosphorylation of extracellular signal-regulated kinase (pERK) leads to activation of the STING pathway, and increased interferon β [84]. This modulates the presentation of Th1 cells and the presentation of both anxiety and depression. Of the inflammatory effects, only neutrophil elastase [85] and PKA/CREB signaling [78] have been specifically linked to the cortex and amygdala, respectively. Other pathways have not been assigned a regional specificity.

Alongside apoptosis and inflammation, the primary injury also triggers immediate changes in oxidative stress, which contribute to psychological dysfunction. This is characterized by dysregulated interactions between glutathione and histone deacetylases (HDACs) [89,90] and increased lipid peroxidation [72]. Activation of HDAC1, HDAC3, and NADPH oxidase 4 (NOX4) [90] influences prostaglandin production in the cortex via microglia [78], which is associated with heightened anxiety and altered glutathione metabolism [89].

Following the primary injury, secondary damage is associated with disrupted neurotransmission and synaptic plasticity, including dysregulated expression of NMDAr and glutamatergic activity [99,110], reduced expression of GABAergic neurons [92], and decreased expression of VEGF receptor, cannabinoid receptors, and catechol-O-methyltransferase (*Comt*) [105]. The increase in NMDAr expression in the hyper-acute phase is followed by a decrease in the chronic phase, mediating the hyper-acute increase in Na^{2+} and Ca^{2+} influx into the cell [93–96,99,105]. This decreases brain-derived neurotrophic factor (BDNF) and nerve growth factor (NGF) production within the hippocampus and promotes the development of anxiety. The changes in NMDAr are further accompanied by bilateral alterations in synaptic glutamatergic a-amino-3-hydroxy5-methyl-4-isoxazolepropionic acid (AMPA; GluA1, GluA2) and GABA-synthetic (GAD65, GAD67) proteins, leading to alterations in synaptic plasticity [110]. Anxiety in TBI is also potentiated by the loss of GABAergic neurons and $GABA_A$ receptor subunits in the basolateral amygdala (BLA), alongside an increase in acetylcholine receptors (α_7-nAChR) [92]. This mediates both a reduction in inhibition within the BLA, as well as an increase in the excitability of principal neurons within the system. Catecholaminergic and endocannabinoid dysregulation also underlies the presentation of psychological dysfunction in TBI [105]. The decrease in *Comt* in the prefrontal cortex mediates a decrease in the release of striatal dopamine, while the decrease in cannabinoid receptor 1 (*Cnr1*) mediates the potentiation of brain damage.

Research examining sex differences in psychological dysfunction following TBI has indicated that estrogen and progesterone modulate the activity of GABA, dopamine, and serotonin systems, pointing to these pathways as potential targets for future investigation [19,119–121]. These findings are further supported by changes in calcitonin gene-related peptide (CGRP) levels, which influence the increase in Nrf2 expression and the reduction of psychological dysfunction observed in female rats [127], as well as alterations in CRF receptor expression [113,117]. Lastly, TBI is associated with increased expression of DNA methyltransferase (DNMT) and histone deacetylase 2 (HDAC2), along with decreased expression of growth arrest and DNA damage 45 (GADD45) [130,131]. Notably, while the BDNF IV promoter shows increased binding of HDAC2 and higher levels of histone H3 lysine 9 acetylation (H3K9ac) in both acute and chronic TBI, the BDNF IX promoter exhibits increased HDAC2 binding and decreased H3K9ac levels [130]. These epigenetic modifications result in reduced expression of BDNF IV and IX, which correlates with the development of anxiety.

Pre-clinical trends in therapeutic intervention development reflect the neurobiological mechanisms that have been assessed, targeting inflammation, oxidative stress, neurotransmission, sex hormones, and stress hormones (Figure 4, Table 3). Neurotransmission appears to be the most frequently targeted mechanism, focusing on the modulation of GABA, NMDAr, glutamate uptake, and the endocannabinoid system [92,95,99,105]. Interestingly, this largely follows the trends established by the treatment of injury-independent psychological dysfunction, which also appears to focus on neurotransmission as a therapeutic intervention. Beyond neurotransmission, however, interventions also target inflammation and/or oxidative stress, via the decrease in lipid peroxidation and inflammatory cytokines [72] or the modulation of glutathione expression [89], PKA/CREB [78], or Sepina3n [85]. Hormonal modulation has also shown a certain degree of promise, with the administration of progesterone decreasing anxiety-like behavior in rats [119], and the blockade of CRFR1 attenuating anxiety-related symptoms and HPA axis reactivity in mice [113].

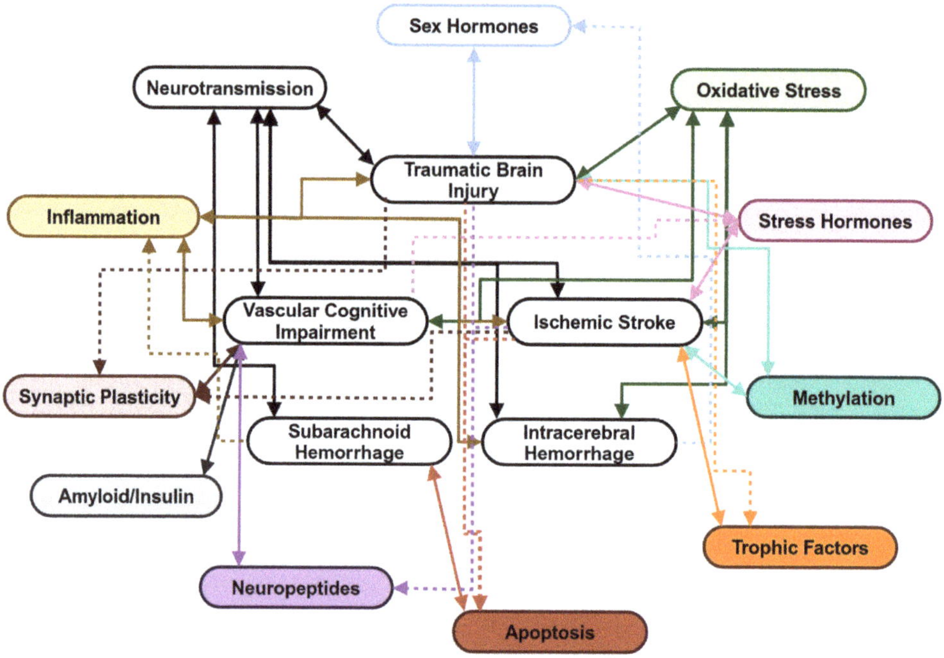

Figure 4. Therapeutic interventions targeting neurobiological mechanisms underlying psychological dysfunction in injury-related contexts. Eleven overarching categories of neurobiological mechanisms have been identified as contributing to psychological dysfunction in the context of brain injury. While there is considerable overlap across different types of brain injury, not all mechanisms have been evaluated in every form of brain injury. Additionally, among the assessed mechanisms, only a subset has been specifically targeted with the aim of influencing psychological function (solid, double-headed arrows = identified mechanisms that have been targeted, others have not been targeted at all for therapeutic intervention (dashed, single-headed arrows (see: Table 3, greyed out rows)).

Table 3. Neurobiological mechanisms targeted for psychological dysfunction in preclinical brain injury models.

Mechanism	Disease/Model Implicated In	Pre-Clinical Interventional Target	References
Neurotransmission	Vascular Cognitive Impairment and Dementia	GABAr	[21,22]
		Serotonin	[102]
	Traumatic Brain Injury	GABAA	[92]
		NMDAr	[93,96,99]
		Glutamate	[95,99]
		Endocannabinoid system	[105]
	Ischemic Stroke	Dopamine	[91]
		Glutamate	[94]
		Serotonin	[103]
	Intracerebral Hemorrhage	Indoleamine 2, 3-dioxygenase and serotonin	[101]
	Subarachnoid Hemorrhage	HDAC2 and glutamate	[97]

Table 3. *Cont.*

Mechanism	Disease/Model Implicated In	Pre-Clinical Interventional Target	References
Inflammation	Vascular Cognitive Impairment and Dementia	Angiotensin system	[128]
	Traumatic Brain Injury	Inflammatory cytokines	[72]
		Sepina3n and Neutrophil elastase	[85]
		PKA/CREB	[78]
	Ischemic Stroke	HDAC	[79]
	Intracerebral Hemorrhage	Inflammatory cytokines	[73]
		TRPV4 channel opening	[74]
Oxidative Stress	Vascular Cognitive Impairment and Dementia	Glutathione expression	[87]
		SOD	[87]
		Angiotensin system	[128]
	Traumatic Brain Injury	Lipid peroxidation	[72]
		Glutathione expression	[89,90]
	Ischemic Stroke	Lipid peroxidation	[82]
		Nrf2	[82]
	Intracerebral Hemorrhage	Nrf2	[81,83]
		SOD and ROS	[73]
Synaptic Plasticity	Vascular Cognitive Impairment and Dementia	TGF-β1	[108]
		BDNF	[21,109]
		Glycogen synthase kinase-3β	[108]
		Synaptophysin	[111]
		Lipid peroxidation	[107]
Sex Hormones	Traumatic Brain Injury	Progesterone	[119]
Stress Hormones	Traumatic Brain Injury	CRFR1 and HPA axis	[113]
	Ischemic Stroke	CRHR1	[82]
	Vascular Cognitive Impairment and Dementia	Glucocorticoid receptors	[112]
Methylation	Traumatic Brain Injury	DNMT and HDAC	[130,131]
	Ischemic Stroke	M6A demethylase and FTO	[129]
Trophic Factors	Ischemic Stroke	BDNF	[114,125]
Apoptosis	Subarachnoid Hemorrhage	p38 MAPK	[66]
Neuropeptides	Vascular Cognitive Impairment and Dementia	Angiotensin system	[128]

2.3. Ischemic Stroke

2.3.1. Clinical Presentation and Clinical Therapeutic Efficacy

Ischemic stroke affects over 75 million individuals worldwide annually, with its incidence steadily increasing [155]. It is characterized by a disruption in cerebral blood flow, leading to an initially reversible loss of function, which subsequently progresses to cerebral infarction [156]. This ischemic event triggers a cascade of pathological processes, starting with electrical dysfunction, followed by membrane disruption, calcium influx, excitotoxicity, reactive oxygen species (ROS) generation, and culminating in cell lysis and death. Psychological dysfunction is a frequent comorbidity following ischemic stroke. Meta-analyses show that between 39% and 52% of stroke patients develop one or more depressive episodes within the first five years post-stroke [157], and up to 30% develop anxiety within the first year [158]. The development of post-stroke depression (PSD) is influenced by multiple factors, including a history of depression and the severity and location of the stroke [159]. PSD is often associated with more severe depressive symptoms than generalized major depressive disorder, and the efficacy of treatments remains inconsistent and moderate [160]. Current pharmacological treatments for PSD include antidepressants, SSRIs, TCAs, CBT, and neuromodulation. However, the most effective treatment approach seems to involve a combination of pharmacological, psychosocial, and stroke-focused interventions, although these interventions tend to show higher rates of attrition due to the significant patient involvement required. Anxiety treatments following stroke typically involve SSRIs, TCAs, MAOIs, psychotherapy, acupuncture, exercise, and CBT. However, the greater emphasis on PSD research has limited the number of randomized controlled trials with placebo controls for anxiety, resulting in low-quality evidence with a high risk of bias regarding the efficacy of therapeutic interventions [26].

2.3.2. Neurobiological Mechanisms and Pre-Clinical Therapeutic Targets

Preclinical studies of ischemic stroke have identified a range of mechanisms that contribute to the psychological dysfunction observed in the post-stroke period. These include apoptosis, neuroinflammation, oxidative stress, disruptions in neurotransmission, alterations in stress hormones, trophic factors, and DNA methylation (Table 1). Clinical research further suggests that synaptic plasticity and neuropeptides may also play a significant role in the development of psychological symptoms (Table 2). Given the extent of the primary injury, it is expected that mediators of this injury include pathways such as Akt1 signaling, the Nrf2 pathway, caspases 3, 8, and 9, HDAC3, and glial activation, all of which have been implicated in PSD [24,67,69,79,82]. Unique from TBI, dysfunction of the Akt/mTOR pathway has been observed to mediate anxiety following ischemic stroke [67], while increased expression of caspases 3, 8, and 9 in the hypothalamus, prefrontal cortex, and hippocampus mediate the presentation of PSD [69]. Separate from the effect of apoptosis, neuroinflammation, and oxidative stress, there is also a reduction in the m6A demethylase fat mass and obesity-associated protein (FTO) in the acute phase of ischemic stroke, accompanied by hypermethylation of N6-methyladenosine [129]. This epigenetic modification exerts significant downstream effects, as FTO regulates critical factors such as BDNF, a subunit of the NMDA receptor, myelocytomatosis oncogene, Jun, and oligodendrocyte transcription factor 2. FTO-mediated m6A demethylation is therefore essential for the regulation of neurogenesis, gliogenesis, axonal growth, synaptic plasticity, and stress responses. These changes are reflected in the altered expression of NMDA receptors, decreased serotonin levels, reduced CB1 receptor expression, increased CRH receptor expression, increased NGR, and disruptions in BDNF signaling within the hippocampus [82,100,103,106,114–116,118,125]. Clinically, elevated cytokine levels in saliva, cerebrospinal fluid (CSF), and serum have been associated with the onset of PSD, as have increased ferritin, cortisol, ApoE, substance P, and imbalances in neurotransmitter systems [139,140,142–149,152,153,161].

Much like in TBI, therapeutic targets in ischemic stroke include the overarching categories of neurotransmission, inflammation, oxidative stress, stress hormones, and trophic factors (Figure 4). Again, the most targeted mechanism is neurotransmission, with dopamine, glutamate, and serotonin presenting as promising interventional foci for PSD [91,94,97,103]. Oxidative stress and neuroinflammation are then the second-most targeted, with researchers focusing on HDAC, lipid peroxidation, and Nrf2, for a combined effect on both PSD and anxiety [79,82]. Lastly, BDNF and stress receptors have presented as potential targets for depression and anxiety, respectively [114,116,125].

2.4. Intracerebral Hemorrhage

2.4.1. Clinical Presentation and Clinical Therapeutic Efficacy

ICH accounts for approximately 10 to 15% of all stroke types [162], with an estimated 3.4 million global cases reported in 2019, reflecting a substantial global health burden [163]. ICH is regarded as the most fatal form of stroke and is characterized by the formation of a hematoma within the brain parenchyma, with or without extension of blood into the ventricles. The damage results from both the spread of blood within the brain tissue and ischemia due to impaired cerebral perfusion. Consequently, 15 to 30% of patients experience depression and/or anxiety within the first year following ICH [164,165]. Treatment options include solution-focused approaches, SSRIs, TCAs, and anxiolytics, targeting behavioral modification, serotonin and norepinephrine uptake, and GABA [164,166–168]. While clinical trials have assessed the efficacy of these interventions, comprehensive evaluations have predominantly focused on SSRIs. Although SSRIs are effective in treating depression after ICH, their use is associated with an increased risk of ICH recurrence and poor neurological outcomes, thus limiting their application in certain patients [167,169].

2.4.2. Neurobiological Mechanisms and Pre-Clinical Therapeutic Targets

Psychological dysfunction following ICH has predominantly been studied through the mechanisms of neuroinflammation, oxidative stress, and neurotransmission (Table 1), likely due to the lower incidence rate and the emphasis on primary injury rather than secondary complications. Neuroinflammatory processes, particularly glial activation, dysregulation of PTEN-induced putative kinase 1 (Pink1) and Nrf2 pathways, and cytokine release, have been strongly associated with the development of anxiety and depression after ICH [24,73,76,80,81,83,86]. In fact, Pink1 and Nrf2 have been independently identified as potential upstream mediators following ICH, their downregulation mediating the increased expression of monocyte chemoattractant protein-1, macrophage inflammatory protein, and proinflammatory macrophages, as well as decreased TrkB and BDNF expression. Concurrently, oxidative stress-induced damage is exacerbated by increased lipid peroxidation and a reduction in antioxidative enzymes such as glutathione (GSH), superoxide dismutase (SOD), and glutaminase, as again potentially mediated by Nrf2 [72,74,88]. Additionally, neurotransmission alterations are evident, with a reduction in dopaminergic neurons mediated by α-tubulin, impaired glutamate uptake, decreased Nrf2 activity, and lowered serotonin levels [74,76,81,101,162]. These changes are widespread throughout the brain.

The focus on neuroinflammation, oxidative stress, and neurotransmission is reflected in the therapeutics, which have been assessed pre-clinically for initial efficacy (Figure 4). These include hydrogen gas (H_2) [73], Nrf2 [81,83], TRPV4 channel opening [74], and serotonin [101]. H_2 administration decreases depressive symptomology in murine ICH via the increase in SOD, decrease in ROS, and downregulation of inflammatory cytokines, such as NF-κβ, IL-1β, and IL-6 [73]. The effect is significantly more apparent in female mice and is potentiated via the concurrent administration of estrogen in male mice, indicating a potential adjunctive target. Nrf2 induction via either pharmacological means or by enriched environment also resulted in the reduction of depressive symptomology in mice, presumably via the upregulation of BDNF and downregulation of inflammatory responses, ROS, and glutaminase activity [81,83]. Similarly, TRPV4 inhibition resulted in the reduction of astrocytic activation, lipid peroxidation, and total SOD, demonstrating a decrease in the anxiogenic effects of ICH [74]. Lastly, in terms of neurotransmission, inhibiting indoleamine 2, and 3-dioxygenase to prevent the downregulation of serotonin levels following ICH decreased stress presentation in mice [101].

2.5. Subarachnoid Hemorrhage

2.5.1. Clinical Presentation and Clinical Therapeutic Efficacy

SAH represents a major global health challenge, with an estimated 8 million cases annually and an incidence rate projected to increase in parallel with the aging global population [170]. Meta-analytic studies have reported that approximately 28% and 32% of SAH patients experience depression and anxiety, respectively [171,172]. However, clinical randomized trials in SAH primarily focus on traditional outcomes such as aneurysm recurrence, retreatment, and mortality, rather than addressing psychological dysfunction [173]. As a result, there is limited clarity regarding the specific interventions for psychological dysfunction in this population and the efficacy of these treatments.

2.5.2. Neurobiological Mechanisms and Pre-Clinical Therapeutic Targets

Preclinical investigations have examined psychological dysfunction in the context of apoptosis and neuroinflammation, in conjunction with clinical assessments of stress hormone and neuropeptide levels (Tables 1 and 2). The mediation of anxiety, compulsive-like behaviors, and depressive symptoms through apoptosis is driven by ERK and P2Y14 receptor signaling, particularly within the hypothalamus and cortex, respectively [65,68,69]. ERK signaling mediated by TNFα is actually necessary for hypothalamic apoptosis and the production of anxiety [68]. Depression has also been linked to impaired glutamate uptake in the striatum [97]. Clinically, alterations in neuropeptide and stress hormone levels are observed globally, with these changes manifesting as measurable variations in urine, serum, and cerebrospinal fluid, correlating with psychological dysfunction [141,151]. These include an increase in admission CGRP within the CSF correlating to poorer long-term psychological outcomes [151] and the increase in early cortisol associating with late-stage depression [141]. Reflecting the decreased assessment of mechanisms of psychological dysfunction following SAH, only apoptosis, and neurotransmission have been targeted as potential therapeutic interventional foci in pre-clinical research (Figure 4). To be exact, acute administration of hydrogen gas inhibited the p38 MAPK pathway to prevent apoptosis, and the resultant anxiety and depression [66], while the acute modulation of glutamate uptake via HDAC2 has been found to decrease depression [97].

2.6. Vascular Cognitive Impairment and Dementia

2.6.1. Clinical Presentation and Clinical Therapeutic Efficacy

VCID accounts for approximately 20–40% of all dementia diagnoses and is primarily caused by cardiovascular disease, which is associated with vascular risk factors leading to chronic cerebral hypoperfusion (CCH) [174]. CCH is a common pathophysiological condition, often resulting from the chronic reduction in cerebral blood flow, leaving the brain in a state of prolonged hypoxic ischemia. This condition contributes to progressive neurological and cognitive dysfunction [175]. In addition to cardiovascular disease, CCH can also develop as a long-term consequence of acute brain injuries. In 2018, the global prevalence of VCID was estimated at 50 million individuals, with projections suggesting a threefold increase in the next 25 years, particularly in high-income countries. Large-scale studies of VCID report that at least 80% of patients exhibit some form of neuropsychiatric symptom [176,177]. While many VCID treatments focus on addressing cognitive impairment, rather than neuropsychiatric symptoms, certain interventions, including NMDA antagonists and SSRIs, have been evaluated for their efficacy in mitigating psychological dysfunction, although their effectiveness in VCID remains limited [178,179].

2.6.2. Neurobiological Mechanisms and Pre-Clinical Therapeutic Targets

CCH is characterized by a gradual, chronic development of injury. Although it might be expected that unique mechanisms would contribute to psychological dysfunction in CCH, this is not entirely the case. Apart from amyloid and cerebral insulin metabolism, CCH shares several mechanisms of psychological dysfunction with acute brain injuries (Tables 1 and 2). However, in CCH, the primary focus shifts from apoptosis, neuroinflammation, and oxidative stress to alterations in neurotransmission and synaptic plasticity, particularly within the hippocampus. Multiple aspects of these processes are implicated in the development of anxiety and depression in CCH, including aberrant GABA and glutamate signaling, reduced BDNF, impaired synaptic potentiation, changes in synaptophysin, serotonin, and HCN1, as well as reduced dopamine and KCNQ3 activity [21,22,77,91,102,104,107–109,111]. Uniquely, the decrease in CNPase and synaptophysin in the white matter mediate an impairment of synaptic potentiation and in anxiety and depression [111], alongside the upregulation of HCN1 and downregulation of KCNQ3 in the amygdala [104]. While neuroinflammation and oxidative stress are not overlooked in the study of psychological dysfunction in CCH, only specific factors are typically assessed, including phosphorylation of TrkB, ERK, CREB, antioxidant expression, and lipid peroxidation [87], which reflect previously discussed patterns. Additionally, the upregulation of the angiotensin-converting enzyme/angiotensin II/angiotensin II type 1 receptor (ACE/Ang II/AT1R) axis in the hippocampus, as opposed to the neuroprotective angiotensin-converting enzyme 2/angiotensin-(1–7) (ACE2/Ang-(1–7)) axis results in decreased cerebral blood flow and increased oxidative stress and neuroinflammation, which contributes to the manifestation of anxiety [128].

Notably, CCH is the only model in which a link between psychological dysfunction and increased tau metabolism or decreased insulin metabolism has been identified, potentially reflecting the chronic nature of VCID [122–124]. Amyloid beta peptide (Aβ) deposition induces neuronal damage and has been linked to behavioral and anxiety disorders in CCH. Specifically, β-amyloid precursor protein cleavage enzyme 1 (BACE1), which functions as the β-secretase responsible for processing amyloid precursor protein, is upregulated after cerebral ischemia [123]. This upregulation of BACE1 is associated with increased Aβ levels in the hippocampus [122]. Abnormal hyperphosphorylation of Tau is mediated by reduced levels of tau O-GlcNAcylation, a post-translational modification involving β-N-acetylglucosamine (GlcNAc) [124]. Alongside GlcNAc dysfunction, the downregulation of the insulin signaling pathway components—such as the insulin receptor, IGF-1 receptor, insulin receptor substrate-1, phosphatidylinositide 3-kinases, 3-phosphoinositide-dependent protein kinase-1, protein kinase B, and glycogen synthase kinase-3—has been associated with anxiety-like behavior following cerebral ischemia [124].

Reflecting the differing investigative foci of CCH, therapeutic interventional research has largely focused on the modulation of synaptic plasticity, though neurotransmission, oxidative stress, neuropeptides, and neuroinflammation have also been targets (Figure 4). In regards to synaptic plasticity, specific neurobiological mechanistic targets for anxiety and depression include transforming growth factor β, BDNF, glycogen synthase kinase-3β, synaptophysin, and lipid peroxidation [77,107–109,111]. Neurotransmitter targets include GABA receptors in the amygdala and hippocampus and serotonin within the hippocampus [21,102], while the modulation of SOD and glutathione decreases anxiety-like symptoms [87]. Lastly, modulation of the angiotensin system decreases oxidative stress and neuroinflammation, resulting in the ablation of depressive symptomology [128].

3. Sexually Dimorphic Responses Reveal Potential Targets for Therapeutic Intervention in Psychological Dysfunction Within Injured Brains

Sex serves as an independent modulator of psychological dysfunction symptomatology and treatment efficacy, regardless of the presence of cerebral injury. Epidemiological studies indicate that absent of injury, the female population is more likely to develop depression, anxiety, or PTSD than males [34,180–182]. Similarly, older females show an increased incidence of depression and anxiety following ischemic stroke and SAH, with a comparable trend noted in pediatric TBI [18,20,23,183]. However, in the adult military population, mTBI is linked to a higher prevalence of anxiety in males than females [184]. The divergence in sex differences between pediatric and adult mTBI may arise during adolescence, as females show reduced severity of psychological dysfunction following mTBI during mid-puberty [185]. In pre-clinical observations, female rodents exhibit lower depression and anxiety levels in blast injury and severe TBI [19,117,119,120,127], as well as enhanced responses to therapeutic intervention in ICH [73]. The factors underlying these observed differences are not limited to sex hormones and may represent novel targets for therapeutic intervention.

Sexually dimorphic presentations of psychological dysfunction have been linked to multiple neurobiological mechanisms. In non-injured contexts, estrogen appears to reduce the impact of stress on cognitive function in females [33], yet fluctuations in estrogen levels, particularly during menopause, are associated with an increased risk of developing depression [36]. Under healthy conditions, females also exhibit heightened activation of the locus coeruleus and greater sensitivity to CRF within the region [33–35]. While this may increase susceptibility to hyperarousal in females, it also presents CRF as a potential target for therapeutic intervention. Pre-clinical examinations within blast injury TBI support this supposition, with the observation that variations in CRFR2 between males and females lead to reduced anxiety levels [117]. While expression within the locus coeruleus itself was not measured, it was observed that female animals exhibit decreased CRFR2 in the dentate hilus of the hippocampus following blast TBI, while males exhibit increased expression in the ventral hippocampus and decreased expression in the amygdala. In addition to stress hormones, neuropeptides may be potential targets of therapeutic intervention. In a severe TBI model, female rodents exhibit both decreased anxiety and decreased depression; this is mediated by the action of CGRP [127]. The increased expression of CGRP not only reduces oxidative stress and vascular damage in female brains, but females also exhibit a more pronounced response to CGRP inhibition compared to males. This suggests that female animals have developed adaptive mechanisms reliant on elevated CGRP expression and sensitivity, which, if replicable, could be exploited as a broad therapeutic strategy.

Sex hormones themselves also present an attractive potential therapeutic target. Clinically, the higher female anxiety observed in pediatric TBI is linked to increased extracellular diffusivity, which is not observed in females with adult hormone levels [150,186]. In males, however, decreased testosterone levels in military males following TBI are associated with the development of anxiety [184,187]. Preclinically, estradiol and progesterone have been shown to influence anxiety in ICH and TBI by modulating neurotransmission, oxidative stress, and inflammation [73,119,188]. Estradiol impacts $GABA_A$ receptor levels, serotonin signaling, mitochondrial function, and antioxidant expression [119,188], while in ICH estrogen modulates astrocyte-mediated inflammation, which may influence the development of depression [73]. While there have been mixed results in mortality and neurological outcomes in clinical trials administering estrogen and progesterone in TBI [127], this may be mediated by other factors, namely severity of brain injury, which ought to be taken into account alongside hormone levels. Meta-analyses of preclinical studies indicate that lower TBI severity is associated with poorer outcomes in females, while moderate-to-severe TBI often correlates with improved outcomes in females [20]. This suggests that injury severity may influence the extent of sexual dimorphism and highlights the significant role of injury severity and etiology in determining the response to therapeutic interventions. Additionally, these findings challenge the assertion that the mechanisms of psychological dysfunction are consistent across various types of brain injury [26], despite differences in etiology. Given the observed sexual dimorphism in response to TBI severity, it is crucial to consider both the type and severity of injury when selecting and evaluating therapeutic interventions for psychological dysfunction following brain injury.

4. Future Directions for Enhancing Therapeutic Efficacy in Injury-Dependent Psychological Dysfunction

4.1. Targeting Upstream Mechanisms as a Comprehensive Therapeutic Strategy

Throughout this review, it has become evident that while current therapeutic approaches for psychological dysfunction are diverse and effective in injury-independent contexts, they are not fully effective in addressing injury-dependent psychological dysfunction [58,189–193]. CBT, for example, is a first-line treatment in non-injured individuals with anxiety and depression [194], but its reliance on active patient engagement diminishes its utility in injured contexts [189,190]. Neuromodulation, which concurrently targets multiple neurobiological mechanisms, shows promise but remains underutilized in the clinical treatment of psychological dysfunction following brain injury [58,189,190]. Vagus nerve stimulation, trigeminal nerve stimulation, and transcranial magnetic stimulation, among others, have shown positive effects in treating anxiety and depression in non-injured individuals [58,195–197], however, only transcranial magnetic stimulation has been studied more extensively in injured populations [198,199]. Consequently, in the context of injury-dependent psychological dysfunction, pharmacological interventions remain the primary treatment approach. Many of these treatments, however, exhibit reduced therapeutic efficacy, potentially due to their focus on downstream factors, like serotonin dysregulation, which does not address the factors driving the dysregulation. Targeting multiple underlying mechanisms pharmacologically, however, presents a challenge, as pharmaceutical cocktails can result in potentially harmful drug–drug interactions [200,201]. The prevalence of side effects with such combinations limits their clinical application and increases the necessary testing and validation. Therefore, the identification of a comprehensive, multi-target approach capable of simultaneously modulating the various factors involved in psychological dysfunction may represent a promising avenue for future research.

One approach to targeting the multiple, divergent pathophysiological cascades that contribute to psychological dysfunction may involve focusing on upstream mediators (Figure 5). Nrf2, for instance, has emerged as a promising target in depression research and in the treatment of brain injury [202–206]. As an upstream regulator, Nrf2 modulates key factors involved in the development of psychological dysfunction (Table 1), including neurotransmitters such as serotonin and dopamine, which are commonly targeted by pharmacological therapies, as well as processes like apoptosis, neuroinflammation, and oxidative stress. By concurrently modulating factors involved in both injury progression and the development of psychological dysfunction, Nrf2 emerges as a promising therapeutic target for injury-associated psychological dysfunction. This has already been assessed as a target for PSD, with controlled regulation of Nrf2 activation, via intravenous application of CDDO-Im, resulting in decreased depressive symptomology in a murine PSD model, as well as an increase in autophagic factors [82]. Further, in ICH, Nrf2 induction, via either an enriched environment or the application of TP-500, results in the reduction of depressive symptomology in mice [81,83]. This was mediated via an effect on BDNF and the downregulation of inflammatory responses, ROS, and glutaminase, illustrating the potential of Nrf2 to modulate neurotransmission, neuroinflammation, and oxidative stress. This indicates Nrf2′s promise as an interventional target in different forms of brain injury, addressing multiple factors affecting the presentation of psychological dysfunction.

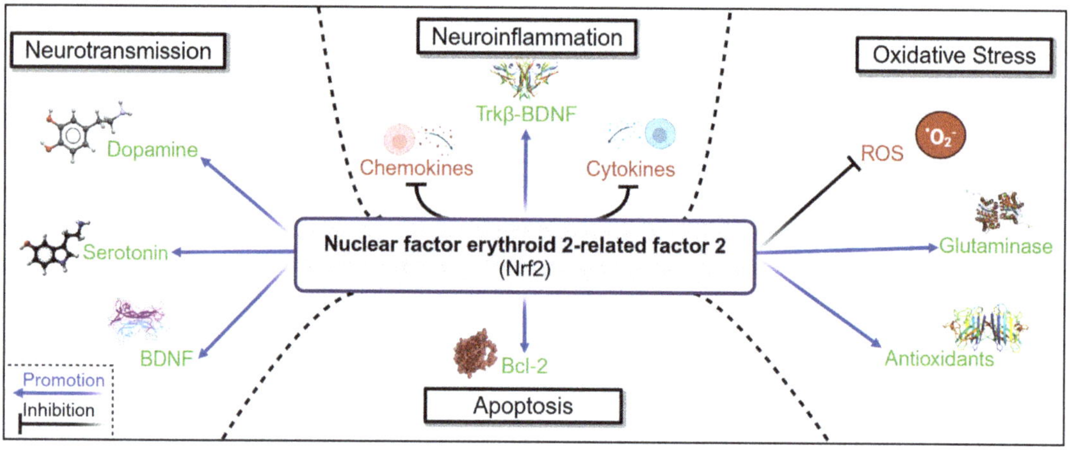

Figure 5. Nrf2 emerges as a promising upstream target for therapeutic intervention in psychological dysfunction associated with brain injury. To effectively address the multiple underlying mechanisms contributing to psychological dysfunction following brain injury, targeting upstream pathways may offer a promising direction for future research. Given its ability to simultaneously modulate neurotransmission, neuroinflammation, oxidative stress, and apoptosis, Nrf2 stands out as a potential target for therapeutic intervention. (Created with Biorender.com).

Nrf2 is not the only upstream mediator that may be a promising target for the treatment of psychological dysfunction. Neuropeptides are multi-focal molecules packaged and co-released with neurotransmitters, the functional implications of which have been studied extensively in numerous reviews [32,207–210]. The neuropeptides connected to psychological dysfunction in non-injured contexts are multitudinous, including CCK, NPY, galanin, oxytocin, and vasopressin [32]. When examining injured contexts, the list increases to include CGRP, angiotensin II, and substance P [127,128,151], while both CCK-8 and NPY are also implicated in PSD [147,152,153]. Together, these molecules have the capacity to mediate changes in oxidative stress, inflammation, the stress axis, vasoreactivity, apoptosis, neurotransmission, and synaptic plasticity [211] (Figure 6). CGRP, for example, can modulate ERK, PI3K/AkT, FoxO3a, CREB, MAPK, BDNF, glial activation, and cytokine release [212]. This is in addition to the activation of Nrf2, which angiotensin II, NPY, and substance P also mediate [127,213,214]. Targeting angiotensin II and the angiotensin system in pre-clinical CCH has already demonstrated an ablation of depressive symptomology, as mediated by a decrease in inflammation and oxidative stress [128]. However, the relative lack of investigation into the therapeutic effect of these molecules belies their apparent promise and indicates the necessity for investigation in the context of injury-induced psychological dysfunction.

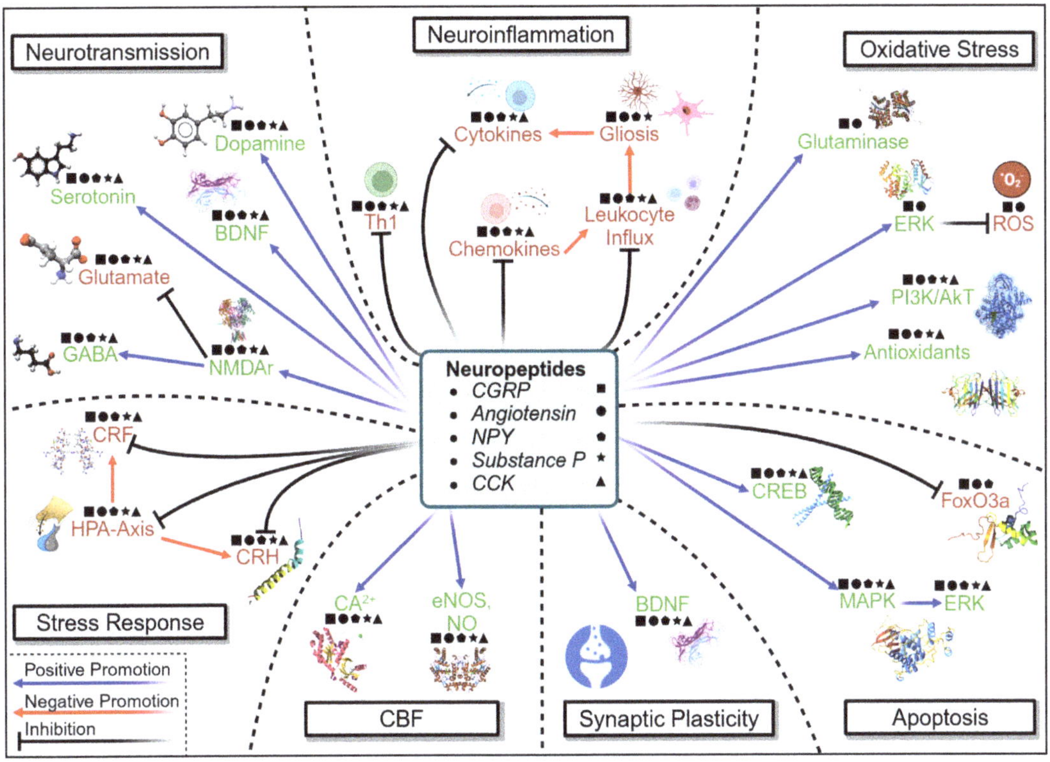

Figure 6. The potential of neuropeptides as promising upstream targets for therapeutic intervention in injury-dependent psychological dysfunction. Neuropeptides function as multifaceted molecules that regulate several processes within the nervous system, including neurotransmission, neuroinflammation, oxidative stress, apoptosis, synaptic plasticity, cerebral blood flow (CBF), and stress responses. These pathways are integral to the development of injury-dependent psychological dysfunction, highlighting the significant potential of neuropeptides as a target for therapeutic intervention. (Each mechanism is annotated with the corresponding neuropeptides that influence it, represented by specific icons: CGRP (square), angiotensin (circle), NPY (pentagon), substance P (star), and CCK (triangle)). (Created with Biorender.com).

As epitomized in the usage of enriched environments to enhance Nrf2 activation and reduce depressive symptomology in ICH, the integration of non-pharmacological therapeutics may offer substantial benefits (Figure 7). These approaches influence multiple neurobiological mechanisms involved in the development of psychological dysfunction following brain injury, including the concurrent modulation of neuroinflammation, neurotransmission, apoptosis, oxidative stress, neuropeptides, and synaptic plasticity [58,215,216], all of which independently represent attractive targets for treatment of psychological dysfunction following injury. Both physical exercise and various forms of neuromodulation have also been shown to induce these effects and may thus be relevant in managing psychological dysfunction. For instance, trigeminal nerve stimulation modulates neuropeptides, including the aforementioned CGRP, NPY, and substance P, neuroinflammation, via the modulation of microglia, leukocytes, and Th1 cells, and neurotransmission, mediated by changes in NMDAr and glutamate and GABA signaling [58]. Currently, it is already under clinical investigation for the treatment of non-traumatic depression, social anxiety disorder, PTSD, and loss of consciousness following brain injury, and is being explored preclinically as a potential treatment for SAH, TBI, and ischemic stroke. Physical exercise, on the other hand, is known to have positive effects on mental health in non-injurious conditions, mediating a 43.2% decrease in self-reported mental health dysfunction among 1.2 million individuals [217]. Not only does it mediate endorphin release, but it also modulates the HPA axis, neurotransmitter activity, cytokines, vagal tone, toll-like receptors, and mitochondria, as well as mirroring effects of CBT via providing a sensation of mastery and a distraction from negative thoughts and ruminations [218]. These examples thus exemplify a theoretical multi-mechanistic strategy for addressing psychological dysfunction following brain injury and illustrate non-pharmacological interventions as a promising area for further experimental investigation, alongside upstream modulators.

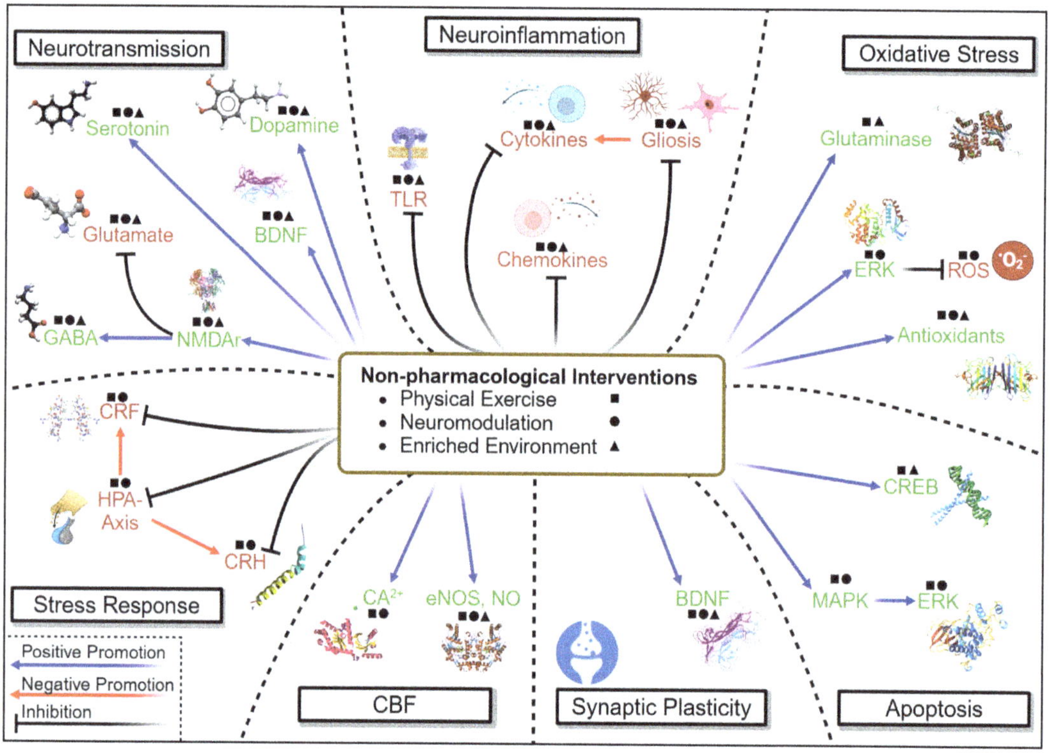

Figure 7. Non-pharmacological interventions as potential therapeutic strategies for alleviating psychological dysfunction following brain injury. Non-pharmacological interventions are inherently multi-mechanistic, enabling them to influence a broad array of factors that contribute to the onset of psychological dysfunction following brain injury. These factors include neurotransmission, neuroinflammation, oxidative stress, apoptosis, synaptic plasticity, cerebral blood flow (CBF), and stress responses, thereby targeting a significant portion of the pathways involved in the development of psychological distress. (Each mechanism is annotated with the interventions known to influence it, represented by specific icons: physical exercise (square), neuromodulation (circle), and an enriched environment (triangle)). (Created with Biorender.com).

4.2. Tailoring Therapeutic Interventions to Address Psychological Dysfunction Based on the Specific Type of Brain Injury

This review reveals that psychological dysfunction following brain injury is often treated similarly to injury-independent psychological dysfunction. While this approach leverages established knowledge, it may not fully address the unique complexities of injury-related psychological dysfunction. For instance, in TBI, both injury severity and age at the time of injury can influence psychological outcomes, with sex-specific differences observed [20,185]. This suggests a pathophysiological cascade influenced by injury severity, which interacts with sex-related factors to differentially affect symptomatology. Therefore, it is unlikely that mechanisms in injury-dependent contexts are identical to those in injury-independent psychological dysfunction [26]. Additionally, the assertion that psychological dysfunction mechanisms are similar across various brain injury types is questionable. For example, NMDAr dysfunction plays a role in psychological dysfunction in both SAH and TBI [93–97,99,105]. However, due to differences in injury development, a single NMDAr-targeting intervention may not be effective across both conditions. In SAH and ICH, NMDAr dysfunction is directly mediated by thrombin, which potentiates NMDAr function, worsening glutamate-induced cell death [219,220]. Thrombin levels remain elevated for up to nine days after SAH, indicating prolonged NMDAr potentiation [221]. In contrast, TBI exhibits distinct temporal patterns, with NMDAr expression initially increasing in the hyper-acute phase but decreasing in the chronic phase [93–96,99,105]. Thus, therapies targeting NMDAr activity may be beneficial for SAH/ICH but harmful in TBI. This highlights the importance of considering injury-specific contexts, as preclinical studies on SAH have primarily focused on neurotransmission and apoptosis [65,66,69,70,97], while clinical conditions also involve neuroinflammation, stress hormones, and neuropeptides [141,151]. This suggests that the assessment of underlying mechanisms of psychological dysfunction is being influenced more by observations from other types of brain injuries, rather than being specifically tailored to the injury in question. Given the diverse etiologies and disease progression patterns [40,44,46,49,51,54](Figure 2), it is unlikely that the brain regions and mechanisms affected are the same across all types of brain injury. This raises concerns about the current research approach, which often evaluates similar mechanisms across different brain injury models (Table 1). As an initial step, we recommend concurrently examining the mechanisms and presentations of psychological dysfunction in multiple models to identify potential discrepancies, which could inform the development of more targeted therapeutic interventions. Additionally, we advocate for considering psychological dysfunction in various brain injuries as separate entities to prevent conflating results between models.

A further complicating factor is the observation that in pre-clinical research, current therapeutic interventions do not appear to be tailored to the state of disease progression (Table 1). Instead, therapies applied during the chronic phase of disease progression target the same mechanisms as those used in the acute phase, focusing on oxidative stress, neuroinflammation, neurotransmission, and DNA methylation (Tables 1 and 3, Figure 4). A review of the relevant studies reveals that interventions were selected based on their demonstrated efficacy in the acute phase of the disease [72,82,103,113,130]. There was no indication that these interventions were specifically designed to address the unique pathophysiological cascades associated with the chronic phase, implying that these approaches treat acute and chronic stages in similar contexts. However, as illustrated by the pattern of NMDAr changes in TBI [93–96,99,105], and the pathological development due to blood presence after SAH and ICH [221], brain injury leads to ongoing pathological changes over time, which are reflected in evolving patterns of psychological dysfunction. Fundamentally, therapeutic interventions following brain injury ought to be tailored to the stage of disease progression, either by adjusting interventions over time as the injury or recovery progresses or by using a single intervention that can be modulated to suit temporal changes in the disease state, such as neuromodulation and physical exercise. In addition to their ability to address multiple underlying mechanisms and their proven safety in clinical settings [58,215,216], neuromodulation offers the unique advantage of modulating treatment intensity, making it adaptable to different injury severities and stages of recovery.

This flexibility provides a potential avenue for managing psychological dysfunction in individuals recovering from brain injuries.

4.3. The Implementation of an Integrative Theoretical Framework to Address the Limitations of Current Approaches That Contribute to Diminished Therapeutic Efficacy

The development and treatment of injury-independent psychological dysfunction is inherently complex. Previous studies on psychological dysfunction and its treatments have primarily focused on symptomatology and clinical presentation, often overlooking the underlying etiological factors, which led to the application of imprecise or ineffective therapies [222]. In response, integrated approaches have emerged that not only address symptoms but also examine environmental and neurobiological factors, alongside personal history, to construct a comprehensive model of disease development [1,222–224]. These frameworks have enhanced the understanding of causal factors, allowing for targeted interventions that address the most relevant elements. For example, the reciprocal interaction model for obsessive–compulsive disorder considers neuropsychological deficits, cognitive-behavioral processes, and the source of the compulsions, enabling the identification of disease subtypes, triggers, and their underlying causes, while facilitating the construction of personalized treatment plans [225]. Similar integrative frameworks for depression have incorporated environmental factors, regional "kindling" by induced stress or trauma, HPA axis activity, stress levels, and personal vulnerability [1,223]. These models not only identify neurobiological targets for treatment (e.g., serotonin or dopamine) but also incorporate non-pharmacological interventions, such as CBT, to address triggers or environmental factors, leading to more effective interventions than those based on symptom-driven treatments alone [222].

Following brain injury, the factors influencing the development and presentation of psychological dysfunction become more intricate, with the added effect of multiple pathophysiological cascades resulting from tissue damage. For instance, the decreased efficacy of dopamine and serotonin in the context of injury in clinical settings [26,134,136,137] may stem from the fact that they are not targeting the underlying factors, such as oxidative stress, neuroinflammation, apoptosis, or dysregulation of hormones and peptides (Table 1), that contribute to the altered levels of dopamine and serotonin. Consequently, simply administering SSRIs, benzodiazepines, or MAOIs potentially allows upstream dysregulation to persist and undermines therapeutic efficacy. However, due to understandable experimental limitations, current preclinical research often fails to adequately address the interacting causal factors (Table 1). Thus, the first step is to identify the most relevant factors to target, akin to strategies in injury-independent contexts [222]. By synthesizing current knowledge into an integrative theoretical framework, it is possible to inform therapeutic interventions and guide future research. This review itself reveals two key observations: (1) there are notable differences between injury-independent and injury-dependent psychological dysfunction, indicating that it is inappropriate to assume the same mechanisms are predominant in both contexts and (2) targeting factors directly associated with psychological dysfunction (e.g., serotonin and dopamine) is insufficient for addressing injury-dependent psychological dysfunction. These insights can directly inform future research, and a more comprehensive theoretical framework will likely raise additional areas for intervention.

In the vein of approaching injury-dependent psychological dysfunction in an integrated manner, it is crucial to: (1) adopt more clinically relevant models, which at least partially integrate the environmental factors and individual susceptibility (Table 1), and (2) examine psychological dysfunction in a manner encompassing the effects of multiple brain regions (Table 1). Currently, preclinical models typically isolate brain injury as the sole stressor [1,222–224]. While it is understandable to initially evaluate therapeutic interventions with brain injury alone as the stressor, the absence of additional stressors in animal models represents a limitation that can be addressed via the inclusion of a chronic stressor post-injury, such as chronic unpredictable mild stress [77], or the usage of animals subjected to maternal deprivation prior to injury [226,227]. An additional potential target for future research is the examination of different brain regions in injury-dependent psychological dysfunction. In brain injury, the location of the initial insult does not necessarily determine where secondary damage will occur, with primary damage to the cortex resulting in subsequent damage in anatomically distant regions, such as the amygdala or thalamus, in TBI [127]. Further, due to brain connectivity, damage in one region may lead to effects in other regions [49]. Therefore, limiting the examination of mechanisms to specific sub-regions may fail to capture critical interactional effects and cross-regional influences. As such, in TBI, research ought to expand beyond the cortex, hippocampus, and amygdala, which currently represent the majority of research in TBI-dependent psychological dysfunction (Table 1). Similarly, in VCID, the focus on the hippocampus, due to its role in cognition [178], largely neglects other brain sub-regions involved in behavior and emotional regulation [178,179]. While examining all brain regions at once is impractical, shifting the focus of psychological dysfunction research away from regions most associated with overt pathophysiology could facilitate the identification of more contextually appropriate interventions.

5. Conclusions

This review provides a comprehensive evaluation of the neurobiological mechanisms and therapeutic strategies related to neuropsychological dysfunction, with a particular emphasis on psychological disturbances arising from brain injuries. It highlights promising therapeutic approaches for addressing injury-dependent psychological dysfunction, identifying Nrf2, neuropeptides, and nonpharmacological therapies as potential multi-mechanistic interventions that act as upstream mediators. These interventions target the complex interactions of various factors contributing to psychological dysfunction in the context of brain injury. Additionally, the review underscores the importance of sexually dimorphic factors, suggesting them as key areas for further investigation and advocating for a deeper exploration of sex-specific patterns to identify additional contributors to psychological dysfunction. The review also identifies significant gaps in current research, particularly the limited consideration of the intricate interactions among causal factors, environmental influences, and individual susceptibilities. This highlights the need for an integrative framework to guide the identification of effective therapeutic targets and promote the use of clinically relevant models. Furthermore, it points to an over-reliance on findings from injury-independent contexts, the assumption that neurobiological mechanisms are consistent across different types of brain injury, and the prevalent focus on acute-phase therapeutic interventions in preclinical research. This approach neglects the distinct pathophysiological cascades associated with various forms and stages of brain injury and overlooks how factors such as the presence or absence of hemorrhage in acute brain injuries can influence the underlying mechanisms of psychological dysfunction. Moreover, it fails to address the chronic phase, during which psychological dysfunction typically emerges. To address these challenges, the review recommends prioritizing research on modulators of psychological dysfunction that reflect injury-specific deviations from non-injury patterns, considering factors such as injury severity and type. It also advocates for tailoring therapeutic strategies according to the stage of injury progression and the specific characteristics of the injury to optimize outcomes. In conclusion, the review emphasizes the need for multi-mechanistic, context-specific approaches to improve therapeutic outcomes and address existing gaps in the understanding of the neurobiological mechanisms of psychological dysfunction across different brain injuries.

Author Contributions: Conceptualization, C.L.; writing—original draft preparation, P.U., T.R. and K.P.; investigation, E.A., V.D.S., A.X. and J.J.; writing—review and editing, K.P. and C.L.; supervision, C.L.; project administration, C.L.; funding acquisition, C.L. All authors have read and agreed to the published version of the manuscript.

Funding: This work is supported in part by the United States Army Medical Research Acquisition Activity (USAMRAA) under award# HT9425-24-1-1017 and the merit-based career enhancement award at the Feinstein Institutes for Medical Research.

Institutional Review Board Statement: Not applicable.

Informed Consent Statement: Not applicable.

Data Availability Statement: There are no data generated for this manuscript.

Acknowledgments: We would like to thank Jake Stiles, Kanheng Lin, Steven Wadolowski, Willians Tambo, and Zain Qureshi for their aid in the initial literature searches.

Conflicts of Interest: The authors declare no conflicts of interest. The funders had no role in the design of the study; in the collection, analyses, or interpretation of data; in the writing of the manuscript; or in the decision to publish the results.

Abbreviations

5-HT	Serotonin
AAP	Amyloid precursor protein
ACE	Angiotensin converting enzyme
ACT	Acceptance and commitment therapy
ACTH	Adrenocorticotropin hormone
AKT	Protein kinase B
AMPA	α-amino-3-hydroxy-5-methyl-4-isoxazolepropionic acid
Ang	Angiotensin
Arg-1	Arginase-1
AT1R	Angiotensin II type 1 receptor
ATP	Adenosine triphosphate
Aβ	Amyloid beta
BACE	β-amyloid precursor protein cleavage enzyme 1
Bax	B-cell lymphoma-2-associated X
Bcl-2	B-cell lymphoma-2
BDNF	Brain-derived neurotrophic factor
BrdU+	Bromodeoxyuridine
CA1	Cornu ammonis 1
Ca2+	Calcium
CB	Endocannabinoid receptor
CBF	Cerebral blood flow
CBT	Cognitive behavioral therapy
CCH	Chronic cerebral hypoperfusion
CCK-8	Cholecystokinin octapeptide
Cd206	Cluster of differentiation 206
Cd86	Cluster of differentiation 86
CGRP	Calcitonin gene-related peptide
CNPase	2′, 3′-cyclic-nucleotide 3′-phosphodiesterase
CNR1	Cannabinoid receptor 1
COMT	Catechol-O-methyl-transferase
CREB	Cyclic-AMP-responsive element-binding protein
CRF-1	Corticotropin-releasing factor 1
CRFR2	Corticotropin-releasing hormone receptor 2

CRH1	Corticotropin-releasing hormone 1
CRHR1	Corticotropin-releasing hormone receptor 1
CSF	Cerebral spinal fluid
CXCL10	C-X-C motif chemokine ligand 10
CXCR3+	C-X-C motif chemokine receptor 3
DCX	Doublecortin
DG	Dentate gyrus
DNA	Deoxyribonucleic acid
DNMT	DNA methyltransferase
ERK	Extracellular signaling-regulated kinase
FOX01	Forkhead box protein O1
FTO	Fat mass and obesity-associated protein
GABA	Gamma-aminobutyric acid
GAD	Generalized anxiety disorder
GAD67	Glutamate decarboxylase 67
GADD45	DNA damage 45
GAP43	Growth-associated protein 43
GIRK/Kir3	G protein-activated inwardly rectifying potassium
GlcNAc	β-N-acetylglucosamine
GLT-1	Glutamate transporter 1
GluA1	Glutamate A1 subunit
GluA2	Glutamate A2 subunit
GPER	G protein-coupled estrogen receptors
Grin1	Glutamate receptor ionotropic N-methyl-D-aspartate subunit 1
GSDMD+	Gasdermin D
GSK-3β	Glycogen synthase kinase-3β
H3K9ac	Histone H3 lysine 9
HCN	Hyperpolarization-activated Cyclic Nucleotide-gated
HCN2	Potassium/sodium hyperpolarization-activated cyclic nucleotide-gated ion channel 2
HDAC	Histone deacetylase
HPA	Hypothalamus-pituitary-pituitary-adrenal
IBA-1	Ionized calcium binding adaptor molecule 1
ICH	Intracerebral hemmorhage
IFN	Interferon
IGF-1	Insulin-like growth factor 1
IL-10	Interleukin-10
IL-1β	Interleukin 1β
IL-6	Interleukin 6
IR	Insulin receptor
IRE1α	Inositol requiring enzyme-1 alpha
IRS-1	Insulin receptor substrate-1
ISI	Inter-Spike Interval
JAK2	Janus kinase 2
JNK	c-Jun N-terminal kinase
K+	Potassium
KCNQ	Potassium channel Kv7
Keap1	Kelch-like ECH-associated protein 1
LTP	Long-term potentiation
m6A	N6-methyladenosine
MAP-2	Microtubule-associated protein
MAPK	Mitogen-activated protein kinase
MCP-1	Monocyte chemoattractant protein-1
MDD	Major depressive disorder
mFPI	Mild fluid percussion injury

MIP-2	Macrophage inflammatory protein-2
miR-25-3p	MicroRNA-25-3p
mRNA	Messenger ribonucleic acid
mTBI	Mild traumatic brain injury
mTOR	Mammalian target of rapamycin
Myc	Myelocytomatosis oncogene
NCAM	Neural cell adhesion molecule
NE	Neutrophil elastase
NF-kβ	Nuclear factor kappa β
NGF	Nerve growth factor
NMDA	N-methyl-D-aspartate
NMDAR	NMDA receptor
NOS2	Nitric oxide synthase 2
NOX4	NADPH oxidase complex
NPY	Neuropeptide Y
NR2A	NMDA receptor subtype 2A
NR2B	NMDA receptor subtype 2B
Nrf2	Nuclear factor erythroid-derived-2-related factor 2
Olig2	Oligodendrocyte transcription factor 2
PARP	Poly (ADP-ribose) polymerase
PDK1	3-phosphoinositide-dependent protein kinase-1
PERK	Protein kinase R-like ER kinase
PI3K	Nrf2-TrkB-phosphoinositide-3-kinase
PINK1	PTEN-induced putative kinase 1
PKA	Protein kinase A
PPF	Paired-pulse facilitation
PPR	Paired-pulse ratio
PSD	Post stroke depression
PSD-93/95	Postsynaptic density protein-93/95
PTSD	Post-traumatic stress disorder
RAS	Renin-angiotensin system
ROS	Reactive oxygen species
SAH	Subarachnoid hemorrhage
SERT	Serotonin transporter
SNRI	Serotonin–norepinephrine reuptake inhibitors
SOD	Superoxide dismutase
SP	Substance P
SSRI	Selective serotonin reuptake inhibitors
STAT1	Signal transducer and activator of transcription
STING	Stimulator of interferon genes
SYP	Synaptophysin
TBI	Traumatic brain injury
TCA	Tricyclic antidepressants
TFRC	Transferrin receptor
TGFβ	Transforming growth factor β
TLR	Toll-like receptor
TNFR1	Tumor necrosis factor receptor
TNFα	Tumor necrosis factor-alpha
TrkB	Tropomycin receptor kinase B
UPR	Unfolded protein response
VCAN	Versican
VEGF	Vascular endothelial growth factor
VCID	Vascular cognitive impairment and dementia
VTA	Ventral tegmental area
α7-nAChR	Alpha-7 nicotinic receptor

References

1. Duval, F.; Lebowitz, B.D.; Macher, J.-P. Treatments in depression. *Dialogues Clin. Neurosci.* **2006**, *8*, 191–206. [CrossRef] [PubMed]
2. Fleminger, S.; Oliver, D.L.; Williams, W.H.; Evans, J. The neuropsychiatry of depression after brain injury. *Neuropsychol. Rehabil.* **2003**, *13*, 65–87. [CrossRef] [PubMed]
3. Naraparedddy, B.R.; Narapareddy, L.; Lin, A.; Wigh, S.; Nanavati, J.; Dougherty, J.; Nowrangi, M.; Roy, D. Treatment of Depression After Traumatic Brain Injury: A Systematic Review Focused on Pharmacological and Neuromodulatory Interventions. *Psychosomatics* **2020**, *61*, 481–497. [CrossRef]
4. Villa, R.F.; Ferrari, F.; Moretti, A. Post-stroke depression: Mechanisms and pharmacological treatment. *Pharmacol. Ther.* **2018**, *184*, 131–144. [CrossRef] [PubMed]
5. Stein, M.B.; Sareen, J. Generalized Anxiety Disorder. *N. Engl. J. Med.* **2015**, *373*, 2059–2068. [CrossRef] [PubMed]
6. Albrecht, J.S.; Barbour, L.; Abariga, S.A.; Rao, V.; Perfetto, E.M. Risk of Depression after Traumatic Brain Injury in a Large National Sample. *J. Neurotrauma* **2019**, *36*, 300–307. [CrossRef]
7. Bartlett, M.; Bulters, D.; Hou, R. Psychological distress after subarachnoid haemorrhage: A systematic review and meta-analysis. *J. Psychosom. Res.* **2021**, *148*, 110559. [CrossRef]
8. Campbell Burton, C.A.; Murray, J.; Holmes, J.; Astin, F.; Greenwood, D.; Knapp, P. Frequency of anxiety after stroke: A systematic review and meta-analysis of observational studies. *Int. J. Stroke* **2013**, *8*, 545–559. [CrossRef]
9. Gould, K.R.; Ponsford, J.L.; Johnston, L.; Schönberger, M. Predictive and associated factors of psychiatric disorders after traumatic brain injury: A prospective study. *J. Neurotrauma* **2011**, *28*, 1155–1163. [CrossRef] [PubMed]
10. Hackett, M.L.; Yapa, C.; Parag, V.; Anderson, C.S. Frequency of depression after stroke: A systematic review of observational studies. *Stroke* **2005**, *36*, 1330–1340. [CrossRef] [PubMed]
11. Scholten, A.C.; Haagsma, J.A.; Cnossen, M.C.; Olff, M.; van Beeck, E.F.; Polinder, S. Prevalence of and Risk Factors for Anxiety and Depressive Disorders after Traumatic Brain Injury: A Systematic Review. *J. Neurotrauma* **2016**, *33*, 1969–1994. [CrossRef]
12. West, R.; Hill, K.; Hewison, J.; Knapp, P.; House, A. Psychological disorders after stroke are an important influence on functional outcomes: A prospective cohort study. *Stroke* **2010**, *41*, 1723–1727. [CrossRef]
13. Frank, D.; Gruenbaum, B.F.; Zlotnik, A.; Semyonov, M.; Frenkel, A.; Boyko, M. Pathophysiology and Current Drug Treatments for Post-Stroke Depression: A Review. *Int. J. Mol. Sci.* **2022**, *23*, 15114. [CrossRef]
14. Liu, Q.; Li, R.; Qu, W.; Li, B.; Yang, W.; Cui, R. Pharmacological and non-pharmacological interventions of depression after traumatic brain injury: A systematic review. *Eur. J. Pharmacol.* **2019**, *865*, 172775. [CrossRef]
15. Segal, Z.; Vincent, P.; Levitt, A. Efficacy of combined, sequential and crossover psychotherapy and pharmacotherapy in improving outcomes in depression. *J. Psychiatry Neurosci.* **2002**, *27*, 281–290.
16. Voineskos, D.; Daskalakis, Z.J.; Blumberger, D.M. Management of Treatment-Resistant Depression: Challenges and Strategies. *Neuropsychiatr. Dis. Treat.* **2020**, *16*, 221–234. [CrossRef]
17. Broussard, J.I.; Acion, L.; De Jesús-Cortés, H.; Yin, T.; Britt, J.K.; Salas, R.; Costa-Mattioli, M.; Robertson, C.; Pieper, A.A.; Arciniegas, D.B.; et al. Repeated mild traumatic brain injury produces neuroinflammation, anxiety-like behaviour and impaired spatial memory in mice. *Brain Inj.* **2018**, *32*, 113–122. [CrossRef]
18. Bushnell, C.D.; Chaturvedi, S.; Gage, K.R.; Herson, P.S.; Hurn, P.D.; Jiménez, M.C.; Kittner, S.J.; Madsen, T.E.; McCullough, L.D.; McDermott, M.; et al. Sex differences in stroke: Challenges and opportunities. *J. Cereb. Blood Flow. Metab.* **2018**, *38*, 2179–2191. [CrossRef]
19. Fox, L. Influences of Steroid Hormones on Psychological Symptoms of Mild Traumatic Brain Injury in Males and Females—ProQuest. 2019. Available online: https://www.proquest.com/openview/eea33ef1ee70b277b8b7f1e572ed3cfd/1?pq-origsite=gscholar&cbl=18750&diss=y (accessed on 31 July 2024).
20. Gupte, R.; Brooks, W.; Vukas, R.; Pierce, J.; Harris, J. Sex Differences in Traumatic Brain Injury: What We Know and What We Should Know. *J. Neurotrauma* **2019**, *36*, 3063–3091. [CrossRef]
21. Lu, Y.; Li, C.; Chen, C.; Luo, P.; Zhou, M.; Li, C.; Xu, X.; Lu, Q.; He, Z.; Guo, L. Activation of GABAB2 subunits alleviates chronic cerebral hypoperfusion-induced anxiety-like behaviours: A role for BDNF signalling and Kir3 channels. *Neuropharmacology* **2016**, *110*, 308–321. [CrossRef]
22. Lu, Y.; Li, C.-J.; Li, W.-J.; Zhou, M.; Liu, L. Depressive-like behaviors induced by chronic cerebral hypoperfusion associate with a dynamic change of GABAB1/B2 receptors expression in hippocampal CA1 region. *Physiol. Behav.* **2022**, *254*, 113887. [CrossRef]
23. Passier, P.E.C.A.; Visser-Meily, J.M.A.; Rinkel, G.J.E.; Lindeman, E.; Post, M.W.M. Determinants of health-related quality of life after aneurysmal subarachnoid hemorrhage: A systematic review. *Qual. Life Res.* **2013**, *22*, 1027–1043. [CrossRef]
24. Zhu, W.; Gao, Y.; Wan, J.; Lan, X.; Han, X.; Zhu, S.; Zang, W.; Chen, X.; Ziai, W.; Hanley, D.F.; et al. Changes in motor function, cognition, and emotion-related behavior after right hemispheric intracerebral hemorrhage in various brain regions of mouse. *Brain Behav. Immun.* **2018**, *69*, 568–581. [CrossRef] [PubMed]
25. Tiel, C.; Sudo, F.K.; Alves, G.S.; Ericeira-Valente, L.; Moreira, D.M.; Laks, J.; Engelhardt, E. Neuropsychiatric symptoms in Vascular Cognitive Impairment: A systematic review. *Dement. Neuropsychol.* **2015**, *9*, 230–236. [CrossRef] [PubMed]

26. Chun, H.-Y.Y.; Newman, R.; Whiteley, W.N.; Dennis, M.; Mead, G.E.; Carson, A.J. A systematic review of anxiety interventions in stroke and acquired brain injury: Efficacy and trial design. *J. Psychosom. Res.* **2018**, *104*, 65–75. [CrossRef]
27. Blokhin, I.O.; Khorkova, O.; Saveanu, R.V.; Wahlestedt, C. Molecular mechanisms of psychiatric diseases. *Neurobiol. Dis.* **2020**, *146*, 105136. [CrossRef]
28. Bouayed, J.; Rammal, H.; Soulimani, R. Oxidative Stress and Anxiety: Relationship and Cellular Pathways. *Oxidative Med. Cell. Longev.* **2009**, *2*, 623654. [CrossRef]
29. Correia, A.S.; Cardoso, A.; Vale, N. Oxidative Stress in Depression: The Link with the Stress Response, Neuroinflammation, Serotonin, Neurogenesis and Synaptic Plasticity. *Antioxidants* **2023**, *12*, 470. [CrossRef] [PubMed]
30. Hu, P.; Lu, Y.; Pan, B.-X.; Zhang, W.-H. New Insights into the Pivotal Role of the Amygdala in Inflammation-Related Depression and Anxiety Disorder. *Int. J. Mol. Sci.* **2022**, *23*, 11076. [CrossRef] [PubMed]
31. Won, E.; Kim, Y.-K. Neuroinflammation-Associated Alterations of the Brain as Potential Neural Biomarkers in Anxiety Disorders. *Int. J. Mol. Sci.* **2020**, *21*, 6546. [CrossRef]
32. Martin, E.I.; Ressler, K.J.; Binder, E.; Nemeroff, C.B. The Neurobiology of Anxiety Disorders: Brain Imaging, Genetics, and Psychoneuroendocrinology. *Clin. Lab. Med.* **2010**, *30*, 865–891. [CrossRef]
33. Bangasser, D.A.; Curtis, A.; Reyes, B.a.S.; Bethea, T.T.; Parastatidis, I.; Ischiropoulos, H.; Van Bockstaele, E.J.; Valentino, R.J. Sex differences in corticotropin-releasing factor receptor signaling and trafficking: Potential role in female vulnerability to stress-related psychopathology. *Mol. Psychiatry* **2010**, *15*, 896–904. [CrossRef]
34. Bangasser, D.A.; Eck, S.R.; Telenson, A.M.; Salvatore, M. Sex differences in stress regulation of arousal and cognition. *Physiol. Behav.* **2018**, *187*, 42–50. [CrossRef]
35. Filkowski, M.M.; Olsen, R.M.; Duda, B.; Wanger, T.J.; Sabatinelli, D. Sex differences in emotional perception: Meta analysis of divergent activation. *NeuroImage* **2017**, *147*, 925–933. [CrossRef] [PubMed]
36. Marzolini, S.; Colella, T.J.F. Factors That Predispose Women to Greater Depressive Symptoms: A Sex-, Age-, and Diagnosis-Matched Cardiac Rehabilitation Cohort. *Can. J. Cardiol.* **2021**, *37*, 382–390. [CrossRef]
37. Fries, G.R.; Saldana, V.A.; Finnstein, J.; Rein, T. Molecular pathways of major depressive disorder converge on the synapse. *Mol. Psychiatry* **2023**, *28*, 284–297. [CrossRef]
38. Zhu, J.-H.; Bo, H.-H.; Liu, B.-P.; Jia, C.-X. The associations between DNA methylation and depression: A systematic review and meta-analysis. *J. Affect. Disord.* **2023**, *327*, 439–450. [CrossRef] [PubMed]
39. Murphy, T.M.; O'Donovan, A.; Mullins, N.; O'Farrelly, C.; McCann, A.; Malone, K. Anxiety is associated with higher levels of global DNA methylation and altered expression of epigenetic and interleukin-6 genes. *Psychiatr. Genet.* **2015**, *25*, 71–78. [CrossRef] [PubMed]
40. Dias, G.P.; Thuret, S. The Hippocampus and Panic Disorder: Evidence from Animal and Human Studies. In *Panic Disorder: Neurobiological and Treatment Aspects*; Nardi, A.E., Freire, R.C.R., Eds.; Springer International Publishing: Cham, Switzerland, 2016; pp. 79–91. ISBN 978-3-319-12538-1.
41. Chen, A.C.; Etkin, A. Hippocampal Network Connectivity and Activation Differentiates Post-Traumatic Stress Disorder From Generalized Anxiety Disorder. *Neuropsychopharmacology* **2013**, *38*, 1889–1898. [CrossRef]
42. Cho, W.-H.; Noh, K.; Lee, B.H.; Barcelon, E.; Jun, S.B.; Park, H.Y.; Lee, S.J. Hippocampal astrocytes modulate anxiety-like behavior. *Nat. Commun.* **2022**, *13*, 6536. [CrossRef] [PubMed]
43. MacQueen, G.; Frodl, T. The hippocampus in major depression: Evidence for the convergence of the bench and bedside in psychiatric research? *Mol. Psychiatry* **2011**, *16*, 252–264. [CrossRef]
44. Roddy, D.W.; Farrell, C.; Doolin, K.; Roman, E.; Tozzi, L.; Frodl, T.; O'Keane, V.; O'Hanlon, E. The Hippocampus in Depression: More Than the Sum of Its Parts? Advanced Hippocampal Substructure Segmentation in Depression. *Biol. Psychiatry* **2019**, *85*, 487–497. [CrossRef]
45. Atmaca, M.; Yildirim, H.; Gurok, M.G.; Akyol, M. Orbito-frontal cortex volumes in panic disorder. *Psychiatry Investig.* **2012**, *9*, 408–412. [CrossRef] [PubMed]
46. Kenwood, M.M.; Kalin, N.H.; Barbas, H. The prefrontal cortex, pathological anxiety, and anxiety disorders. *Neuropsychopharmacology* **2022**, *47*, 260–275. [CrossRef]
47. Pizzagalli, D.A.; Roberts, A.C. Prefrontal cortex and depression. *Neuropsychopharmacology* **2022**, *47*, 225–246. [CrossRef] [PubMed]
48. Wang, H.-Y.; Xu, G.-Q.; Ni, M.-F.; Zhang, C.-H.; Li, X.-L.; Chang, Y.; Sun, X.-P.; Zhang, B.-W. Neural basis of implicit cognitive reappraisal in panic disorder: An event-related fMRI study. *J. Transl. Med.* **2021**, *19*, 304. [CrossRef] [PubMed]
49. LeDoux, J. The amygdala. *Curr. Biol.* **2007**, *17*, R868–R874. [CrossRef]
50. Schindler, S.; Schmidt, L.; Stroske, M.; Storch, M.; Anwander, A.; Trampel, R.; Strauß, M.; Hegerl, U.; Geyer, S.; Schönknecht, P. Hypothalamus enlargement in mood disorders. *Acta Psychiatr. Scand.* **2019**, *139*, 56–67. [CrossRef] [PubMed]
51. Fischer, S. The hypothalamus in anxiety disorders. *Handb. Clin. Neurol.* **2021**, *180*, 149–160. [CrossRef]

52. Wang, H.; Cui, W.; Chen, W.; Liu, F.; Dong, Z.; Xing, G.; Luo, B.; Gao, N.; Zou, W.-J.; Zhao, K.; et al. The laterodorsal tegmentum-ventral tegmental area circuit controls depression-like behaviors by activating ErbB4 in DA neurons. *Mol. Psychiatry* **2023**, *28*, 1027–1045. [CrossRef] [PubMed]
53. Tong, Q.; Cui, X.; Xu, H.; Zhang, X.; Hu, S.; Huang, F.; Xiao, L. D1 receptor-expressing neurons in ventral tegmental area alleviate mouse anxiety-like behaviors via glutamatergic projection to lateral septum. *Mol. Psychiatry* **2023**, *28*, 625–638. [CrossRef] [PubMed]
54. Morris, L.S.; Mehta, M.; Ahn, C.; Corniquel, M.; Verma, G.; Delman, B.; Hof, P.R.; Jacob, Y.; Balchandani, P.; Murrough, J.W. Ventral tegmental area integrity measured with high-resolution 7-Tesla MRI relates to motivation across depression and anxiety diagnoses. *Neuroimage* **2022**, *264*, 119704. [CrossRef]
55. Carl, E.; Witcraft, S.M.; Kauffman, B.Y.; Gillespie, E.M.; Becker, E.S.; Cuijpers, P.; Van Ameringen, M.; Smits, J.A.J.; Powers, M.B. Psychological and pharmacological treatments for generalized anxiety disorder (GAD): A meta-analysis of randomized controlled trials. *Cogn. Behav. Ther.* **2020**, *49*, 1–21. [CrossRef] [PubMed]
56. Lipsman, N.; Sankar, T.; Downar, J.; Kennedy, S.H.; Lozano, A.M.; Giacobbe, P. Neuromodulation for treatment-refractory major depressive disorder. *Can. Med. Assoc. J.* **2014**, *186*, 33–39. [CrossRef] [PubMed]
57. Nava-Ruelas, R.; Jarde, A.; Elsey, H.; Siddiqi, K.; Todowede, O.; Zavala, G.; Siddiqi, N. Pharmacological and psychological interventions for depression in people with tuberculosis. *Cochrane Database Syst. Rev.* **2021**, *2021*, CD014848. [CrossRef]
58. Powell, K.; Lin, K.; Tambo, W.; Saavedra, A.P.; Sciubba, D.; Al Abed, Y.; Li, C. Trigeminal nerve stimulation: A current state-of-the-art review. *Bioelectron. Med.* **2023**, *9*, 30. [CrossRef] [PubMed]
59. Witt, K.G.; Hetrick, S.E.; Rajaram, G.; Hazell, P.; Taylor Salisbury, T.L.; Townsend, E.; Hawton, K. Pharmacological interventions for self-harm in adults. *Cochrane Database Syst. Rev.* **2021**, *2021*, CD013669. [CrossRef]
60. Perrault, G.; Morel, E.; Sanger, D.J.; Zivkovic, B. Differences in pharmacological profiles of a new generation of benzodiazepine and non-benzodiazepine hypnotics. *Eur. J. Pharmacol.* **1990**, *187*, 487–494. [CrossRef] [PubMed]
61. Dale, E.; Bang-Andersen, B.; Sánchez, C. Emerging mechanisms and treatments for depression beyond SSRIs and SNRIs. *Biochem. Pharmacol.* **2015**, *95*, 81–97. [CrossRef]
62. Carradori, S.; Secci, D.; Petzer, J.P. MAO inhibitors and their wider applications: A patent review. *Expert. Opin. Ther. Pat.* **2018**, *28*, 211–226. [CrossRef]
63. Kandola, A.; Ashdown-Franks, G.; Hendrikse, J.; Sabiston, C.M.; Stubbs, B. Physical activity and depression: Towards understanding the antidepressant mechanisms of physical activity. *Neurosci. Biobehav. Rev.* **2019**, *107*, 525–539. [CrossRef] [PubMed]
64. Bystritsky, A. Treatment-resistant anxiety disorders. *Mol. Psychiatry* **2006**, *11*, 805–814. [CrossRef] [PubMed]
65. Nanegrungsunk, D.; Ragozzino, M.E.; Xu, H.-L.; Haselton, K.J.; Paisansathan, C. Subarachnoid hemorrhage in C57BL/6J mice increases motor stereotypies and compulsive-like behaviors. *Neurol. Res.* **2021**, *43*, 239–251. [CrossRef] [PubMed]
66. Song, J.-H.; Jia, H.-Y.; Shao, T.-P.; Liu, Z.-B.; Zhao, Y.-P. Hydrogen gas post-conditioning alleviates cognitive dysfunction and anxiety-like behavior in a rat model of subarachnoid hemorrhage. *Exp. Ther. Med.* **2021**, *22*, 1121. [CrossRef] [PubMed]
67. Yang, J.; Pan, Y.; Li, X.; Wang, X. Atorvastatin attenuates cognitive deficits through Akt1/caspase-3 signaling pathway in ischemic stroke. *Brain Res.* **2015**, *1629*, 231–239. [CrossRef]
68. Ma, L.; Jiang, Y.; Dong, Y.; Gao, J.; Du, B.; Liu, D. Anti-TNF-alpha antibody attenuates subarachnoid hemorrhage-induced apoptosis in the hypothalamus by inhibiting the activation of Erk. *Neuropsychiatr. Dis. Treat.* **2018**, *14*, 525–536. [CrossRef] [PubMed]
69. Liu, S.; Han, S.; Dai, Q.; Li, S.; Li, J. BICAO-induced ischaemia caused depressive-like behaviours and caspase-8/-9-dependent brain regional neural cell apoptosis in mice. *Stroke Vasc. Neurol.* **2018**, *3*, 1–8. [CrossRef] [PubMed]
70. Kanamaru, H.; Zhu, S.; Dong, S.; Takemoto, Y.; Huang, L.; Sherchan, P.; Suzuki, H.; Tang, J.; Zhang, J.H. UDP-Glucose/P2Y14 Receptor Signaling Exacerbates Neuronal Apoptosis After Subarachnoid Hemorrhage in Rats. *Stroke* **2024**, *55*, 1381–1392. [CrossRef] [PubMed]
71. Tweedie, D.; Milman, A.; Holloway, H.W.; Li, Y.; Harvey, B.K.; Shen, H.; Pistell, P.J.; Lahiri, D.K.; Hoffer, B.J.; Wang, Y.; et al. Apoptotic and behavioral sequelae of mild brain trauma in mice. *J. Neurosci. Res.* **2007**, *85*, 805–815. [CrossRef] [PubMed]
72. Afzal, M.; Kazmi, I.; Quazi, A.M.; Khan, S.A.; Zafar, A.; Al-Abbasi, F.A.; Imam, F.; Alharbi, K.S.; Alzarea, S.I.; Yadav, N. 6-Shogaol Attenuates Traumatic Brain Injury-Induced Anxiety/Depression-like Behavior via Inhibition of Oxidative Stress-Influenced Expressions of Inflammatory Mediators TNF-α, IL-1β, and BDNF: Insight into the Mechanism. *ACS Omega* **2022**, *7*, 140–148. [CrossRef]
73. An, P.; Zhao, X.-C.; Liu, M.-J.; You, Y.-Q.; Li, J.-Y. Gender-based differences in neuroprotective effects of hydrogen gas against intracerebral hemorrhage-induced depression. *Neurochem. Int.* **2022**, *153*, 105276. [CrossRef]
74. An, P.; Zhao, X.-C.; Liu, M.-J.; You, Y.-Q.; Li, J.-Y.; Gong, H.-S. Dexmedetomidine Alleviates Intracerebral Hemorrhage-Induced Anxiety-Like Behaviors in Mice Through the Inhibition of TRPV4 Opening. *Front. Pharmacol.* **2022**, *13*, 852401. [CrossRef]

75. Tao, X.; Wu, S.; Tang, W.; Li, L.; Huang, L.; Mo, D.; Liu, C.; Song, T.; Wang, S.; Wang, J.; et al. Alleviative effects of foraging exercise on depressive-like behaviors in chronic mild stress-induced ischemic rat model. *Brain Inj.* **2022**, *36*, 127–136. [CrossRef] [PubMed]
76. Yang, Y.; Zhang, K.; Zhong, J.; Wang, J.; Yu, Z.; Lei, X.; Chen, X.; Quan, Y.; Xian, J.; Chen, Y.; et al. Stably maintained microtubules protect dopamine neurons and alleviate depression-like behavior after intracerebral hemorrhage. *Sci. Rep.* **2018**, *8*, 12647. [CrossRef]
77. Tan, Z.; Qiu, J.; Zhang, Y.; Yang, Q.; Yin, X.; Li, J.; Liu, G.; Li, H.; Yang, G. Tetramethylpyrazine Alleviates Behavioral and Psychological Symptoms of Dementia Through Facilitating Hippocampal Synaptic Plasticity in Rats with Chronic Cerebral Hypoperfusion. *Front. Neurosci.* **2021**, *15*, 646537. [CrossRef]
78. Xie, L.-L.; Li, S.-S.; Fan, Y.-J.; Qi, M.-M.; Li, Z.-Z. Melatonin alleviates traumatic brain injury-induced anxiety-like behaviors in rats: Roles of the protein kinase A/cAMP-response element binding signaling pathway. *Exp. Ther. Med.* **2022**, *23*, 248. [CrossRef] [PubMed]
79. Zhu, H.; Guo, Y.; Huang, A.; Shen, H.; Chen, Y.; Song, J.; Guan, A.; Wu, L.; Wang, H.; Deng, B. HDAC3-Regulated PGE2 Production by Microglia Induces Phobic Anxiety Susceptibility After Stroke and Pointedly Exploiting a Signal-Targeted Gamma Visual Stimulation New Therapy. *Front. Immunol.* **2022**, *13*, 845678. [CrossRef]
80. Li, J.; Li, J.; Guo, E.; Wang, Y.; Yang, M.; Huo, H.; Shi, Y.; Zhao, L. Pink1 deficiency enhances neurological deficits and inflammatory responses after intracerebral hemorrhage in mice. *Neurotherapeutics* **2024**, *21*, e00317. [CrossRef] [PubMed]
81. Jia, P.; Wang, J.; Ren, X.; He, J.; Wang, S.; Xing, Y.; Chen, D.; Zhang, X.; Zhou, S.; Liu, X.; et al. An enriched environment improves long-term functional outcomes in mice after intracerebral hemorrhage by mechanisms that involve the Nrf2/BDNF/glutaminase pathway. *J. Cereb. Blood Flow. Metab.* **2023**, *43*, 694–711. [CrossRef] [PubMed]
82. Liu, H.; Zhang, Y.; Hou, X.; Zhu, C.; Yang, Q.; Li, K.; Fan, L.; Zhang, X.; Jiang, X.; Jin, X.; et al. CRHR1 antagonist alleviated depression-like behavior by downregulating p62 in a rat model of post-stroke depression. *Exp. Neurol.* **2024**, *378*, 114822. [CrossRef]
83. Ren, H.; Han, R.; Liu, X.; Wang, L.; Koehler, R.C.; Wang, J. Nrf2-BDNF-TrkB pathway contributes to cortical hemorrhage-induced depression, but not sex differences. *J. Cereb. Blood Flow. Metab.* **2021**, *41*, 3288–3301. [CrossRef]
84. Sen, T.; Saha, P.; Gupta, R.; Foley, L.M.; Jiang, T.; Abakumova, O.S.; Hitchens, T.K.; Sen, N. Aberrant ER Stress Induced Neuronal-IFNβ Elicits White Matter Injury Due to Microglial Activation and T-Cell Infiltration after TBI. *J. Neurosci.* **2020**, *40*, 424–446. [CrossRef] [PubMed]
85. Ma, X.; Niu, X.; Zhao, J.; Deng, Z.; Li, J.; Wu, X.; Wang, B.; Zhang, M.; Zhao, Y.; Guo, X.; et al. Downregulation of Sepina3n Aggravated Blood-Brain Barrier Disruption after Traumatic Brain Injury by Activating Neutrophil Elastase in Mice. *Neuroscience* **2022**, *503*, 45–57. [CrossRef] [PubMed]
86. Zhou, M.; Liu, Y.-W.-Y.; He, Y.-H.; Zhang, J.-Y.; Guo, H.; Wang, H.; Ren, J.-K.; Su, Y.-X.; Yang, T.; Li, J.-B.; et al. FOXO1 reshapes neutrophils to aggravate acute brain damage and promote late depression after traumatic brain injury. *Mil. Med. Res.* **2024**, *11*, 20. [CrossRef]
87. Pujari, R.R.; Vyawahare, N.S.; Thakurdesai, P.A. Neuroprotective and antioxidant role of *Phoenix dactylifera* in permanent bilateral common carotid occlusion in rats. *J. Acute Dis.* **2014**, *3*, 104–114. [CrossRef]
88. Yan, Q.; Zhang, H.; Hui, K.; Shao, Q. Cordycepin Ameliorates Intracerebral Hemorrhage Induced Neurological and Cognitive Impairments Through Reducing Anti-Oxidative Stress in a Mouse Model. *J. Stroke Cerebrovasc. Dis.* **2022**, *31*, 106199. [CrossRef]
89. Tchantchou, F.; Miller, C.; Goodfellow, M.; Puche, A.; Fiskum, G. Hypobaria-Induced Oxidative Stress Facilitates Homocysteine Transsulfuration and Promotes Glutathione Oxidation in Rats with Mild Traumatic Brain Injury. *J. Cent. Nerv. Syst. Dis.* **2021**, *13*, 1179573520988193. [CrossRef]
90. Lu, Y.; Chen, Y.; Xu, S.; Wei, L.; Zhang, Y.; Chen, W.; Liu, M.; Zhong, C. HDAC inhibitor attenuates rat traumatic brain injury induced neurological impairments. *Heliyon* **2023**, *9*, e18485. [CrossRef]
91. Fan, Y.; Wang, Y.; Ji, W.; Liu, K.; Wu, H. Exercise preconditioning ameliorates cognitive impairment and anxiety-like behavior via regulation of dopamine in ischemia rats. *Physiol. Behav.* **2021**, *233*, 113353. [CrossRef] [PubMed]
92. Almeida-Suhett, C.P.; Prager, E.M.; Pidoplichko, V.; Figueiredo, T.H.; Marini, A.M.; Li, Z.; Eiden, L.E.; Braga, M.F.M. Reduced GABAergic Inhibition in the Basolateral Amygdala and the Development of Anxiety-Like Behaviors after Mild Traumatic Brain Injury. *PLoS ONE* **2014**, *9*, e102627. [CrossRef]
93. Beitchman, J.A.; Griffiths, D.R.; Hur, Y.; Ogle, S.B.; Bromberg, C.E.; Morrison, H.W.; Lifshitz, J.; Adelson, P.D.; Thomas, T.C. Experimental Traumatic Brain Injury Induces Chronic Glutamatergic Dysfunction in Amygdala Circuitry Known to Regulate Anxiety-Like Behavior. *Front. Neurosci.* **2019**, *13*, 1434. [CrossRef]
94. Frank, D.; Kuts, R.; Tsenter, P.; Gruenbaum, B.F.; Grinshpun, Y.; Zvenigorodsky, V.; Shelef, I.; Natanel, D.; Brotfain, E.; Zlotnik, A.; et al. The effect of pyruvate on the development and progression of post-stroke depression: A new therapeutic approach. *Neuropharmacology* **2019**, *155*, 173–184. [CrossRef]

95. Frank, D.; Gruenbaum, B.F.; Shelef, I.; Zvenigorodsky, V.; Severynovska, O.; Fleidervish, I.; Knyazer, B.; Frenkel, A.; Zlotnik, A.; Kofman, O.; et al. Blood glutamate scavenging as a novel glutamate-based therapeutic approach for post-traumatic brain injury anxiety and social impairment. *Transl. Psychiatry* **2023**, *13*, 1–9. [CrossRef]
96. Fromm, L.; Heath, D.L.; Vink, R.; Nimmo, A.J. Magnesium attenuates post-traumatic depression/anxiety following diffuse traumatic brain injury in rats. *J. Am. Coll. Nutr.* **2004**, *23*, 529S–533S. [CrossRef]
97. Tao, K.; Cai, Q.; Zhang, X.; Zhu, L.; Liu, Z.; Li, F.; Wang, Q.; Liu, L.; Feng, D. Astrocytic histone deacetylase 2 facilitates delayed depression and memory impairment after subarachnoid hemorrhage by negatively regulating glutamate transporter-1. *Ann. Transl. Med.* **2020**, *8*, 691. [CrossRef] [PubMed]
98. Day, N.L.; Carle, M.S.; Floyd, C.L. Post-injury administration of a combination of memantine and 17β-estradiol is protective in a rat model of traumatic brain injury. *Neurochem. Int.* **2017**, *111*, 57–68. [CrossRef]
99. Sönmez, A.; Sayın, O.; Gürgen, S.G.; Çalişir, M. Neuroprotective effects of MK-801 against traumatic brain injury in immature rats. *Neurosci. Lett.* **2015**, *597*, 137–142. [CrossRef]
100. Zhang, X.; Shi, Y.; Fan, T.; Wang, K.; Zhan, H.; Wu, W. Analysis of Correlation Between White Matter Changes and Functional Responses in Post-stroke Depression. *Front. Aging Neurosci.* **2021**, *13*, 728622. [CrossRef] [PubMed]
101. Ohnishi, M.; Akagi, M.; Kotsuki, M.; Yonemura, S.; Aokawa, H.; Yamashita-Ibara, M.; Yokofujita, O.; Maehara, S.; Hata, T.; Inoue, A. Indoleamine 2, 3-dioxygenase is responsible for low stress tolerance after intracerebral hemorrhage. *PLoS ONE* **2023**, *18*, e0273037. [CrossRef]
102. Xiao, W.; Li, J.; Gao, X.; Yang, H.; Su, J.; Weng, R.; Gao, Y.; Ni, W.; Gu, Y. Involvement of the gut-brain axis in vascular depression via tryptophan metabolism: A benefit of short chain fatty acids. *Exp. Neurol.* **2022**, *358*, 114225. [CrossRef]
103. Zahrai, A.; Vahid-Ansari, F.; Daigle, M.; Albert, P.R. Fluoxetine-induced recovery of serotonin and norepinephrine projections in a mouse model of post-stroke depression. *Transl. Psychiatry* **2020**, *10*, 334. [CrossRef] [PubMed]
104. Zhou, M.; Li, Y.; Lin, K.; Luo, P.; Liu, W. Chronic Cerebral Hypoperfusion-induced Dysregulations of Hyperpolarization- activated Cyclic Nucleotide-gated, KCNQ and G Protein-coupled Inwardly Rectifying Potassium Channels Correlated with Susceptibility and Unsusceptibility to Anxiety Behaviors. *Curr. Neurovasc Res.* **2022**, *19*, 398–405. [CrossRef]
105. Bhatt, D.; Hazari, A.; Yamakawa, G.R.; Salberg, S.; Sgro, M.; Shultz, S.R.; Mychasiuk, R. Investigating the cumulative effects of Δ9-tetrahydrocannabinol and repetitive mild traumatic brain injury on adolescent rats. *Brain Commun.* **2020**, *2*, fcaa042. [CrossRef]
106. Wang, S.; Sun, H.; Liu, S.; Wang, T.; Guan, J.; Jia, J. Role of hypothalamic cannabinoid receptors in post-stroke depression in rats. *Brain Res. Bull.* **2016**, *121*, 91–97. [CrossRef] [PubMed]
107. Meftahi, G.H.; Bayat, M.; Zarifkar, A.H.; Akbari, S.; Borhani Haghighi, A.; Naseh, M.; Yousefi Nejad, A.; Haghani, M. Treatment with edaravone improves the structure and functional changes in the hippocampus after chronic cerebral hypoperfusion in rat. *Brain Res. Bull.* **2021**, *174*, 122–130. [CrossRef] [PubMed]
108. Hooshmandi, E.; Akbari, S.; Pandamooz, S.; Ghobadi, M.; Ghasemi, R.; Maghsoudi, N.; Rai, S.N.; Borhani-Haghighi, A.; Salehi, M.S.; Azarpira, N.; et al. Combined use of hair follicle stem cells and CEPO (carbamylated erythropoietin)-Fc in a rat model of chronic cerebral hypoperfusion: A behavioral, electrophysiological, and molecular study. *Behav. Brain Res.* **2023**, *454*, 114655. [CrossRef]
109. Li, Y.; Qu, C.; Song, H.; Li, T.; Zheng, J.; Wu, L.; Yan, N.; Xu, L.; Qu, C.; Zhang, J. Enriched environment priors to TET1 hippocampal administration for regulating psychiatric behaviors via glial reactivity in chronic cerebral hypoperfusion models. *J. Affect. Disord.* **2022**, *310*, 198–212. [CrossRef]
110. Hoffman, A.N.; Watson, S.; Fanselow, M.S.; Hovda, D.A.; Giza, C. Region-Dependent Modulation of Neural Plasticity in Limbic Structures Early after Traumatic Brain Injury. *Neurotrauma Rep.* **2021**, *2*, 200–213. [CrossRef]
111. Liu, B.; Zhao, G.; Jin, L.; Shi, J. Nicotinamide Improves Cognitive Function in Mice with Chronic Cerebral Hypoperfusion. *Front. Neurol.* **2021**, *12*, 596641. [CrossRef]
112. Lee, S.R.; Choi, B.; Paul, S.; Seo, J.-H.; Back, D.B.; Han, J.-S.; Choi, D.-H.; Kwon, K.J.; Shin, C.Y.; Lee, J.; et al. Depressive-like behaviors in a rat model of chronic cerebral hypoperfusion. *Transl. Stroke Res.* **2015**, *6*, 207–214. [CrossRef] [PubMed]
113. Kosari-Nasab, M.; Sadeghi, T.; Bashiri, H.; Shokouhi, G.; Salari, A.-A. The blockade of corticotropin-releasing factor 1 receptor attenuates anxiety-related symptoms and hypothalamus-pituitary-adrenal axis reactivity in mice with mild traumatic brain injury. *Behav. Pharmacol.* **2019**, *30*, 220–228. [CrossRef] [PubMed]
114. Ai, N.-H.; Kaori, K.; Kotaro, T.; Takuya, W.; Hikari, I.; Risako, F.; Shutaro, K.; Katsunori, I.; Kaori, K.; Takuya, W.; et al. Extract of Yokukansan improves anxiety-like behavior and increases serum brain-derived neurotrophic factor in rats with cerebral ischemia combined with amyloid-42 peptide. *J. Tradit. Chin. Med.* **2019**, *39*, 50–58. [PubMed]
115. Barra de la Tremblaye, P. A Role for CRH and HPA Activation in the Regulation of Plasticity Signaling, Neuroinflammation and Emotional/Mnesic Behavior Following Global Cerebral Ischemia in Rats. University of Ottawa (Canada). 2016. Available online: https://ruor.uottawa.ca/server/api/core/bitstreams/9f87317c-7743-44ad-a43f-8f5f49528f31/content (accessed on 1 August 2024).

116. Khan, S. Mediating Role of Corticotropin-Releasing Hormone (CRH) in Anxiety and Ischemia: Behavioural and Physiological Correlates, University of Ottawa (Canada). 2003. Available online: http://hdl.handle.net/10393/29051 (accessed on 2 August 2024).
117. Russell, A.L.; Handa, R.J.; Wu, T.J. Sex-Dependent Effects of Mild Blast-induced Traumatic Brain Injury on Corticotropin-releasing Factor Receptor Gene Expression: Potential Link to Anxiety-like Behaviors. *Neuroscience* **2018**, *392*, 1–12. [CrossRef]
118. Fernandes, J.G.d.A.; Victoria, E.C.G.; Toscano, E.C.d.B.; Ferreira, R.N.; Silva, D.G.; Oliveira, B.d.S.; Mourão, A.M.; Pedroso, V.S.P.; Teixeira, A.L.; de Miranda, A.S.; et al. High levels of NGF during anxiety-like behavior in a murine model of brain ischemic stroke. *Neurol. Psychiatry Brain Res.* **2020**, *38*, 114–120. [CrossRef]
119. Fox, L.C.; Scholl, J.L.; Palmer, G.M.; Forster, G.L.; Watt, M.J. Sex differences in the effects of mild traumatic brain injury and progesterone treatment on anxiety-like behavior and fear conditioning in rats. *Exp. Neurol.* **2023**, *365*, 114415. [CrossRef] [PubMed]
120. Nykamp, T. The Effect of Gonadal Steroids Following Repeat Mild Traumatic Brain Injury on Anxiety-Like Behaviors in Rats. Master's Thesis, 2021. Available online: https://ecommons.luc.edu/luc_theses/4394 (accessed on 1 January 2021).
121. McQueen, J.K.; Wilson, H.; Sumner, B.E.; Fink, G. Serotonin transporter (SERT) mRNA and binding site densities in male rat brain affected by sex steroids. *Brain Res. Mol. Brain Res.* **1999**, *63*, 241–247. [CrossRef] [PubMed]
122. Abdul-Muneer, P.M.; Saikia, B.B.; Bhowmick, S. Synergistic effect of mild traumatic brain injury and alcohol aggravates neuroinflammation, amyloidogenesis, tau pathology, neurodegeneration, and blood-brain barrier alterations: Impact on psychological stress. *Exp. Neurol.* **2022**, *358*, 114222. [CrossRef]
123. Zhiyou, C.; Yong, Y.; Shanquan, S.; Jun, Z.; Liangguo, H.; Ling, Y.; Jieying, L. Upregulation of BACE1 and beta-amyloid protein mediated by chronic cerebral hypoperfusion contributes to cognitive impairment and pathogenesis of Alzheimer's disease. *Neurochem. Res.* **2009**, *34*, 1226–1235. [CrossRef]
124. Zhao, Y.; Gu, J.; Dai, C.; Liu, Q.; Iqbal, K.; Liu, F.; Gong, C.-X. Chronic cerebral hypoperfusion causes decrease of O-GlcNAcylation, hyperphosphorylation of tau and behavioral deficits in mice. *Front. Aging Neurosci.* **2014**, *6*, 10. [CrossRef]
125. Kim, Y.R.; Kim, H.N.; Hong, K.W.; Shin, H.K.; Choi, B.T. Anti-depressant effects of phosphodiesterase 3 inhibitor cilostazol in chronic mild stress-treated mice after ischemic stroke. *Psychopharmacology* **2016**, *233*, 1055–1066. [CrossRef] [PubMed]
126. Baykara, B.; Cetin, F.; Baykara, B.; Aksu, I.; Dayi, A.; Kiray, M.; Sisman, A.R.; Ozdemir, D.; Arda, M.N.; Uysal, N. Anxiety caused by traumatic brain injury correlates to decreased prefrontal cortex VEGF immunoreactivity and neuron density in immature rats. *Turk. Neurosurg.* **2012**, *22*, 604–610. [CrossRef] [PubMed]
127. Li, C.; Ajmal, E.; Alok, K.; Powell, K.; Wadolowski, S.; Tambo, W.; Turpin, J.; Barthélemy, E.; Al-Abed, Y.; LeDoux, D. CGRP as a potential mediator for the sexually dimorphic responses to traumatic brain injury. *Biol. Sex. Differ.* **2024**, *15*, 44. [CrossRef] [PubMed]
128. Feng, P.; Wu, Z.; Liu, H.; Shen, Y.; Yao, X.; Li, X.; Shen, Z. Electroacupuncture Improved Chronic Cerebral Hypoperfusion-Induced Anxiety-Like Behavior and Memory Impairments in Spontaneously Hypertensive Rats by Downregulating the ACE/Ang II/AT1R Axis and Upregulating the ACE2/Ang-(1-7)/MasR Axis. *Neural Plast.* **2020**, *2020*, 9076042. [CrossRef] [PubMed]
129. Chokkalla, A.K.; Jeong, S.; Mehta, S.L.; Davis, C.K.; Morris-Blanco, K.C.; Bathula, S.; Qureshi, S.S.; Vemuganti, R. Cerebroprotective Role of N6-Methyladenosine Demethylase FTO (Fat Mass and Obesity-Associated Protein) After Experimental Stroke. *Stroke* **2023**, *54*, 245–254. [CrossRef] [PubMed]
130. Jadhav, G.; Dudhabhate, B.B.; Kokare, D.M.; Sakharkar, A.J. Gut Microbiota Regulates Epigenetic Remodelling in the Amygdala: A Role in Repeated Mild Traumatic Brain Injury (rMTBI)-Induced Anxiety. *Mol. Neurobiol.* **2024**, *61*, 9892–9914. [CrossRef]
131. Sagarkar, S.; Bhamburkar, T.; Shelkar, G.; Choudhary, A.; Kokare, D.M.; Sakharkar, A.J. Minimal traumatic brain injury causes persistent changes in DNA methylation at BDNF gene promoters in rat amygdala: A possible role in anxiety-like behaviors. *Neurobiol. Dis.* **2017**, *106*, 101–109. [CrossRef]
132. Feigin, V.L.; Nichols, E.; Alam, T.; Bannick, M.S.; Beghi, E.; Blake, N.; Culpepper, W.J.; Dorsey, E.R.; Elbaz, A.; Ellenbogen, R.G.; et al. Global, regional, and national burden of neurological disorders, 1990–2016: A systematic analysis for the Global Burden of Disease Study 2016. *Lancet Neurol.* **2019**, *18*, 459–480. [CrossRef] [PubMed]
133. Howlett, J.R.; Nelson, L.D.; Stein, M.B. Mental Health Consequences of Traumatic Brain Injury. *Biol. Psychiatry* **2022**, *91*, 413–420. [CrossRef]
134. Kreitzer, N.; Ancona, R.; McCullumsmith, C.; Kurowski, B.G.; Foreman, B.; Ngwenya, L.B.; Adeoye, O. The Effect of Antidepressants on Depression After Traumatic Brain Injury: A Meta-analysis. *J. Head. Trauma. Rehabil.* **2019**, *34*, E47–E54. [CrossRef]
135. Plantier, D.; Luauté, J.; SOFMER Group. Drugs for behavior disorders after traumatic brain injury: Systematic review and expert consensus leading to French recommendations for good practice. *Ann. Phys. Rehabil. Med.* **2016**, *59*, 42–57. [CrossRef]
136. Slowinski, A.; Coetzer, R.; Byrne, C. Pharmacotherapy Effectiveness in Treating Depression After Traumatic Brain Injury: A Meta-Analysis. *J. Neuropsychiatry Clin. Neurosci.* **2019**, *31*, 220–227. [CrossRef]

137. Wheaton, P.; Mathias, J.L.; Vink, R. Impact of pharmacological treatments on cognitive and behavioral outcome in the postacute stages of adult traumatic brain injury: A meta-analysis. *J. Clin. Psychopharmacol.* **2011**, *31*, 745–757. [CrossRef]
138. Mckee, A.C.; Daneshvar, D.H. The neuropathology of traumatic brain injury. *Handb. Clin. Neurol.* **2015**, *127*, 45–66. [CrossRef] [PubMed]
139. Zhanina, M.Y.; Druzhkova, T.A.; Yakovlev, A.A.; Vladimirova, E.E.; Freiman, S.V.; Eremina, N.N.; Guekht, A.B.; Gulyaeva, N.V. Development of Post-Stroke Cognitive and Depressive Disturbances: Associations with Neurohumoral Indices. *Curr. Issues Mol. Biol.* **2022**, *44*, 6290–6305. [CrossRef] [PubMed]
140. Kim, J.-M.; Kang, H.-J.; Kim, J.-W.; Bae, K.-Y.; Kim, S.-W.; Kim, J.-T.; Park, M.-S.; Cho, K.-H. Associations of Tumor Necrosis Factor-α and Interleukin-1β Levels and Polymorphisms with Post-Stroke Depression. *Am. J. Geriatr. Psychiatry Off. J. Am. Assoc. Geriatr. Psychiatry* **2017**, *25*, 1300–1308. [CrossRef]
141. Kofler, M.; Schiefecker, A.J.; Gaasch, M.; Sperner-Unterweger, B.; Fuchs, D.; Beer, R.; Ferger, B.; Rass, V.; Hackl, W.; Rhomberg, P.; et al. A reduced concentration of brain interstitial amino acids is associated with depression in subarachnoid hemorrhage patients. *Sci. Rep.* **2019**, *9*, 2811. [CrossRef] [PubMed]
142. Zhu, L.; Han, B.; Wang, L.; Chang, Y.; Ren, W.; Gu, Y.; Yan, M.; Wu, C.; Zhang, X.Y.; He, J. The association between serum ferritin levels and post-stroke depression. *J. Affect. Disord.* **2016**, *190*, 98–102. [CrossRef] [PubMed]
143. Zhang, Z.; Mu, J.; Li, J.; Li, W.; Song, J. Aberrant apolipoprotein E expression and cognitive dysfunction in patients with poststroke depression. *Genet. Test. Mol. Biomark.* **2013**, *17*, 47–51. [CrossRef]
144. Fuying, Z.; Yingying, Y.; Shining, Z.; Kezhong, Z.; Yanyan, S.; Xuemei, Z.; Yin, C.; Jinxia, C.; Huajie, L.; Yanfeng, W.; et al. Novel susceptibility genes were found in a targeted sequencing of stroke patients with or without depression in the Chinese Han population. *J. Affect. Disord.* **2019**, *255*, 1–9. [CrossRef]
145. Guo, W.Y.; Zhang, Z.H.; Mu, J.L.; Liu, D.; Zhao, L.; Yao, Z.Y.; Song, J.G. Relationship between 5-HTTLPR polymorphism and post-stroke depression. *Genet. Mol. Res.* **2016**, *15*, 1. [CrossRef]
146. Liang, J.; Yue, Y.; Jiang, H.; Geng, D.; Wang, J.; Lu, J.; Li, S.; Zhang, K.; Wu, A.; Yuan, Y. Genetic variations in the p11/tPA/BDNF pathway are associated with post stroke depression. *J. Affect. Disord.* **2018**, *226*, 313–325. [CrossRef] [PubMed]
147. Zhang, X.; Wang, C.-B.; Duan, L.-H.; Long, J.-J.; Xiao, P.; Wang, Y.-L.; Zhang, X.-H.; Liu, Q.-Q. Correlation research of serum substance P, CCK-8, and 5-HT values with depression levels in stroke survivors. *Eur. Rev. Med. Pharmacol. Sci.* **2023**, *27*, 1248–1254. [CrossRef]
148. Cheng, S.-Y.; Zhao, Y.-D.; Li, J.; Chen, X.-Y.; Wang, R.-D.; Zeng, J.-W. Plasma levels of glutamate during stroke is associated with development of post-stroke depression. *Psychoneuroendocrinology* **2014**, *47*, 126–135. [CrossRef]
149. Geng, L.-Y.; Qian, F.-Y.; Qian, J.-F.; Zhang, Z.-J. The combination of plasma glutamate and physical impairment after acute stroke as a potential indicator for the early-onset post-stroke depression. *J. Psychosom. Res.* **2017**, *96*, 35–41. [CrossRef] [PubMed]
150. Braga, M.F.M.; Juranek, J.; Eiden, L.E.; Li, Z.; Figueiredo, T.H.; de Araujo Furtado, M.; Marini, A.M. GABAergic circuits of the basolateral amygdala and generation of anxiety after traumatic brain injury. *Amino Acids* **2022**, *54*, 1229–1249. [CrossRef]
151. Bründl, E.; Proescholdt, M.; Störr, E.-M.; Schödel, P.; Bele, S.; Höhne, J.; Zeman, F.; Brawanski, A.; Schebesch, K.-M. Endogenous calcitonin gene-related peptide in cerebrospinal fluid and early quality of life and mental health after good-grade spontaneous subarachnoid hemorrhage-a feasibility series. *Neurosurg. Rev.* **2021**, *44*, 1479–1492. [CrossRef]
152. Baranowska, B.; Kochanowski, J.; Grudniak, M.; Wolinska-Witort, E.; Kalisz, M.; Bik, W. Plasma NPY concentrations in women with acute ischemic stroke. *Neuro Endocrinol. Lett.* **2013**, *34*, 124–128.
153. Yue, Y.; Jiang, H.; Yin, Y.; Zhang, Y.; Liang, J.; Li, S.; Wang, J.; Lu, J.; Geng, D.; Wu, A.; et al. The Role of Neuropeptide Y mRNA Expression Level in Distinguishing Different Types of Depression. *Front. Aging Neurosci.* **2016**, *8*, 323. [CrossRef] [PubMed]
154. Barnett, N.; Ljubic, M.; Chung, J.; Capizzi, A. Testosterone and neurobehavioral outcomes in special operations forces military with multiple mild traumatic brain injury. *NeuroRehabilitation* **2024**, *55*, 271–279. [CrossRef] [PubMed]
155. Pu, L.; Wang, L.; Zhang, R.; Zhao, T.; Jiang, Y.; Han, L. Projected Global Trends in Ischemic Stroke Incidence, Deaths and Disability-Adjusted Life Years From 2020 to 2030. *Stroke* **2023**, *54*, 1330–1339. [CrossRef]
156. Feske, S.K. Ischemic Stroke. *Am. J. Med.* **2021**, *134*, 1457–1464. [CrossRef]
157. Ayerbe, L.; Ayis, S.; Wolfe, C.D.A.; Rudd, A.G. Natural history, predictors and outcomes of depression after stroke: Systematic review and meta-analysis. *Br. J. Psychiatry* **2013**, *202*, 14–21. [CrossRef] [PubMed]
158. Rafsten, L.; Danielsson, A.; Sunnerhagen, K.S. Anxiety after stroke: A systematic review and meta-analysis. *J. Rehabil. Med.* **2018**, *50*, 769–778. [CrossRef]
159. Guo, J.; Wang, J.; Sun, W.; Liu, X. The advances of post-stroke depression: 2021 update. *J. Neurol.* **2022**, *269*, 1236–1249. [CrossRef]
160. Medeiros, G.C.; Roy, D.; Kontos, N.; Beach, S.R. Post-stroke depression: A 2020 updated review. *Gen. Hosp. Psychiatry* **2020**, *66*, 70–80. [CrossRef]
161. Zhang, X.-F.; Zou, W.; Yang, Y. Effects of IL-6 and cortisol fluctuations in post-stroke depression. *J. Huazhong Univ. Sci. Technol. Med. Sci.* **2016**, *36*, 732–735. [CrossRef]

162. Wang, Q.; Qi, Y.; Li, Y.; Yan, Z.; Wang, X.; Ma, Q.; Tang, C.; Liu, X.; Wei, M.; Zhang, H. Psychiatric traits and intracerebral hemorrhage: A Mendelian randomization study. *Front. Psychiatry* **2023**, *13*, 1049432. [CrossRef]
163. Sun, T.; Yuan, Y.; Wu, K.; Zhou, Y.; You, C.; Guan, J. Trends and patterns in the global burden of intracerebral hemorrhage: A comprehensive analysis from 1990 to 2019. *Front. Neurol.* **2023**, *14*, 1241158. [CrossRef]
164. Scopelliti, G.; Casolla, B.; Boulouis, G.; Kuchcinski, G.; Moulin, S.; Leys, D.; Hénon, H.; Cordonnier, C.; Pasi, M. Long-term anxiety in spontaneous intracerebral hemorrhage survivors. *Int. J. Stroke* **2022**, *17*, 1093–1099. [CrossRef] [PubMed]
165. Suswati, I.; Rahayu; Maulida, A.P.; Prasetio, A.D. Managing Mental Disorders in Intracranial Hemorrhage (ICH) Patients: A Case Study on The Importance of Early Recognition and Intervention. *Brain Hemorrhages* **2024**, *5*, 42–45. [CrossRef]
166. Douros, A.; Ades, M.; Renoux, C. Risk of Intracranial Hemorrhage Associated with the Use of Antidepressants Inhibiting Serotonin Reuptake: A Systematic Review. *CNS Drugs* **2018**, *32*, 321–334. [CrossRef] [PubMed]
167. Kubiszewski, P.; Sugita, L.; Kourkoulis, C.; DiPucchio, Z.; Schwab, K.; Anderson, C.D.; Gurol, M.E.; Greenberg, S.M.; Viswanathan, A.; Rosand, J.; et al. Association of Selective Serotonin Reuptake Inhibitor Use After Intracerebral Hemorrhage with Hemorrhage Recurrence and Depression Severity. *JAMA Neurol.* **2020**, *78*, 1–8. [CrossRef] [PubMed]
168. Song, Y. Observation on the Effect of Solution-Focused Approach Combined with Family Involvement in WeChat Platform Management on Inpatients with Intracerebral Hemorrhage. *J. Health Eng.* **2022**, *2022*, 9951374. [CrossRef]
169. Liu, L.; Fuller, M.; Behymer, T.P.; Ng, Y.; Christianson, T.; Shah, S.; King, N.K.K.; Woo, D.; James, M.L. Selective Serotonin Reuptake Inhibitors and Intracerebral Hemorrhage Risk and Outcome. *Stroke* **2020**, *51*, 1135–1141. [CrossRef]
170. Hoh, B.L.; Ko, N.U.; Amin-Hanjani, S.; Chou, S.H.-Y.; Cruz-Flores, S.; Dangayach, N.S.; Derdeyn, C.P.; Du, R.; Hänggi, D.; Hetts, S.W.; et al. 2023 Guideline for the Management of Patients with Aneurysmal Subarachnoid Hemorrhage: A Guideline From the American Heart Association/American Stroke Association. *Stroke* **2023**, *54*, e314–e370. [CrossRef]
171. Tang, W.K.; Wang, L.; Kwok Chu Wong, G.; Ungvari, G.S.; Yasuno, F.; Tsoi, K.K.F.; Kim, J.S. Depression after Subarachnoid Hemorrhage: A Systematic Review. *J. Stroke* **2020**, *22*, 11–28. [CrossRef]
172. Tang, W.K.; Wang, L.; Tsoi, K.K.; Kim, J.M.; Lee, S.-J.; Kim, J.S. Anxiety after subarachnoid hemorrhage: A systematic review and meta-analysis: Anxiety in subarachnoid hemorrhage. *J. Affect. Disord. Rep.* **2021**, *3*, 100060. [CrossRef]
173. Diestro, J.D.B.; Vyas, M.; Jung, Y.; Kishibe, T.; Leochico, C.; Espiritu, A.; Dorotan, M.K.; Dimal, N.; Omar, A.T.; Sienes, A.; et al. Long-term neuropsychiatric complications of aneurysmal subarachnoid hemorrhage: A narrative review. *J. Neurointerv Surg.* **2024**, 1–8. [CrossRef]
174. Rundek, T.; Tolea, M.; Ariko, T.; Fagerli, E.A.; Camargo, C.J. Vascular Cognitive Impairment (VCI). *Neurotherapeutics* **2022**, *19*, 68–88. [CrossRef]
175. He, Y.; Chen, X.; Wu, M.; Hou, X.; Zhou, Z. What type of cell death occurs in chronic cerebral hypoperfusion? A review focusing on pyroptosis and its potential therapeutic implications. *Front. Cell Neurosci.* **2023**, *17*, 1073511. [CrossRef]
176. Chiu, P.-Y.; Liu, C.-H.; Tsai, C.-H. Neuropsychiatric manifestations in vascular cognitive impairment patients with and without dementia. *Acta Neurol. Taiwan.* **2007**, *16*, 86–91.
177. Gupta, M.; Dasgupta, A.; Khwaja, G.A.; Chowdhury, D.; Patidar, Y.; Batra, A. The profile of behavioral and psychological symptoms in vascular cognitive impairment with and without dementia. *Ann. Indian. Acad. Neurol.* **2013**, *16*, 599–602. [CrossRef] [PubMed]
178. Levine, D.A.; Langa, K.M. Vascular cognitive impairment: Disease mechanisms and therapeutic implications. *Neurotherapeutics* **2011**, *8*, 361–373. [CrossRef]
179. Linh, T.T.D.; Hsieh, Y.-C.; Huang, L.-K.; Hu, C.-J. Clinical Trials of New Drugs for Vascular Cognitive Impairment and Vascular Dementia. *Int. J. Mol. Sci.* **2022**, *23*, 11067. [CrossRef] [PubMed]
180. Christiansen, D.M.; McCarthy, M.M.; Seeman, M.V. Where Sex Meets Gender: How Sex and Gender Come Together to Cause Sex Differences in Mental Illness. *Front. Psychiatry* **2022**, *13*, 856436. [CrossRef] [PubMed]
181. Kessler, R.C.; Petukhova, M.; Sampson, N.A.; Zaslavsky, A.M.; Wittchen, H.-U. Twelve-month and lifetime prevalence and lifetime morbid risk of anxiety and mood disorders in the United States. *Int. J. Methods Psychiatr. Res.* **2012**, *21*, 169–184. [CrossRef]
182. Tolin, D.F.; Foa, E.B. Sex differences in trauma and posttraumatic stress disorder: A quantitative review of 25 years of research. *Psychol. Bull.* **2006**, *132*, 959–992. [CrossRef] [PubMed]
183. Lapchak, P.A.; Boitano, P.D. Reflections on Neuroprotection Research and the Path Toward Clinical Success. In *Neuroprotective Therapy for Stroke and Ischemic Disease*; Lapchak, P.A., Zhang, J.H., Eds.; Springer International Publishing: Cham, Switzerland, 2017; pp. 3–71. ISBN 978-3-319-45345-3. [CrossRef]
184. Ciarlone, S.L.; Statz, J.K.; Goodrich, J.A.; Norris, J.N.; Goforth, C.W.; Ahlers, S.T.; Tschiffely, A.E. Neuroendocrine function and associated mental health outcomes following mild traumatic brain injury in OEF-deployed service members. *J. Neurosci. Res.* **2020**, *98*, 1174–1187. [CrossRef]
185. Wade, S.L.; Kaizar, E.E.; Narad, M.E.; Zang, H.; Kurowski, B.G.; Miley, A.E.; Moscato, E.L.; Aguilar, J.M.; Yeates, K.O.; Taylor, H.G.; et al. Behavior Problems following Childhood TBI: The Role of Sex, Age, and Time since Injury. *J. Head. Trauma. Rehabil.* **2020**, *35*, E393–E404. [CrossRef]

186. Juranek, J.; Johnson, C.P.; Prasad, M.R.; Kramer, L.A.; Saunders, A.; Filipek, P.A.; Swank, P.R.; Cox, C.S.; Ewing-Cobbs, L. Mean diffusivity in the amygdala correlates with anxiety in pediatric TBI. *Brain Imaging Behav.* **2012**, *6*, 36–48. [CrossRef]
187. Barton, D.J.; Kumar, R.G.; McCullough, E.H.; Galang, G.; Arenth, P.M.; Berga, S.L.; Wagner, A.K. Persistent Hypogonadotropic Hypogonadism in Men After Severe Traumatic Brain Injury: Temporal Hormone Profiles and Outcome Prediction. *J. Head. Trauma. Rehabil.* **2016**, *31*, 277–287. [CrossRef] [PubMed]
188. Nilsen, J. Estradiol and neurodegenerative oxidative stress. *Front. Neuroendocr.* **2008**, *29*, 463–475. [CrossRef]
189. Gartlehner, G.; Dobrescu, A.; Chapman, A.; Toromanova, A.; Emprechtinger, R.; Persad, E.; Affengruber, L.; Pieh, C.; Klerings, I.; Wagner, G. Nonpharmacologic and Pharmacologic Treatments of Adult Patients with Major Depressive Disorder: A Systematic Review and Network Meta-analysis for a Clinical Guideline by the American College of Physicians. *Ann. Intern. Med.* **2023**, *176*, 196–211. [CrossRef] [PubMed]
190. Lee, H.J.; Stein, M.B. Update on treatments for anxiety-related disorders. *Curr. Opin. Psychiatry* **2023**, *36*, 140. [CrossRef] [PubMed]
191. Ruberto, V.L.; Jha, M.K.; Murrough, J.W. Pharmacological Treatments for Patients with Treatment-Resistant Depression. *Pharmaceuticals* **2020**, *13*, 116. [CrossRef]
192. Scangos, K.W.; Khambhati, A.N.; Daly, P.M.; Makhoul, G.S.; Sugrue, L.P.; Zamanian, H.; Liu, T.X.; Rao, V.R.; Sellers, K.K.; Dawes, H.E.; et al. Closed-loop neuromodulation in an individual with treatment-resistant depression. *Nat. Med.* **2021**, *27*, 1696–1700. [CrossRef]
193. Tafet, G.E.; Nemeroff, C.B. Pharmacological Treatment of Anxiety Disorders: The Role of the HPA Axis. *Front. Psychiatry* **2020**, *11*, 443. [CrossRef] [PubMed]
194. Hofmann, S.G.; Asnaani, A.; Vonk, I.J.J.; Sawyer, A.T.; Fang, A. The Efficacy of Cognitive Behavioral Therapy: A Review of Meta-analyses. *Cogn. Ther. Res.* **2012**, *36*, 427–440. [CrossRef]
195. Aaronson, S.T.; Sears, P.; Ruvuna, F.; Bunker, M.; Conway, C.R.; Dougherty, D.D.; Reimherr, F.W.; Schwartz, T.L.; Zajecka, J.M. A 5-Year Observational Study of Patients with Treatment-Resistant Depression Treated with Vagus Nerve Stimulation or Treatment as Usual: Comparison of Response, Remission, and Suicidality. *Am. J. Psychiatry* **2017**, *174*, 640–648. [CrossRef]
196. Austelle, C.W.; Cox, S.S.; Wills, K.E.; Badran, B.W. Vagus nerve stimulation (VNS): Recent advances and future directions. *Clin. Auton. Res.* **2024**, *34*, 529–547. [CrossRef]
197. Guo, Z.-P.; Liao, D.; Chen, L.; Wang, C.; Qu, M.; Lv, X.-Y.; Fang, J.-L.; Liu, C.-H. Transcutaneous Auricular Vagus Nerve Stimulation Modulating the Brain Topological Architecture of Functional Network in Major Depressive Disorder: An fMRI Study. *Brain Sci.* **2024**, *14*, 945. [CrossRef]
198. Calderone, A.; Cardile, D.; Gangemi, A.; De Luca, R.; Quartarone, A.; Corallo, F.; Calabrò, R.S. Traumatic Brain Injury and Neuromodulation Techniques in Rehabilitation: A Scoping Review. *Biomedicines* **2024**, *12*, 438. [CrossRef]
199. Philip, N.S.; Ramanathan, D.; Gamboa, B.; Brennan, M.C.; Kozel, F.A.; Lazzeroni, L.; Madore, M.R. Repetitive Transcranial Magnetic Stimulation for Depression and Posttraumatic Stress Disorder in Veterans with Mild Traumatic Brain Injury. *Neuromodulation* **2023**, *26*, 878–884. [CrossRef] [PubMed]
200. Holm, J.; Eiermann, B.; Eliasson, E.; Mannheimer, B. A limited number of prescribed drugs account for the great majority of drug-drug interactions. *Eur. J. Clin. Pharmacol.* **2014**, *70*, 1375–1383. [CrossRef] [PubMed]
201. Tornio, A.; Filppula, A.M.; Niemi, M.; Backman, J.T. Clinical Studies on Drug-Drug Interactions Involving Metabolism and Transport: Methodology, Pitfalls, and Interpretation. *Clin. Pharmacol. Ther.* **2019**, *105*, 1345–1361. [CrossRef]
202. Brandes, M.S.; Gray, N.E. NRF2 as a Therapeutic Target in Neurodegenerative Diseases. *ASN Neuro* **2020**, *12*, 1759091419899782. [CrossRef] [PubMed]
203. Cuadrado, A.; Rojo, A.I.; Wells, G.; Hayes, J.D.; Cousin, S.P.; Rumsey, W.L.; Attucks, O.C.; Franklin, S.; Levonen, A.-L.; Kensler, T.W.; et al. Therapeutic targeting of the NRF2 and KEAP1 partnership in chronic diseases. *Nat. Rev. Drug Discov.* **2019**, *18*, 295–317. [CrossRef]
204. Dinkova-Kostova, A.T.; Copple, I.M. Advances and challenges in therapeutic targeting of NRF2. *Trends Pharmacol. Sci.* **2023**, *44*, 137–149. [CrossRef]
205. Zhang, Q.; Zhang, J.; Mo, J. Pharmacological Modulations of Nrf2 and Therapeutic Implications in Aneurysmal Subarachnoid Hemorrhage. *Molecules* **2023**, *28*, 1747. [CrossRef]
206. Zuo, C.; Cao, H.; Song, Y.; Gu, Z.; Huang, Y.; Yang, Y.; Miao, J.; Zhu, L.; Chen, J.; Jiang, Y.; et al. Nrf2: An all-rounder in depression. *Redox Biol.* **2022**, *58*, 102522. [CrossRef]
207. Barrera, G.; Echevarria, D.J.; Poulin, J.-F.; Laforest, S.; Drolet, G.; Morilak, D.A. One for all or one for one: Does co-transmission unify the concept of a brain galanin "system" or clarify any consistent role in anxiety? *Neuropeptides* **2005**, *39*, 289–292. [CrossRef] [PubMed]
208. Gysling, K.; Forray, M.I.; Haeger, P.; Daza, C.; Rojas, R. Corticotropin-releasing hormone and urocortin: Redundant or distinctive functions? *Brain Res. Rev.* **2004**, *47*, 116–125. [CrossRef]

209. Honkaniemi, J.; Pelto-Huikko, M.; Rechardt, L.; Isola, J.; Lammi, A.; Fuxe, K.; Gustafsson, J.-Å.; Wikström, A.-C.; Hökfelt, T. Colocalization of Peptide and Glucocorticoid Receptor Immunoreactivities in Rat Central Amygdaloid Nucleus. *Neuroendocrinology* **2008**, *55*, 451–459. [CrossRef]
210. Palkovits, M. Stress-induced expression of co-localized neuropeptides in hypothalamic and amygdaloid neurons. *Eur. J. Pharmacol.* **2000**, *405*, 161–166. [CrossRef]
211. Russo, A.F. Overview of Neuropeptides: Awakening the Senses? *Headache* **2017**, *57* (Suppl. 2), 37–46. [CrossRef]
212. Lin, K.; Stiles, J.; Tambo, W.; Ajmal, E.; Piao, Q.; Powell, K.; Li, C. Bimodal functions of calcitonin gene-related peptide in the brain. *Life Sci.* **2024**, *359*, 123177. [CrossRef] [PubMed]
213. Chu, Z.; Zhu, L.; Zhou, Y.; Yang, F.; Hu, Z.; Luo, Y.; Li, W.; Luo, F. Targeting Nrf2 by bioactive peptides alleviate inflammation: Expanding the role of gut microbiota and metabolites. *Crit. Rev. Food Sci. Nutr.* **2024**, 1–20. [CrossRef] [PubMed]
214. Powell, K.; Wadolowski, S.; Tambo, W.; Kim, D.; Jacob, A.; Wang, P.; Li, C. *Endogenous Cgrp Activates Nrf2 Signaling Via Non-Electrophilic Mechanisms*; Social Science Research Network: Rochester, NY, USA, 2024; Available online: https://papers.ssrn.com/abstract=4996292 (accessed on 6 December 2024).
215. Nash, C.; Powell, K.; Lynch, D.G.; Hartings, J.A.; Li, C. Nonpharmacological modulation of cortical spreading depolarization. *Life Sci.* **2023**, *327*, 121833. [CrossRef] [PubMed]
216. White, T.G.; Powell, K.; Shah, K.A.; Woo, H.H.; Narayan, R.K.; Li, C. Trigeminal Nerve Control of Cerebral Blood Flow: A Brief Review. *Front. Neurosci.* **2021**, *15*, 649910. [CrossRef]
217. Chekroud, S.R.; Gueorguieva, R.; Zheutlin, A.B.; Paulus, M.; Krumholz, H.M.; Krystal, J.H.; Chekroud, A.M. Association between physical exercise and mental health in 1·2 million individuals in the USA between 2011 and 2015: A cross-sectional study. *Lancet Psychiatry* **2018**, *5*, 739–746. [CrossRef]
218. Mikkelsen, K.; Stojanovska, L.; Polenakovic, M.; Bosevski, M.; Apostolopoulos, V. Exercise and mental health. *Maturitas* **2017**, *106*, 48–56. [CrossRef]
219. Gingrich, M.B.; Junge, C.E.; Lyuboslavsky, P.; Traynelis, S.F. Potentiation of NMDA receptor function by the serine protease thrombin. *J. Neurosci.* **2000**, *20*, 4582–4595. [CrossRef]
220. Shlobin, N.A.; Har-Even, M.; Itsekson-Hayosh, Z.; Harnof, S.; Pick, C.G. Role of Thrombin in Central Nervous System Injury and Disease. *Biomolecules* **2021**, *11*, 562. [CrossRef]
221. Qian, Y.; Wang, J.; Chen, J.; Lin, W.; Shen, H.; Fang, Y.; Yu, W. Multifaceted role of thrombin in subarachnoid hemorrhage: Focusing on cerebrospinal fluid circulation disorder. *Exp. Neurol.* **2025**, *383*, 115036. [CrossRef]
222. Pizzagalli, D.A. Depression, Stress, and Anhedonia: Toward a Synthesis and Integrated Model. *Annu. Rev. Clin. Psychol.* **2014**, *10*, 393–423. [CrossRef]
223. Beck, A.T.; Bredemeier, K. A Unified Model of Depression: Integrating Clinical, Cognitive, Biological, and Evolutionary Perspectives. *Clin. Psychol. Sci.* **2016**, *4*, 596–619. [CrossRef]
224. Vega-Zelaya, L.; Pastor, J. The Network Systems Underlying Emotions: The Rational Foundation of Deep Brain Stimulation Psychosurgery. *Brain Sci.* **2023**, *13*, 943. [CrossRef] [PubMed]
225. Kalanthroff, E.; Wheaton, M.G. An Integrative Model for Understanding Obsessive-Compulsive Disorder: Merging Cognitive Behavioral Theory with Insights from Clinical Neuroscience. *J. Clin. Med.* **2022**, *11*, 7379. [CrossRef]
226. Čater, M.; Majdič, G. How early maternal deprivation changes the brain and behavior? *Eur. J. Neurosci.* **2022**, *55*, 2058–2075. [CrossRef]
227. Đorović, Đ.; Lazarevic, V.; Aranđelović, J.; Stevanović, V.; Paslawski, W.; Zhang, X.; Velimirović, M.; Petronijević, N.; Puškaš, L.; Savić, M.M.; et al. Maternal deprivation causes CaMKII downregulation and modulates glutamate, norepinephrine and serotonin in limbic brain areas in a rat model of single prolonged stress. *J. Affect. Disord.* **2024**, *349*, 286–296. [CrossRef]

MDPI AG
Grosspeteranlage 5
4052 Basel
Switzerland
Tel.: +41 61 683 77 34

Cells Editorial Office
E-mail: cells@mdpi.com
www.mdpi.com/journal/cells

www.ingramcontent.com/pod-product-compliance
Lightning Source LLC
LaVergne TN
LVHW070455061126
851858LV00011B/244